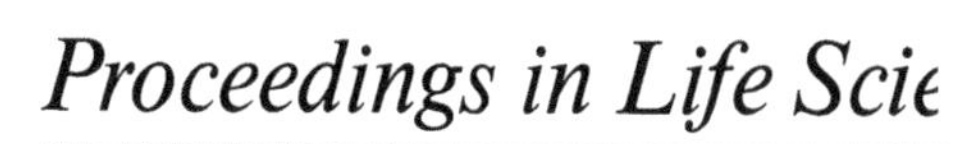
Proceedings in Life Scie

Structure and Function of Haemocyanin

Edited by J. V. Bannister

With 195 Figures

Springer-Verlag
Berlin Heidelberg New York 1977

Dr. Joe V. Bannister, The University of Malta, Department of Physiology and Biochemistry, Msida/Malta

Cover motif: Fig. 8, p. 127

ISBN-13: 978-3-642-66681-0 e-ISBN-13: 978-3-642-66679-7
DOI: 10.1007/ 978-3-642-66679-7

Library of Congress Cataloging in Publication Data. European Molecular Biology International Workshop, 5th, University of Malta, 1976. Structure and function of haemocyanin. (Proceedings in life sciences). Proceedings of the 5th of a series of meetings; proceedings of the 1st are entered under title: Physiology and biochemistry of haemocyanins. Includes index. 1. Hemocyanin–Congresses. I. Bannister, Joe V., 1945-. II. Title. QP99.3.H4E95.1976. 612'.0154'6. 77-2773.

Softcover reprint of the hardcover 1st edition 1977

This book is dedicated to
Professor Jeffries Wyman

Preface

Haemocyanin was first recognised as a respiratory pigment by P. Bert in 1867. Over the years the haemocyanins have attracted attention as macromolecules and copper-containing respiratory proteins. The early functional studies of A.C. Redfield (Biol. Rev. 9, 176, 1934) and the ultracentrifuge work of I.B. Ericksson-Quensel and T. Svedberg (Biol. Bull. 71, 498, 1936) come easily to mind. In recent years haemocyanin studies have come to the forefront with the work of the Ghirettis (Padova), R. Lontie (Louven), E.F.J. Van Bruggen (Groningen) and Joe and Celia Bonaventura (Beaufort) to mention but a few of the number of able investigators coming to this field from diverse disciplines. It is hoped that this book presents a fair cross section of current workers and work on haemocyanin. It is the second collection of haemocyanin studies to be published after a Workshop on the Structure and Function of Haemocyanin.

The first haemocyanin meeting was held by Professors F. "Ghiro" Ghiretti and Anna Ghiretti-Magaldi in Naples in 1966. Subsequent meetings have been held at Groningen (1970), Louven (1971) and Padova (1974).

Current interest in haemocyanin can be judged by the scope and breadth of the papers from the Malta meeting and presented here. In organising the Malta meeting I would like to thank Dr. John Tooze of the European Molecular Biology Organisation and my friend and colleague, Professor Maurizio Brunori for their continuous help and support.

This book would not have been published without the direct intervention of Dr. Konrad Springer. I thank Professor R.J.P. Williams FRS for his lucid review on haemocyanin and for editorial help. Finally I would like to thank my brother Professor W.H. Bannister for his continuous encouragement and Miss V. Vitale for valuable assistance in the preparation of the manuscripts.

Msida, Summer 1977 J.V. BANNISTER

Contents

List of Participants

Angela Anastasi
Dept of Physiology and Biochemistry, University of Malta, Msida, Malta

Joe V. Bannister
Dept. of Physiology and Biochemistry, University of Malta, Msida, Malta

William H. Bannister
Dept. of Physiology and Biochemistry, University of Malta, Msida, Malta

Donatella Barra
Institute of Biological Chemistry, University of Rome, Rome, Italy

Joe Bonaventura
Duke University Marine Laboratory, Beaufort, North Carolina, USA

Celia Bonaventura
Duke University Marine Laboratory, Beaufort, North Carolina, USA

Francesco Bossa
Institute of Biological Chemistry, University of Camerino, Macerata, Italy

Maurizio Brunori
Institute of Chemistry, University of Rome, Rome, Italy

Tony E. Cass
Inorganic Chemistry Department, University of Oxford, Oxford, Great Britain

Marc De Ley
Laboratorium voor Biochemie Katholieke Universiteit te Leuven, Leuven, Belgium

H. David Ellerton
Department of Chemistry, Victoria University, Wellington, New Zealand

Alphonse Galdes
Dept. of Physiology and Biochemistry, University of Malta, Msida, Malta

Francesco Ghiretti
Institute of Animal Biology, CNR Centre for the Physiology and Biochemistry of Haemocyanin, University of Padova, Padova, Italy

Anna Ghiretti-Magaldi
Institute of Animal Biology, CNR Centre for the Physiology and Biochemistry of Haemocyanin, University of Padova, Padova, Italy

Constant Gielens
Laboratorium voor Biochemie, Katholieke Universiteit te Leuven, Leuven, Belgium

Roderick L. Hall
Dept. of Biochemistry University of Leeds, Leeds, Great Britain

H. Allen O. Hill
Inorganic Chemistry Department, University of Oxford, Oxford, Great Britain

George M. Hughes
Research Units for Comparative Animal Respiration, University of Bristol, Bristol, Great Britain

Giulio Jori
Institute of Organic Chemistry, University of Padova, Padova, Italy

Harm. A. Kuiper
Biochemisch Laboratorium, Rijksuniversiteit, Groningen, The Netherlands

Jean Lamy
Laboratoire de Biochemie, Facultè de Médicine, Tours, France

Norman Li
Dept. of Chemistry, Duquesne University, Pennsylvania, USA

Bernt Linzen
Zoologisches Institut, Universität München, Munich, W. Germany

Renate Loewe
Zoologisches Institut, Universität München, Munich, W. Germany

Rene Lontie
Laboratorium voor Biochemie, Katholieke Universiteit te Leuven, Leuven, Belgium

Warner Love
Dept. of Biophysics, The Johns Hopkins University, Maryland, USA

Jennifer Mallia
Dept. of Physiology and Biochemistry, University of Malta, Msida, Malta

Giselle Preaux
Laboratorium voor Biochemie, Katholieke Universiteit te Leuven, Leuven, Belgium

Fernanda Ricchelli
Institute of Animal Biology, CNR Centre for the Physiology and Biochemistry of Haemocyanin, University of Padova, Padova Italy

Austen Riggs
Dept. of Zoology, University of Texas at Austin, Texas, USA

Benedetto Salvato
Institute of Animal Biology, CNR Centre for the Physiology and Biochemistry of Haemocyanin, University of Padova, Padova, Italy

Taede Sminia
Biologisch Laboratorium, Vrije Universiteit, Amsterdam, The Netherlands

Laura Tallandini
Institute of Animal Biology, CNR Centre for the Physiology and Biochemistry of Haemocyanin, University of Padova, Padova, Italy

Ruurd Torensma
Biochemisch Laboratorium, Rijksuniversiteit, Groningen, The Netherlands

Ernst Van Bruggen
Biochemisch Laboratorium, Rijksuniversiteit, Groningen, The Netherlands

A. Van der Berg
Biochemisch Laboratorium, Rijksuniversiteit, Groningen, The Netherlands

Henk Van der Deen
Biochemisch Laboratorium, Rijksuniversiteit, Groningen, The Netherlands

Lies Van Schaick
Biochemisch Laboratorium, Rijksuniversiteit, Groningen, The Netherlands

Raphael Witters
Laboratorium voor Biochemie, Katholieke Universiteit te Leuven, Leuven, Belgium

Edward J. Wood
Dept. of Biochemistry, University of Leeds, Leeds, Great Britain

Introduction

Problems Presented by Haemocyanins

R. J. P. WILLIAMS

In the following articles the reader will find detailed review material covering the work of most if not all of those who participate in the study of haemocyanins. As each article has at least one specialist as an author and as specialists are notoriously anxious both to state detail correctly and to provide personal justification for their activities, the overall impression of this book could be lost or at least difficult to see. It is for this reason that I have been asked to present an introduction to the various papers. An attempt will be made to keep above controversy but as the man in the middle I shall be thought to be biased (by all sides) no matter what I write.

The review material is divided in the same sections as the book:

1. Haemocyanin (a) the full protein
   (b) the subunits
   (c) physical properties
2. Haemocyanin as a copper complex
3. Reactions of haemocyanin (a) thermodynamics
   (b) kinetics
4. Evolutionary studies
5. Physiology

Haemocyanin as a Protein, General Characteristics. Haemocyanin is a very large protein in all species and a major difficulty in its study may arise from heterogeneity. There is no reason to suppose that haemocyanins are simpler than haemoglobins so that iso-proteins within species as well as species differences will abound. It is only in the next phase of study that the true complexities of this protein will be revealed and maybe we shall see then why it is that the haemocyanins have persisted in such a few species in marked contrast with the haemoglobins.

There is as yet no detailed X-ray structure for the subunits but we have the exceedingly fine electron microscope studies. It is to these studies which we must return when we analyse cooperativity.

Subunits. It is generally agreed that the basic subunit of arthropod haemocyanins has a molecular weight of approx. 70,000 containing two copper atoms. It is a single polypeptide chain but it is not clear how many different types of subunit are present i.e. α, β, γ units due to gene duplication.

If the consensus of evidence is believed there is another subunit in molluscan haemocyanins of some 50,000 molecular weight. Thus there are two series of haemocyanins at least and we shall be faced with a more difficult evolutionary problem than that in the haemoglobins where the basic unit of monomeric, tetrameric and polymeric species has a molecular weight of 17,000 and in the hemerythrins where monomers and octamers have a subunit of about 15,000 molecular weight. There is no evidence for a marked difference in the active site of the subunits in the two series of oxygen carriers.

The subunits are packed in particular arrays and these arrays are stabilised by the additional binding of calcium and magnesium ions. The dissociation/association equilibrium is made complex by protein heterogeneity. The action of the calcium ions is uncertain but they do not have strong binding sites and the calcium is probably in fast exchange. A good guess is that the site contains two/three carboxylate residues.

Subunit Structure. Apart from the copper site there are not known to be any unusual protein features in the subunits. The work of sequence analysis has begun by the study of cleavage reactions and it is known that the proteins are glycoproteins. The work of Ghiretti on the composition of the subunits is suggestive and it will be especially interesting to compare 70,000 and 50,000 subunits.

Haemocyanin as a Copper Complex. The general view of the protein copper as a copper(I) has not been shaken and the belief that the copper(II) in the oxygen complex which is therefore a peroxide complex $Cu(II)-O_2^{2-}-Cu(II)$ seems to be accepted. Much greater difficulty surrounds the discussion of ligands. The supposition that three (four) imidazoles (histidines) bind the copper is based on rather flimsy evidence from (1) photo-oxidation (see the papers of groups in Padova and Louvain and from ESR data on nitric oxide complexes. The work also reveals the presence of a nearby tryptophan); (2) photo-electron spectroscopy established the oxidation states but adds little about the ligand. Perhaps an ENDOR study using the methaemocyanin would be more successful for it will be many years before the crystallographers will get to grips with this protein.

It might be well to remember that copper(I) appears in these proteins not zinc(II) and yet zinc(II) is found in carbonic anhydrase (three histidine ligands). Thus the present author would suppose that a special geometry is required which would have one or all of the characteristics so as to accept Cu(I) and exclude Zn(II): (1) A "hole" size between at least two of the histidine which fits Cu(I) (Cu-N distance 1.0 Å) and not Zn(II) (Zn-N distance 0.7 Å). (2) A linear geometry of two histidines (sp) or a trigonal geometry of three histidines (sp^2) but not a disposition of three histidines approaching tetrahedral geometry (sp^3) for the latter favours zinc(II) and the former copper(I). (3) A protein environment of low dielectric constant so that insertion of a unit positive charge (Cu^+) is favoured over a divalent ion (Zn^{2+}).

Given the above characteristics the site geometry will change on reaction:

$$2\ Cu(I) + O_2 \rightarrow 2\ Cu(II)\ O_2^{2-}$$

and so will the Cu-N distances. Thus protein cooperativity could rest on a reaction very like that seen in haemoglobins.

It is of interest to compare the evidence about copper sites in different copper proteins and to ask about their functional significance. There are a growing variety of these sites (Table 1).

Now while it is relatively easy to interpret the EPR and spectral data on small molecule Cu(II) complexes very great difficulty has been experienced with close structural prediction from physical data in all the cases in the Table. It is my opinion today that it may be possible to get a rough approximation to structure from such spectral data but it is quite impossible to get details of the geometry around the copper or the precise nature of the ligands. I illustrate this with the known example of superoxide dismutase. Here the X-ray crystal structure shows

Table 1. Copper sites in enzymes

Enzyme	Site	Ligands proposed
Laccase Caeruloplasmin Plastocyanin	'Blue' site copper type I	$(N)_2$, RS^-
Laccase Caeruloplasmin	EPR copper type II	$(N)_{2-4}$
Lacasse Caeruloplasmin	Copper dimer	?
Galactose oxidase	'Pink' copper	Cu(III) ?
Superoxide dismutase	Cu(II)	$(N)_3(N^-)$ (known)
Tyrosinase	?	$(N)_2$
Haemocyanin	Cu(I)	$(N)_{2-4}$

that there are four imidazole (histidine) nitrogen atoms bound to copper(II) and that one of these is an imidazole anion. The rhombic symmetry seen in the EPR is only a small distortion from tetragonality. The metal is open-sided to one side and would not appear to be blocked by water. With the early knowledge of the spectra and the EPR signals it was clear that the copper(II) was in a site close to tetragonal with three/four nitrogen ligands and readily underwent substitution. However, the very general nature of this description from the physical properties is in some contrast with the curious and particular structural answer found by X-ray crystallographic analysis and this difference leads us to the inevitable conclusion that if the detail is important for function, then functions can never be understood except on the basis of a very detailed structure, i.e., present-day spectroscopy will fail.

Two questions now arise in the case of superoxide dismutase (1) even when the structure is available, can we understand function? (2) If the answer to question (1) is 'No' then we must ask "what other information is required?" I am inclined to answer 'No' to question (1) for two reasons. Although I can see the reason for the open-sided Cu(II) in superoxide dismutase, (compare haemoglobin and the general ideas of the entatic state in the catalysis of reactions involving substitution at metal ion sites) I fail to understand the function of the imidazole anion. Looking at the reaction of the enzyme more closely the copper in superoxide dismutase could cycle between Cu(II) and Cu(I), in the step

$$O_2^{\cdot -} \rightarrow O_2$$

or Cu(II) and Cu(III) in the step

$$O_2^{\cdot -} \rightarrow O_2^{2-}$$

The latter would be greatly assisted by the imidazole anion which would bind Cu(III) very well as in some peptide complexes. In this way superoxide dismutase would fit into a group of enzymes with galactose oxidase if Hamilton's views of this enzyme are correct.

However, if the first step above is required and Cu(I) is a redox state on the reaction path then we must expect a considerable rearrangement of the coordination sphere, see above, and may be a protonation of the imidazole anion. Such steps demand a much more dynamic active site than than provided from the exact structure seen by X-rays and would have to have a mobile structure around the coppers. It is here that the methods outlined by Hill and his group became important for NMR procedures permit not only structural analysis but also provide the parameters of motion. Again, speaking for myself, I believe that the dynamics of protein structure will become blatantly apparent and of overwhelming importance in the next few years.

Returning to Table 1 it is not yet obvious why the different copper sites have different coordination spheres. All we may say clearly is that simple electron-transfer sites should not be designed for substitution reactions and that reaction centres using O_2, H_2O_2 or $O_2^{\cdot}$ should be open-sided. This is observed. But can the inorganic chemist say how to make a complex for the special catalyses which the individual enzymes of Table 1 carry out? Could the chemist design a catalyst which would turn over from Cu(I) to Cu(III) giving little Cu(II)? Perhaps in the case of enzymes or functional proteins we must know the outline structure before we have a chance of mimicking the reactions which biology carries out. Thus biology may be used to lead investigations in catalysis by man. Certainly the precise design of the copper site of haemocyanin will be as full of subtle features as we know already to exist in haemoglobin.

Binding Thermodynamics (Haemocyanin). The copper site is designed to bind oxygen. It must bind as few other small molecules as possible and perhaps it should not even retain water in the absence of oxygen. There are several simple ways of controlling ligand binding but as oxygen is a small molecule, the one of outstanding importance is steric constraint. We see from the work of Li and collaborators that the constraints differ from one haemocyanin species to another. Parallel observations have been made amongst the haem/oxygen carriers.

The ligands which can displace oxygen do not do so in a simple reversible fashion. Even the thiourea-type ligands which displace oxygen directly undergo subsequent reaction. Particularly interesting are the reactions of azide and fluoride for they displace oxygen as peroxide leaving complexes of Cu(II), or Cu(III) (?). Now there is a very useful parallel here with the copper sites of caeruloplasmin. In this protein and in laccase there are three types of copper, Table 1. Ignoring the "blue" copper which is not attacked by ligands we have type II copper(II) and the copper dimer. It has been my contention for some time that this dimeric copper site could react with anions. The work of Lontie and collaborators shows that in haemocyanins a copper(II) pair binds azide to give an absorption band at around 350 nm. Undoubtedly there is some reaction with contaminating monomeric Cu(II) but this reaction with the dimer closely parallels that of laccase. If it is possible to make a true Cu(II) methaemocyanin then the anion binding of the site could be of great interest for the oxygen reactions of laccase and caeruloplasmin. (Both our and Malmstrom's group have also stressed the probability of the close proximity of type II copper and the copper dimer in the active site of the laccase). Recently new EPR signals have been seen by Vanngard and Malmstrom for laccase and they and workers in Japan have recognised new intermediate redox states of the oxygen reaction. What are the steps in the reactions of anions with oxyhaemocyanin which lead to oxidation of the copper(I) to give methaemocyanin? A low temperature study is called for to see if the known intermediates of laccase occur in these reactions.

If there is a true Bohr effect, or a cooperative effect or other allosteric phenomena, in haemocyanins then the same problems of triggering and communication exist as in the haemoglobins but unlike iron(II), copper(I) has no spin state changes. Thus it is *redox-state* changes in different proteins that must be compared. In the above we indicated how the redox switch of copper could cause steric changes which would run through the proteins.

Ligand Kinetics. In the discussion of consecutive reaction constants, either equilibrium or kinetic constants, I do not feel that I can add anything of real significance to the summary by Wyman who was a pioneer and remains a leader in the field. My one caveat is the fear that for huge assemblies such as the haemocyanins the analysis into individual constants will fail as the number of constants is too large. Haemoglobin has proved difficult enough. Again it may be that in haemocyanins we are dealing with several polymeric species $(Hm)_x$ where x could be numbers such as 10, 20, 30 and 40. The structures of these units are different so that the individual sites of copper pairs could change in their physicochemical properties with composition. I do not see how thorough experimental analysis of such a system is possible. It may well be possible, however, to describe the equilibrium for the x = 10, 20, 30, 40 units separately but we shall fail to describe the internal workings of these polymers. I therefore wonder if we may not be forced to revert to a qualitative bulk description and not a molecular one in these cases. It may be that a (mini) phase rule study would be a better approach.

Physiology. It is not yet the custom for chemists to involve themselves with physiological problems but I believe the time is close when this will no longer be true. The haemocyanins may present a favourable case. With the advent of high power methods for following individual elements in very small volumes, e.g., using scanning electron microscopes with analytical devices, it becomes possible to visualise storage regions for elements. Following copper in the whole biological systems is then possible. This conference did not hear of much of this type of work and the contributions in physiology are numerically slight. I hope that this will change in the course of the next few years.

For those who are interested in copper proteins and not just haemocyanin there is also a problem of the general handling of copper by biological systems.

Hemocyanin as a Protein

Heterogeneity of *Panulirus interruptus* Hemocyanin

A. A. VAN DEN BERG, W. GAASTRA, AND H. A. KUIPER

Abstract

Panulirus interruptus hemocyanin is heterogeneous both at the level of undissociated and dissociated protein. It is possible to fractionate dissociated protein into three components with different chromatographic and electrophoretic properties. Investigations on the structural differences probably cause the observed heterogeneity.

Introduction

Hemocyanins show differences in electrophoretic behavior for both the undissociated and dissociated molecules (Wood et al., 1968; Siezen and van Driel, 1973; Carpenter and van Holde, 1973; Loehr and Mason, 1973; Murray and Jeffrey, 1974; Sugita and Sekiguchi, 1975). Sullivan et al. (1974) separated native hemocyanin from *Limulus polyphemus* into at least five different fractions by DEAE-Sephadex chromatography. The amino acid composition, the oxygen equilibrium and the kinetics of oxygen dissociation of the fractions were different (Bonaventura et al., 1975). *P. interruptus* hemocyanin exhibits heterogeneity in electrophoretic mobility in the presence of sodium dodecyl sulphate (Kuiper et al., 1975). Preliminary experiments by Bonaventura (personal communication) showed for the monomers of *Panulirus* hemocyanin chromatographic heterogeneity on DEAE-Sephadex which is also reflected in differences in electrophoretic mobility. This study reports the separation of native subunits into two distinct fractions. We also report a series of investigations characterizing the fractionated protein. Furthermore, electrophoretic heterogeneity of undissociated protein was detected.

Isolation, Fractionation, and Electrophoresis of *Panulirus interruptus* Hemocyanin

Hemolymph of *P. interruptus* was collected and obtained as described by Kuiper et al. (1975). In order to remove clotted material, the hemolymph was centrifuged for 20 min at 1370 g. For removal of a contaminating carotenoid a partial ammonium sulphate fractionation was carried out (Kuiper et al., 1975). After this treatment the hemolymph was dialyzed against 0.052 M Tris-glycine buffer pH 8.9 containing 10 mM EDTA. Under

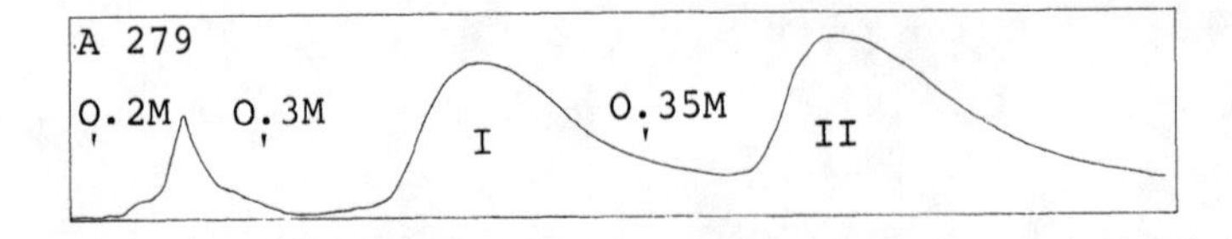

Fig. 1. Uvicord pattern of fractionation of *Panulirus* hemocyanin on a DEAE-Sephadex A-50 column (50 × 3.6 cm), equilibrated in 0.052 M Tris-glycine buffer pH 8.9 containing 10 mM EDTA. The sodium chloride concentration was raised stepwise as in figure. Fraction I was eluted at 0.3 M, fraction II at 0.35 M sodium chloride

these circumstances hemocyanin is present in its monomeric form (Fig. 1). The column was eluted stepwise; at 0.2 M NaCl some remaining carotenoid was eluted, at 0.3 M NaCl hemocyanin fraction I and at 0.35 M NaCl fraction II were eluted.

Polyacrylamide gel electrophoresis, both with and without sodium dodecyl sulphate, (Shapiro et al., 1967; Weber and Osborn, 1969) showed for unfractionated hemocyanin three bands of roughly equal intensity (Fig. 2). Fraction I showed one band corresponding to the middle band of the

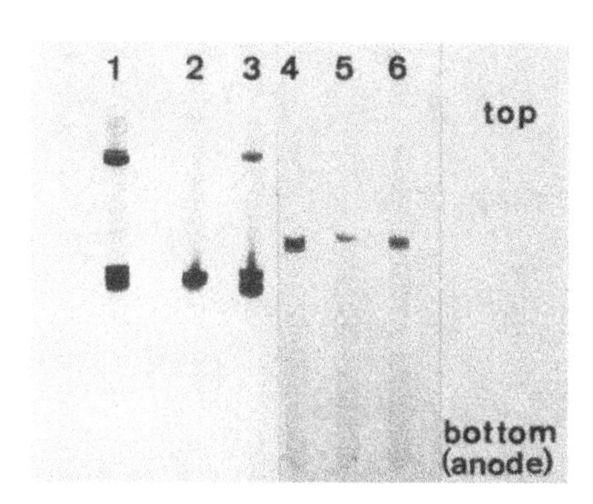

Fig. 2. Polyacrylamide gel electrophoresis of *Panulirus* hemocyanin. Experimental conditions in absence of detergent: 0.052 M Tris-glycine buffer pH 8.9. *1*: unfractionated hemocyanin; *2* and *3*: fraction I and II, respectively (see Fig. 1). In the upper parts of gels, material was often present, which may have been due to aggregation of protein in sucrose. Experimental conditions in presence of detergent: 0.02 M Tris-acetate buffer pH 8.0, ionic strength 0.05 containing 0.1% sodium dodecyl sulphate. *4*: unfractionated hemocyanin; *5* and *6*: fraction I and II, respectively (see Fig. 1)

original mixture, while the bands of fraction II correspond to the outer bands (Fig. 2). The components of fraction II (II^a and II^b) were separated by preparative gel electrophoresis as shown in Figures 3 and 4.

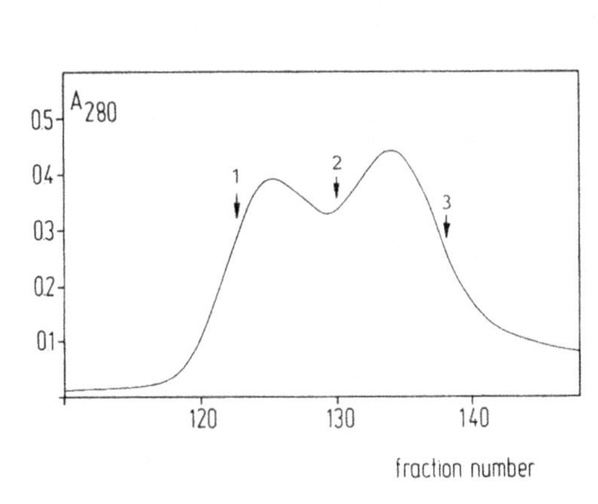

Fig. 3

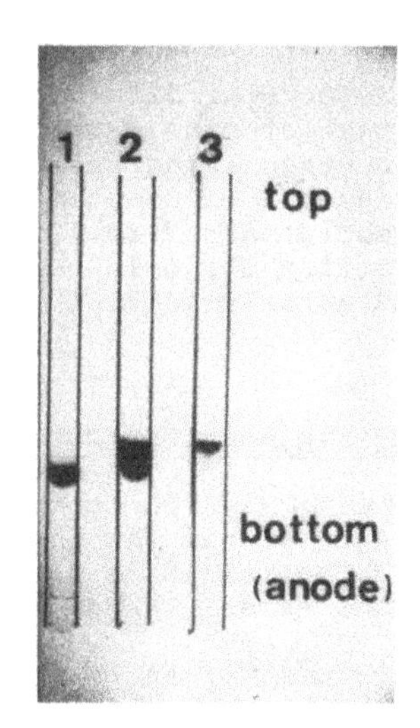

Fig. 4

Fig. 3. Elution profile of preparative polyacrylamide gel electrophoresis of fraction II. 50 mg were subjected to electrophoresis on a 5.5% gel (5 × 15 cm) in 0.052 Tris-glycine buffer pH 8.9 at 100 V and 30 mA for 20 h. Elution was performed at a rate of 30 ml/h with the same buffer. Fractions of 3 ml were collected

Fig. 4. Analysis of elution profile shown in Figure 3 by analytical polyacrylamide gel electrophoresis. *Numbers* correspond to those in Figure 3

The hemocyanin fractions were stored at -18°C after lyophilization in the presence of sucrose at a weight ratio 1:2.5. Before further use the sucrose was removed by dialysis.

Native undissociated hemocyanin, too, was heterogeneous. Three components in different quantities were observed. Reassociated unfractionated hemocyanin (dialysis from 0.052 M Tris-glycine, pH 8.9 to 0.05 M Tris-HCl, pH 7.6 containing 10 mM $CaCl_2$) showed a different electrophoretic pattern, consisting of two bands of equal intensity and a third, less

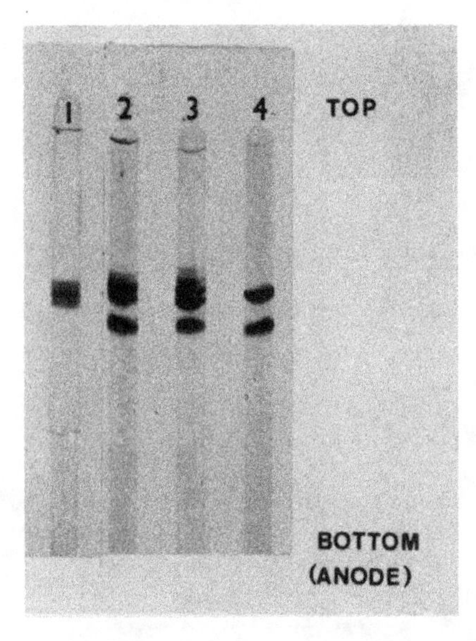

Fig. 5. Polyacrylamide gel electrophoresis of undissociated and reassociated hemocyanin. Experimental conditions: 0.025 M Tris-HCl containing 10 mM $CaCl_2$, ionic strength 0.055, pH 7.6. *1*: undissociated protein; *2*: reassociated unfractionated protein; *3*: reassociated protein from fraction II, *4*: reassociated protein from fraction I. For the reassociation experiments bovine serum albumin was used as an internal standard (protein bands with the highest mobility)

intense one (Fig. 5). This may be caused by the fact that approximately 85% of the material gave rise to whole molecules, while 15% remained as 5S components, as was determined by sedimentation studies. Hemocyanin, reassociated from fraction I was homogeneous. The fact that in the case of reassociated unfractionated hemocyanin and hemocyanin reassociated from Fraction II, only three to four bands were observed, indicates that hexameric molecules of certain compositions are preferentially formed.

Possible Explanations for the Observed Heterogeneity

Genetic Variability. The observed heterogeneity is not due to genetic variability within the population, since hemocyanin isolated from a single animal exhibits the same heterogeneity upon chromatography and electrophoresis.

Amino Acid Analysis. Amino acid analysis of the two fractions obtained from *Panulirus* hemocyanin gave results identical to the results shown in Table 1. From these results it is already clear that the differences between the fractions I and II of *Panulirus* hemocyanin are not as pronounced as found with other hemocyanins (i.e. *Limulus* hemocyanin. Bonaventura et al., 1975).

Automatic Edman Degradation. Eleven steps of Edman degradation with a Beckman 890 C automatic sequence analyzer of both fractions unequivocally showed one amino acid at each position (Fig. 6).

Table 1. Amino acid composition of hemocyanin from *P. interruptus*

Amino acid	Weight %	Amino acid residue 75,000 MW
Aspartic acid	14.52	94.5
Threonine	4.61	34.2
Serine	4.07	35.1
Glutamic acid	11.64	67.5
Proline	3.89	30
Glycine	3.64	47.7
Alanine	3.32	35.1
Half-cystine	0.72	5.4
Valine	5.03	38.1
Methionine	2.78	15.9
Isoleucine	5.43	36
Leucine	8.00	53.1
Tyrosine	6.18	28.5
Phenylalanine	7.57	38.7
Lysine	5.14	30.9
Histidine	7.31	39.9
Arginine	6.85	33

These data are average values from 3 hydrolysis times: 24, 48 and 72 h. The values are normalized for the 48 h hydrolysis. Threonine and serine values are extrapolated to zero time. Valine and isoleucine values are given for the 72-h hydrolysis.

Asp-Ala-Leu-Gly-Thr-Gly-Asn-Ala-Gln-Lys-Gln-
5 10

Fig. 6. The NH_2-terminal sequence of *Panulirus* hemocyanin

Carbohydrate Analysis. Since *Panulirus* hemocyanin is known to be a glycoprotein (Kuiper et al., 1975), one of the possibilities for the observed heterogeneity is a difference in carbohydrate content of the fractions. The total amount of carbohydrate is approximately 1% (w/w). Table 2 shows the results of a carbohydrate analysis of fraction I and II, as

Table 2. Fraction I contained three peaks which have not been identified yet. The molecular weight of hemocyanin is 450,000

Sugar (mol/mol hemocyanin)	Fraction I	Fraction II
Mannose	22.6	22.7
N-acetyl-glucosamine	9.7	8.6
Fucose	-	2.5
Glucose	(13.8)	(5.0)

determined by gas chromatography after methanolysis of the protein (Kamerling et al., 1975). Since the carbohydrate analysis of both fractions gives similar results and since no charged sugar residues are present, a different carbohydrate content or composition is not likely to cause the observed heterogeneity. The differences in glucose content is probably due to incomplete removal of sucrose during dialysis. In the gas-chromatogram of fraction I three peaks with retention times not corresponding to those of any known sugar were found. According to the positions of these peaks it is very likely that they are due to lipid material.

Ouchterlony Immunodiffusion. Ouchterlony immunodiffusion tests (Welling et al., 1976) using antiserum against fraction I, gave no differences between the fractions I, II and II[a] under circumstances of associated or dissociated protein. No spur formation was observed (Fig. 7). This probably means, as found in similar studies with other, though smaller proteins, that the possible difference in amino acid sequence is less then 10% (Prager and Wilson, 1971; Welling et al., 1976).

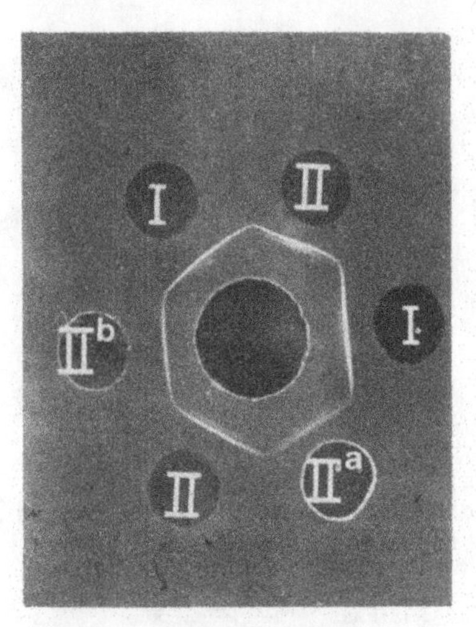

Fig. 7. Immunodiffusion study in which undiluted rabbit antiserum directed against fraction I was developed against fraction I, fraction II from the DEAE-Sephadex chromatography and components II[a] and II[b] from fraction II, isolated by preparative gel electrophoresis

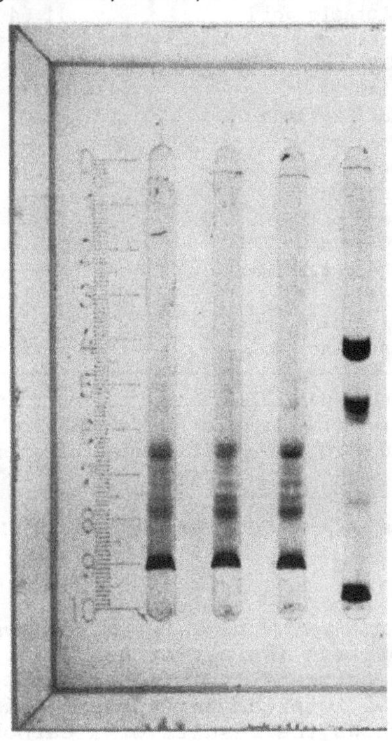

Fig. 8

Fig. 8. Polyacrylamide gel electrophoresis in presence of sodium dodecyl sulphate of fragments obtained after cleavage with cyanogen bromide. *1*: unfractionated hemocyanin; *2*: fraction I; *3*: fraction II; *4*: reference proteins: bovine serum albumin, ovalbumin, chymotrypsinogen A, cytochrome c

Cyanogen Bromide Fragments. Polyacrylamide gel electrophoresis in the presence of sodium dodecyl sulphate of fragments obtained after cyanogen bromide cleavage, revealed no differences between fraction I and II and the unfractionated hemocyanin (Fig. 8). About eleven protein bands were observed with molecular weights varying from 15,000 to 45,000, indicating that the cleavage had been incomplete. Since the protein contains sixteen methionine residues, seventeen fragments with an average MW of 4500 were expected. The cyanogen bromide cleavage was performed in 0.1 N HCl instead of 70% formic acid, since incubation of hemocyanin in 70% formic acid gives rise to many aspecific cleavages.

Isolation of Peptides

Isolation of peptides from *Panulirus* hemocyanin proceeds with many complications. We have tried to isolate tryptic peptides from citraconylated hemocyanin. Chromatography of these peptides on DEAE-cellulose (DE-32 Whatman) after decitraconylation always gives one broad peak at the end of the salt gradient system even when buffers with 8 M urea or 6 M guanidine-HCl are used. This is probably due to incomplete removal of the citraconyl groups.

At the moment we are trying to isolate cyanogen bromide and tryptic fragments from carboxymethylated hemocyanin, but until now no pure peptides could be obtained because of strong aggregation of the peptides in the solvents used for gel filtration and chromatography.

Acknowledgements. We want to thank Drs. G. Gerwig, B.L. Schut and J.F.G. Vliegenthart from the Laboratory for Organic Chemistry, Utrecht, for performing the carbohydrate analysis and Dr. C.H. Monfoort from the laboratory for Physiological Chemistry, Utrecht, for the determination of the amino acid compositions. We thank Dr. G.W. Welling for preparing antiserum, and are much indebted to Mr. H. Assink, Mr. H.J.J. Blik, Mr. F.v.d. Graaf and Mr. J. v.d. Laan for contributing to this study. We would also like to thank Dr. J. Bonaventura from the Duke University Marine Laboratory at Beaufort, North Carolina, U.S.A. for helpful advice. Part of this work has been carried out under auspices of the Netherlands Foundation for Chemical Research (S.O.N.) with financial aid from the Netherlands Organization for the Advancement of Pure Research (Z.W.O.).

References

Bonaventura, J., Bonaventura, C., Sullivan, B.: Hemoglobins and hemocyanins: Comparative aspects of structure and function. J. Exptl. Zool. 194, 155-174 (1975)

Carpenter, D.E., van Holde, K.E.: Amino Acid composition, amino-terminal analysis, and subunit structure of *Cancer magister* hemocyanin. Biochemistry 12, 2231-2238 (1973)

Kamerling, J.P., Gerwig, G.J., Vliegenthart, J.F.G., Clamp, J.R.: Characterization by gas-liquid chromatography - mass spectrometry and proton-magnetic resonance spectroscopy of pentrimethylsilyl methyl glycosides obtained in the methanolysis of glycoproteins and glycopeptides. Biochem. J. 151, 491-495 (1975)

Kuiper, H.A., Gaastra, W., Beintema, J.J., Van Bruggen, E.F.J., Schepman, A.M.H., Drenth, J.: Subunit composition, X-ray diffraction, amino acid analysis and oxygen binding behaviour of *Panulirus interruptus* hemocyanin. J. Mol. Biol. 99, 619-629 (1975)

Loehr, J.S., Mason, H.S.: Dimorphism of *Cancer Magister* hemocyanin subunits. Biochem. Biophys. Res. Commun. 51, 741-745 (1973)

Murray, A.C., Jeffrey, P.D.: Hemocyanin from the Australian fresh water crayfish *Cherax destructor*. Subunit heterogeneity. Biochemistry 13, 3667-3671 (1974)

Prager, E.M., Wilson, A.C.: The dependence of immunological cross-reactivity upon sequence resemblance among lysozymes. J. Biol. Chem. 246, 7010-7017 (1971)

Shapiro, A.L., Vinuela, E., Maizal, S.V.: Molecular weight estimation of polypeptide chains by electrophoresis in SDS polyacrylamide gels. Biochem. Biophys. Res. Commun. 28, 815-820 (1967)

Siezen, R.J., van Driel, R.: Structure and properties of hemocyanins VIII. Microheterogeneity of α-hemocyanin of *Helix pomatia*. Biochem. Biophys. Acta 295, 131-139 (1973)

Sugita, H., Sekiguchi, K.: Heterogeneity of the minimum functional unit of hemocyanins from the spider (*Argiope bruennichii*), the scropion (*Heterometrus* sp.), and the horseshoe crab (*Tachypleus tridentatus*). J. Biochem. 78, 713-718 (1975)

Sullivan, B., Bonaventura, J., Bonaventura, C.: Functional differences in the multiple hemocyanins of the horseshoe crab, *Limulus polyphemus* L. Proc. Natl. Acad. Sci. 71, 2558-2562 (1974)

Weber, K., Osborn, M.: The reliability of molecular weight determinations by dodecyl sulfate-polyacrylamide gel electrophoresis. J. Biol. Chem. 244, 4406-4412 (1969)

Welling, G.W., Groen, G., Beintema, J.J., Emmens, M., Schroder, F.P.: Immunologic comparison of pancreatic ribonucleases. Immunochemistry (1976) (in press)

Wood, E.J., Salisbury, C.M., Formosa, N., Bannister, W.H.: An electrophoretic and immunologic study of *Murex trunculus* hemocyanin. Comp. Biochem. Physiol. 26, 345-351 (1968)

Structure, Dissociation and Reassembly of *Limulus polyphemus* Hemocyanin

W. G. SCHUTTER, E. F. J. VAN BRUGGEN, J. BONAVENTURA, C. BONAVENTURA, AND B. SULLIVAN

Introduction

Hemocyanin of the Xiphosuran horseshoe crab, *Limulus polyphemus*, occurs as a 3.3 × 10^6 dalton protein (van Holde and van Bruggen, 1971). The native multimer can be dissociated into a mixture of at least five subunit fractions whose functional properties differ from one another (Sullivan et al., 1974). The molecular weight of these subunits appears to be 66-70 × 10^3 and each appears to be an unique polypeptide (Bonaventura et al., 1975; Sullivan et al., 1976; Bonaventura et al., 1976).

This paper is a preliminary report on an electron microscopic study of *L. polyphemus* hemocyanin. We studied the intact 60S structure, its pH dependent dissociation and the influence of NaCl and EDTA on this dissociation, its subunits (fractions I to V respectively) and the reassembly products from separate subunit fractions and mixtures of fractions.

We found that none of the separated subunit fractions are capable of forming an intact 60S structure by themselves. They rather appear to act in concert to reassemble the original 48-meric structure.

Materials and Methods

L. polyphemus hemocyanin and its subunits were prepared as described previously (Sullivan et al., 1974).

The influence of pH on the 60S structure was studied by diluting the blood 200 times with 0.05 M Tris-HCl buffer of pH 5.5, 6.5, 7.5, 8.5 and 9.5 respectively, eventually in the presence of 1 M NaCl and/or 0.01 M EDTA. Observations were made after 15 min standing at 20°C.

Reassembly from subunit fractions or their mixtures was initially performed by dialysis during at least 48 h at 20°C versus a solution containing 0.5 M NaCl and 5 mM $CaCl_2$ buffered by 0.02 M Tris/HCl at pH 7.0. Later, better results were obtained with a two-step procedure, firstly dialysis during 48 h at 20°C versus "EDTA-buffer" (= 0.01 M EDTA and 0.02 M Tris-HCl, pH 7.0) followed by dialysis during 48 h at 20°C versus "Ca-buffer" (= 0.01 M $CaCl_2$ and 0.02 M Tris-HCl, pH 7.0).

Electron microscopy of single molecules, dissociation intermediates, subunits and reassembly products was done using negative staining methods as described earlier (Siezen and van Bruggen, 1974).

Results and Discussion

Appearance of 60S Structure. The intact 50S structure (Fig. 1a) measures about 250 Å across. The different molecular profiles are very characteristic and were described earlier (van Bruggen, 1964; Wibo, 1966). The molecules are observed as (sometimes slightly pentagonal) circles,

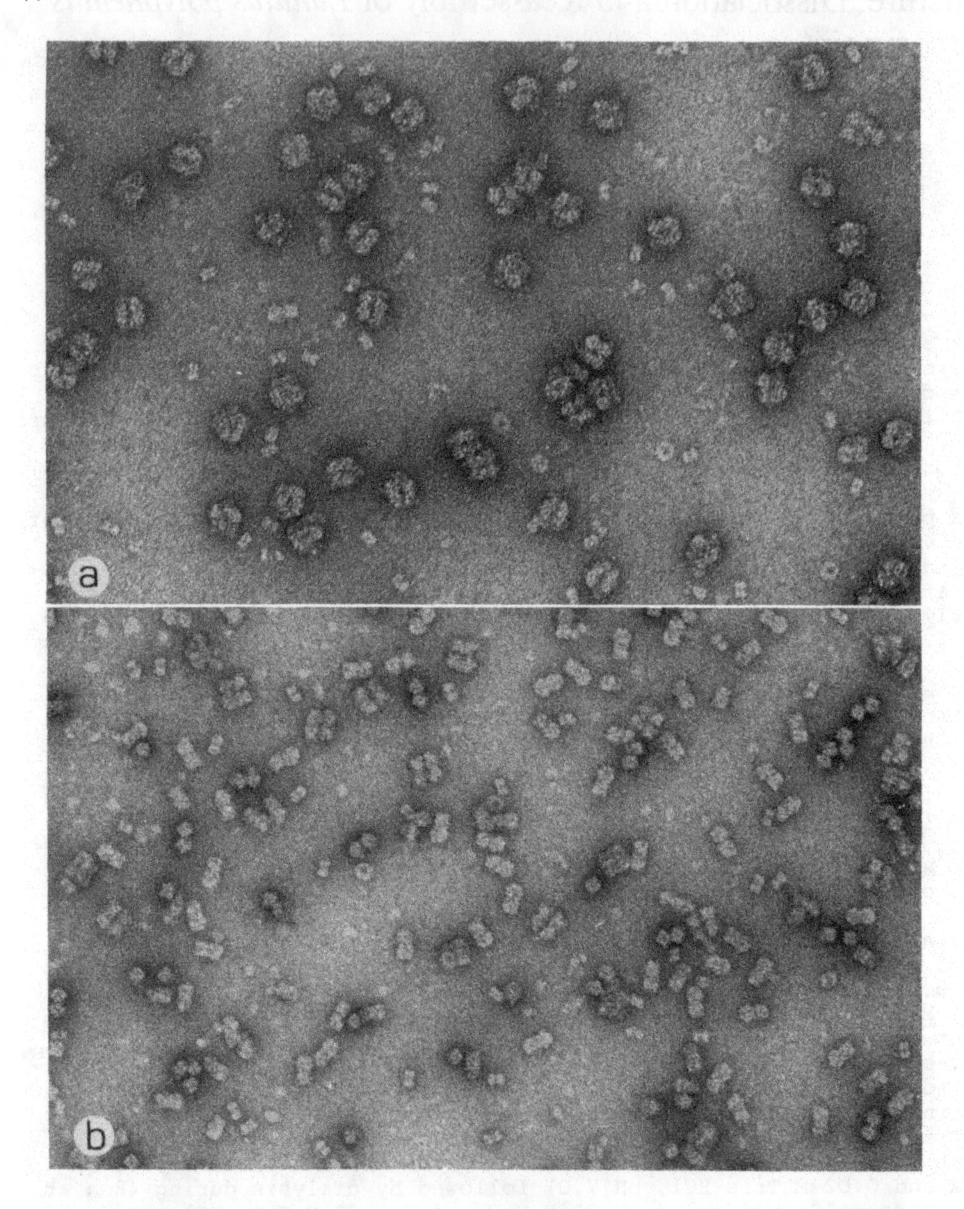

Fig. 1. (a) 60S structure of *L. polyphemus* hemocyanin, (b) different dissociation intermediates

as squares with a gap parallel to one of the sides or as squares with a substructure of four smaller squares.

In addition, *L. polyphemus* blood contains hemagglutinin molecules (Marchalonis and Edelman, 1968; Fernandez-Moran et al., 1968) that are observed as 115 Å circles and as 115 × 60 Å rectangles. We want to remark that during the separation of the subunits we always obtain a fraction with no extinction at 340 nm (Sullivan et al., 1974). We found recently that this fraction consists mainly of these ringshaped hemagglutinin structures.

Dissociation of the 60S Structure. An electron micrograph of different dissociation intermediates is presented on Figure 1b. The 16S units are observed as hexagonal or square profiles measuring about 115 Å across. Dimers of 16S units are seen as a combination of a hexagon and a square with one side in common. Two of these dimers in a probably antiparallel orientation form the very characteristic "bridged" tetramer of four 16S units. Dimeric structures of two 115 Å-squares with one corner in common are interpreted as the side-view of a bridged tetramer.

The results of the pH-dependent dissociation and the influence of 1 M NaCl or 0.01 M EDTA on this dissociation are summarized in Table 1.

Table 1. pH-dependent dissociation of *L. polyphemus* hemocyanin; influence of 1 M NaCl or 0.01 M EDTA on this situation

0.05 M Tris-HCl	200 × diluted blood	+ 1 M NaCl	+ 0.01 M EDTA
pH 5.5	8	8 (4) (2) (1)	(4) 2 1
6.5	8	8	(6) 4 (2)
7.5	8 (4)	8	4 (2)
8.5	(8) 4 (2) (1)	8	0
9.5	(6) 4 2 1	8 (4)	0

The observations were made after 15 min standing at 20°C.
8 = octamer of 16 S units = 8 × 6-mer. 6 = hexamer of 16S units = 6 × 6-mer etc.
0 = 5S subunit = monomer. () = present in small amount

Dissociation normally occurs above pH 8.0. 1 M NaCl stabilizes, while 0.01 M EDTA promotes dissociation finally resulting at pH 9 in a complete transition into 5S subunits. This was already known as the best condition for "stripping" (Sullivan et al., 1974).

Structure of Subunits. The 5S subunits of stripped *L. polyphemus* blood are observed as rounded, triangular or rectangular profiles of about 65 Å diameter (Fig. 2a). Figures 2b-f present the electron microscopy of the separated fractions I to V. The fractionated subunits show the same profiles as their mixture. The subunits of fraction V have some tendency to aggregation.

Reassembly from Subunits. (a) Reassembly by 0.5 M NaCl, 5 mM $CaCl_2$ and 0.02 M Tris-HCl, pH 7.0. Reassembly of an unfractionated mixture of 5S subunits, obtained by stripping, produces about 20% 60S structures, about 60% tetramers of 16S units and the remainder as dimers and smaller submultiples (Fig. 3a). The reassembled 60S structures are morphologically indistinguishable from undissociated 60S molecules.

Roughly the same result was obtained with an artificial mixture of the fractions I + II + III + IV + V (Fig. 4a).

Reassembly experiments with separate fractions are shown on Figures 3b-f, while the results obtained with artificial one-fraction-missing mixtures are presented on Figure 4b-f.

Table 2 summarizes all the reassambly experiments. At this stage the following conclusions can be drawn:

1. None of the separate subunit fractions are capable of forming an intact 60S structure on its own. Reassembly to the level of 16S units is possible with fraction II and fraction III.

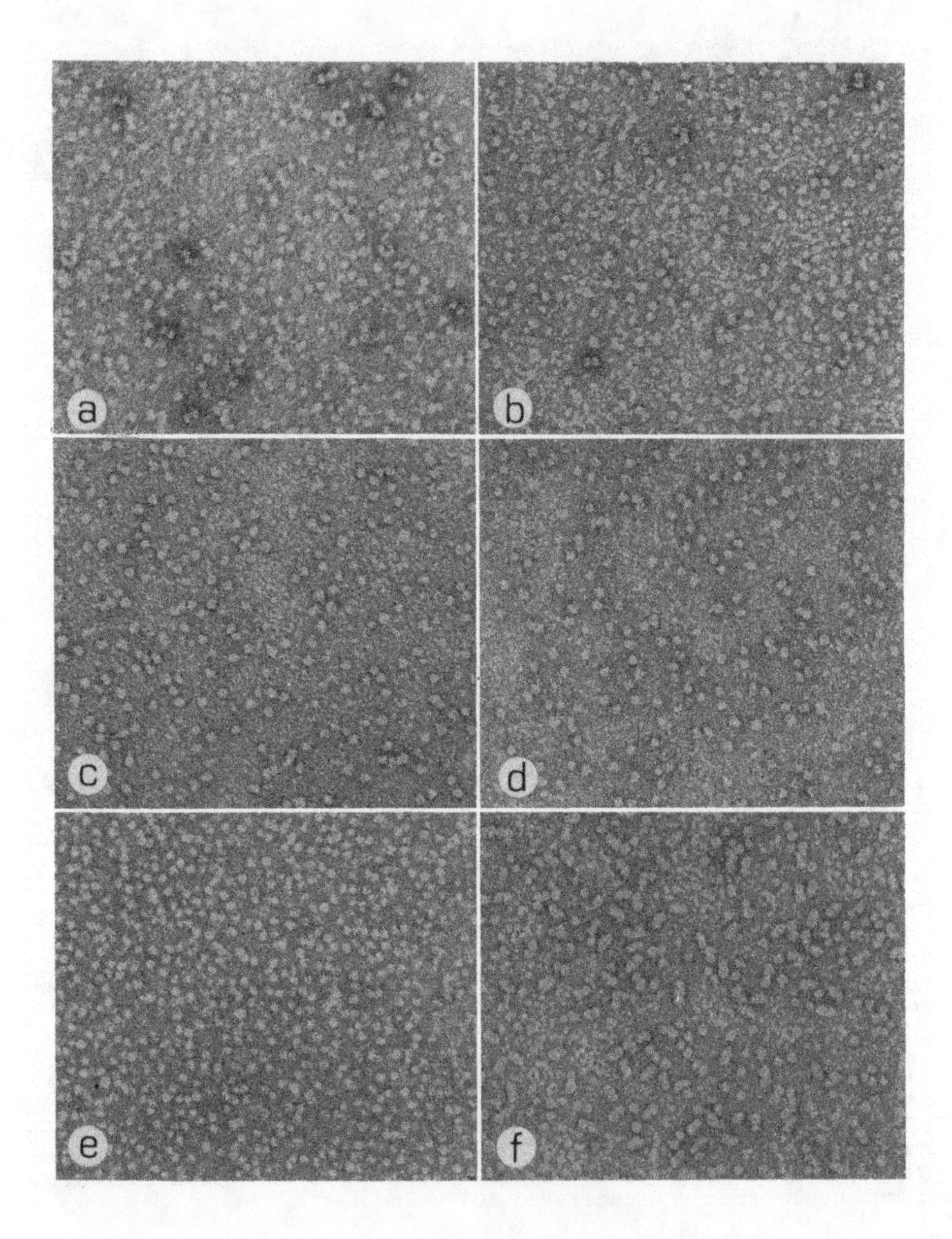

Fig. 2. (a) Unfractionated 5S subunits of stripped *L. polyphemus* hemocyanin. (b) 5S subunits of fraction I, (c) 5S subunits of fraction II, (d) 5S subunits of fraction III, (e) 5S subunits of fraction IV, (f) 5S subunits of fraction V

2. Partial reassembly to intact 60S structures occurs with unfractionated subunits or with their total artificial mixture or with a mixture of fractions II + III + IV + V.

3. Absence of fraction III or IV prevents aggregation above the level of tetramers of 16S.

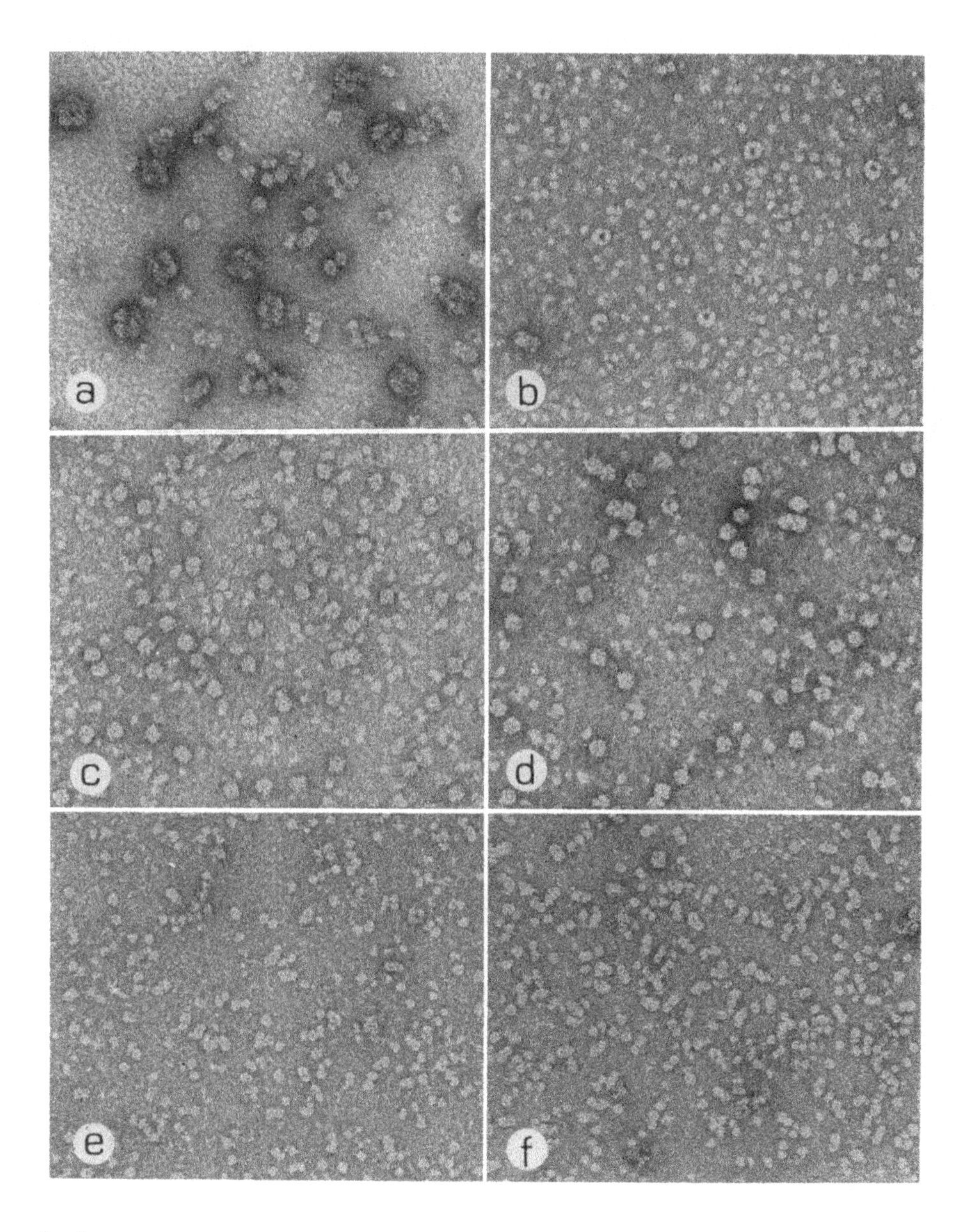

Fig. 3a-f. Reassembly products from 5S subunits by 0.5 M NaCl, 5 mM $CaCl_2$ and 0.02 M Tris-HCl, pH 7.0. (a) Reassembly of unfractionated mixture, (b) reassembly of fraction I, (c) reassembly of fraction II, (d) reassembly of fraction III, (e) reassembly of fraction IV, (f) reassembly of fraction V

4. Absence of fraction II or V stops aggregation at the level of 16S units.

Additional experiments, in which the assembly order of the fractions was varied, were difficult to interpret. It is clear, however, that fraction V plays an essential role in the formation of octamers of

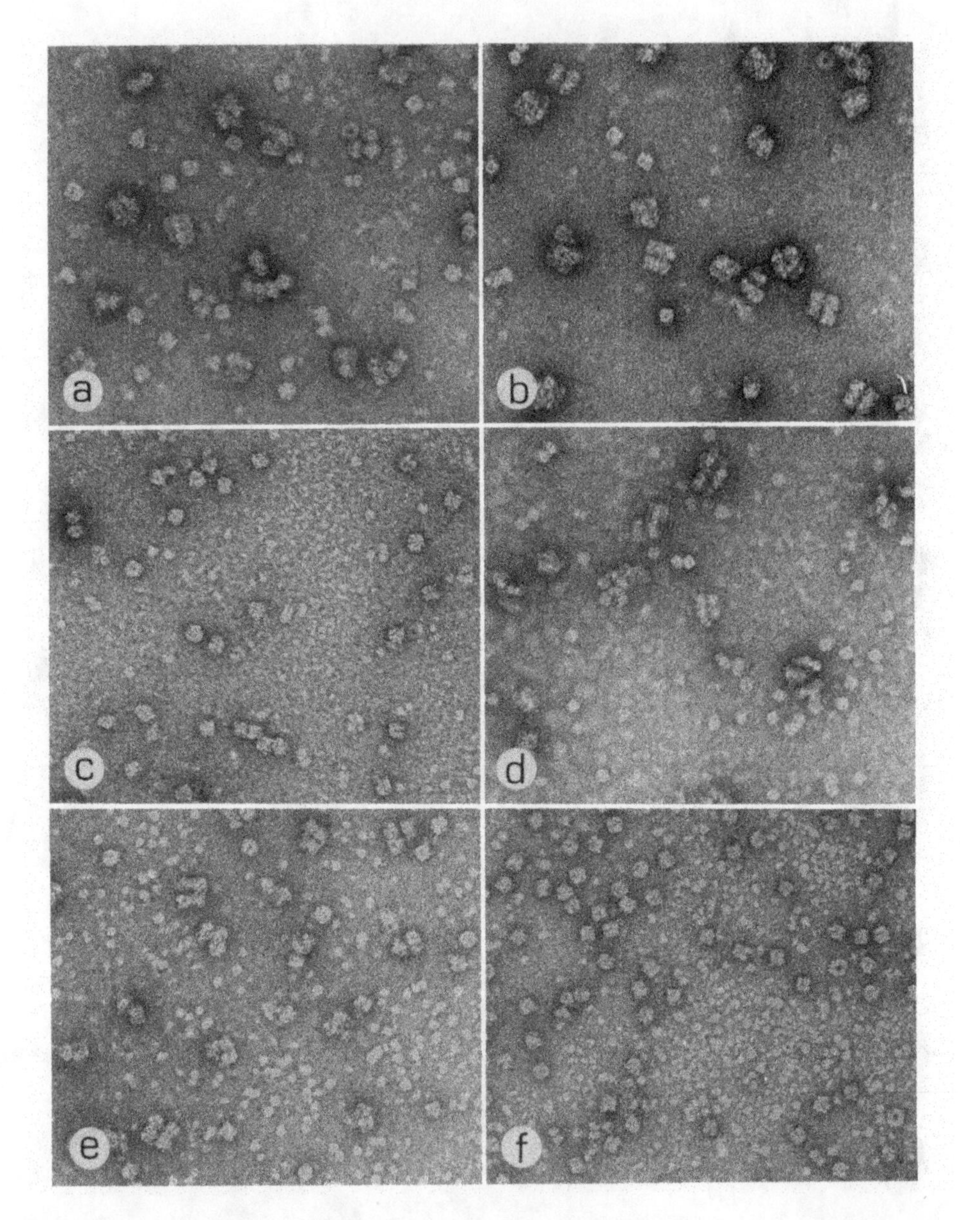

Fig. 4a-f. Reassembly products from different mixtures of the fractionated 5S subunits by 0.5 M NaCl, 5 mM $CaCl_2$ and 0.02 M Tris-HCl, pH 7.0. (a) Mixture of fractions I + II + III + IV + V, (b) mixture of fractions II + III + IV + V, (c) mixture of fractions I + III + IV + V, (d) mixture of fractions I + II + IV + V, (e) mixture of fractions I + II + III + V, (f) mixture of fractions I + II + III + IV

16S units. Further we found that fraction IV can also make 16S units provided the Ca^{2+}-concentration is raised above 5 mM.

(b) Two-step Reassembly by EDTA- and Ca-buffer. Reassembly of an unfractionated, freshly prepared mixture of 5S subunits by dialysis against EDTA-buffer produces almost quantitatively tetramers of 16S units (Fig. 5a). Se-

Table 2. Reassembly products from 5S subunits by 0.5 M NaCl, 5 mM $CaCl_2$ and 0.02 M Tris-HCl, pH 7.0

Reassembly of:	Reassembly		Products			
Unfractionated mixture	O		(1)	(2)	4	8
Fraction I	O	O-O				
" II	(O)		1			
" III	(O)		1			
" IV	O					
" V	O	O-O	(1)			
I + II + III + IV + V	O		1	(2)	4	8
II + III + IV + V	O		1	(2)	4	8
I + III + IV + V	(O)		1			
I + II + IV + V	O		1	(2)	4	
I + II + III + V	O		1	(2)	4	
I + II + III + IV	(O)		1			

8 = 8 × 6-mer; 6 = 6 × 6-mer etc.
O = 5S subunit; O-O = aggregate of 5S subunits; () = present in small amount.

quential dialysis versus Ca-buffer produces octamers of 16S units in a very high yield (Fig. 5b).

A preliminary study showed these molecules to be morphologically and functionally indistinguishable from the original 60S structure.

A Problem of Aging. The last experiments indicated that there occurs some "aging" with stripped hemocyanin during storage in the cold room. This aging makes the reassembly more difficult. We do not know yet the reason for it, neither did we look at its eventual reversibility.

Conclusion

Careful sedimentation velocity and equilibrium ultracentrifugation by Johnson (1973) indicate that the 60S *L. polyphemus* hemocyanin structure dissociates into well-defined units with the following molecular weights:

60S structure MW 3,320,000 ± 4%
36S structure MW 1,705,000 ± 1%
25S structure MW 852,000 ± 3%

Further we know the molecular weight of the 5S subunits to be about 70,000.

Together with our morphological knowledge this brings us to the following model:

1. The 5S structure is a roughly spherical protein particle with a diameter of about 65 Å. There exist at least five and probably more different types of these particles.

2. The 16S structure is a hexamer built from six 5S particles. The structure is observed as a hexagonal or square profile with a diameter of about 115 Å. The relationship with other Arthropod hemocyanins makes a trigonal antiprismatic arrangement (pointgroup symmetry 32) for the 5S subunits within the 16S unit very probable (Schepman, 1975). We do not know how the different 5S components are distributed within one 16S unit. It is clear, however, from the weight ratios of the different components that there must be at least two different types of 16S units.

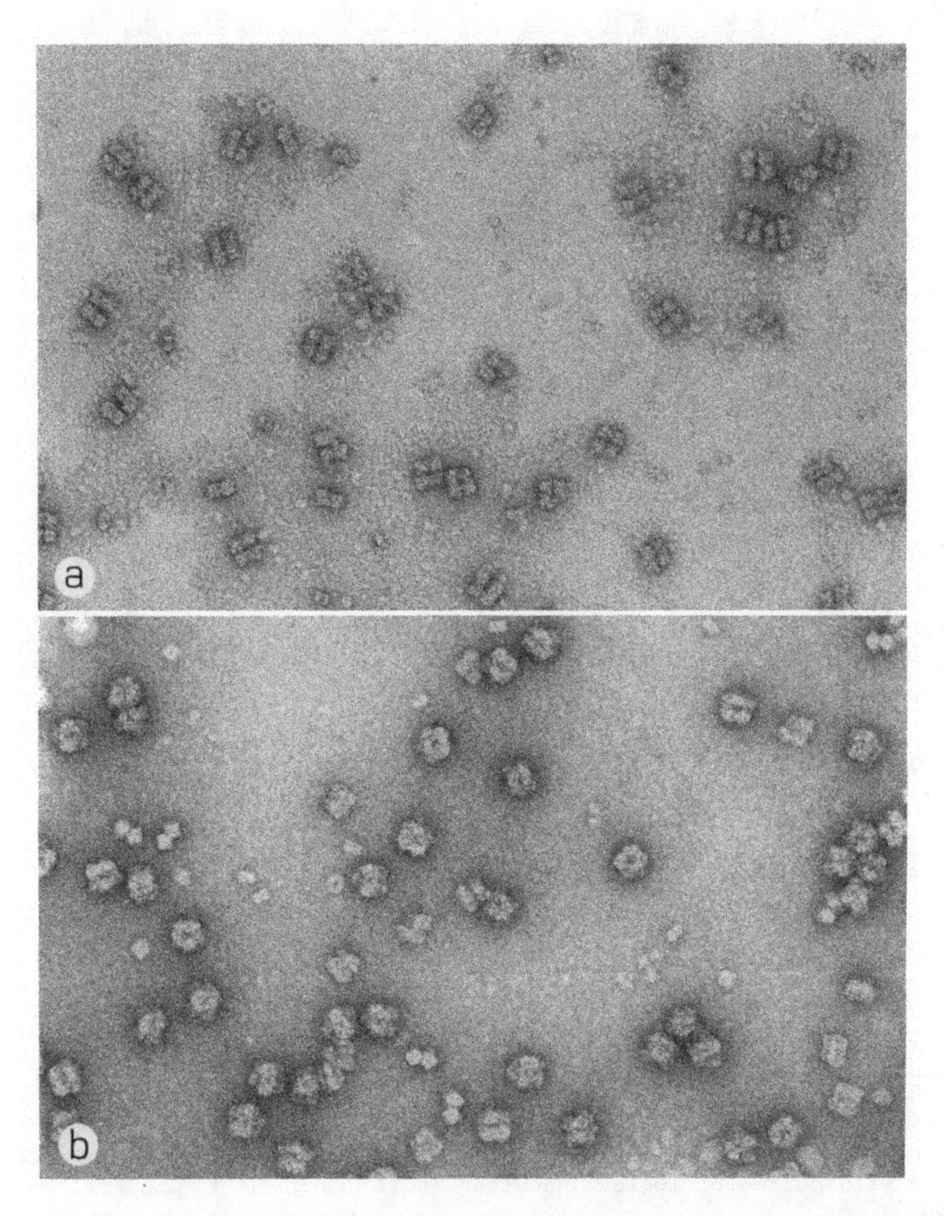

Fig. 5a and b. Two-step reassembly of an unfractionated, freshly prepared mixture of 5S subunits. (a) First step products after dialysis versus EDTA-buffer. (b) Second step products after sequential dialysis versus Ca-buffer

3. The 25S structure is assembled from two 16S structures in such a way that this projection shows a hexagonal and a square profile having one side in common.

4. The 36S structure is the "bridged" tetramer formed by the antiparallel linkage of two 25S structures.

5. The 60S structure is formed by the stacking of two bridged tetramers in a probably staggered arrangement. The forces between the tetramers within the 60S structure are stronger than the forces between 60S structures. This can be concluded from the observation of the 60S structure as a "closed", single molecule with no strong tendency for further stacking.

Experiments that are under way at this moment will hopefully give us an answer on the individual role played by each different 5S component in the assembly process. We already know about the essential role of fraction V and the Ca^{2+}-sensitivity of the hexamerization of fraction IV. However, more detailed information is needed to arrive at an unambiguous model. The paper represents paper number 3 on structure-function relationships in *Limulus* hemocyanin.

Acknowledgments. This work was supported by a grant from the United States National Science Foundation, the National Institute of Health, an EMBO-travel grant and by the Netherlands Foundation for Chemical Research (S.O.N.) with financial aid from the Netherlands Organization for the Advancement of Pure Research (Z.W.O.). J.B. is an established investigator of the American Heart Association.

We would like to thank Annelies van den Berg for her technical assistance, Klaas Gilissen for printing the monographs and Ans van Rijsbergen for typing the manuscript.

References

Bonaventura, J., Bonaventura, C., Sullivan, B.: Hemoglobins and hemocyanins: Comparative aspects of structure and function. J. Exptl. Zool. 194, 155-174 (1975)

Bonaventura, J., Bonaventura, C., Sullivan, B.: Non-heme oxygen transport proteins. In: Oxygen and Physiological Function. Jöbsis, F. (ed.). Dallas, Texas: Professional Information Library, 1976 (in press)

Bruggen, E.F.J. van: Electron microscopy of *Limulus polyphemus* hemocyanin. In: Proc. 3rd Europ. Reg. Conf. Elec. Micr. Titlbach, M. (ed.). Prague: Publ. House Czech. Acad. Sci., 1964, Vol. B, pp. 57-58

Fernandez-Moran, H., Marchalonis, J.J., Edelman, G.M.: Electron microscopy of a hemagglutinin from *Limulus polyphemus*. J. Mol. Biol. 32, 467-469 (1968)

Johnson, M.: Subunit Structure of *Limulus* hemocyanin. PhD. Thesis, Univ. Connecticut 1973

Marchalonis, J.J., Edelman, G.M.: Isolation and characterization of a hemagglutinin from *Limulus polyphemus*. J. Mol. Biol. 32, 453-465 (1968)

Schepman, A.M.H.: X-ray diffraction and electron microscopy of *Panulirus interruptus* hemocyanin. Ph.D. thesis, Groningen 1975

Siezen, R.J., Bruggen, E.F.J. van: Electron microscopy of dissociation products of *Helix pomatia* α-hemocyanin. Quaternary structure. J. Mol. Biol. 90, 77-89 (1974)

Sullivan, B., Bonaventura, J., Bonaventura, C.: Functional differences in the multiple hemocyanins of the horseshoe crab *Limulus polyphemus L.* Proc. Natl. Acad. Sci. 71, 2558-2562 (1974)

Sullivan, B., Bonaventura, J., Bonaventura, C.: Hemocyanin of the horseshoe crab *Limulus polyphemus*. I. Characterization of the isolated components. J. Biol. Chem. (1976) (in press)

Van Holde, K.E., Bruggen, E.F.J. van: The hemocyanins. In: Biological Macromolecules Series. Timasheff, S.N., Fasman, G.D. (eds.). New York: Marcel Dekker, 1971, Vol. V, pp. 1-53

Wibo, M.: Recherches sur les hemocyanines des Arthropodes: constantes de sedimentation et aspects morphologiques. Mémoire présenté au concours des bourses de voyage. 1966

Association Equilibria of *Callianassa* Hemocyanin

K. E. VAN HOLDE, D. BLAIR, N. ELDRED, AND F. ARISAKA

Introduction

The hemocyanins exhibit some of the most complex patterns of quaternary organization known among proteins. In arthropod hemocyanins at least, the individual polypeptide chains are fairly small (about 75,000 daltons). Yet they assemble, via a series of hierarchical levels of structural complexity, into objects which weigh, in some instances, millions of daltons (van Holde and van Bruggen, 1971). It has been recognized since Svedberg's time that hemocyanins could be caused to dissociate to various levels of subunit structure by modifications in the solution environment (Eriksson-Quensel and Svedberg, 1936). In some cases such dissociation appeared to be at least partially reversible. However, it has become apparent in more recent studies that many of these supposedly "reversible" dissociation processes are complicated in a peculiar way: they do not appear to follow the mass action law. In other words, a dilution or concentration of what is by other criteria an equilibrating mixture does not yield the expected shift in composition. Such behavior has been observed, by DiGiamberardino (1967), Konings et al. (1969), and Siezen and van Driel (1973).

The explanation advanced by a number of authors is based on an assumed microheterogeneity in the hemocyanin: there are believed to be multiple forms of the protein, each capable of an abrupt association-dissociation reaction in response to a change of, say, pH. The composition found at any pH then reflects not a normal equilibrium mixture, but a mixture of wholly associated and wholly dissociated microspecies. Quite strong evidence to support this hypothesis has been presented by Siezen and van Driel (1973).

There has recently been direct evidence for the postulated microheterogeneity. Sullivan et al. (1974), have described the separation of multiple polypeptide chain components from *Limulus* hemocyanin. Murray and Jeffrey (1974) report a somewhat similar situation in a crayfish hemocyanin. Loehr and Mason (1973) and Carpenter and van Holde (1973) have identified two polypeptide chain species in *Cancer magister* hemocyanin by SDS gel electrophoresis. Lamy et al. (1973) report multiple 5S components in scorpion hemocyanin.

The existence of microheterogeneity in dissociation behavior should, as Siezen and van Driel (1973) have pointed out, complicate or even vitiate any quantitative analysis of the thermodynamics of the dissociation association reaction. Until recently, only one example has been known of a hemocyanin in which such reactions may be wholly reversible. This is the lobster hemocyanin, studied by Kegeles and co-worker Morimoto (Morimoto and Kegeles, 1971). For this reason we were most interested to discover, some years ago, that the hemocyanin of the ghost shrimp, *Callianassa californiensis,* seemed to exhibit true reversibility in association and dissociation. Furthermore, this hemocyanin is somewhat unusual among arthropod hemocyanins in forming several orders of increasingly complex molecular assemblies (Miller and van Holde, 1974; Roxby et al., 1974).

We have continued studies of this hemocyanin, and shall review here published results (Roxby et al., 1974; Miller and van Holde, 1974; and Blair and van Holde, 1976), and a number of hitherto unpublished results. The aim of this paper will be to explore the data available on the *Callianassa* system, with regard to evidence for reversibility, microheterogeneity, and the nature of the subunit interactions. We preface further discussion with a brief review of the stoichoimetry and quaternary structure.

Stoichiometry and Quaternary Structure

By appropriate choice of solvent composition, pH, and temperature, it is possible to find conditions under which the various orders of structure exhibited by *Callianassa* hemocyanin are individually stabilized as homogeneous components. It is then possible, using such techniques as sedimentation equilibrium and SDS gel electrophoresis, to establish the molecular weights, and hence stoichiometry, of these structures. The results of such studies are summarized in Table 1. Most of these data are from Roxby et al. (1974) but some more recent results have been included. It can be seen that the *Callianassa* hemocyanin can exist in a series of aggregation states, representing successively higher levels of organization. We discuss the detailed properties of each in turn:

Table 1. Observed association states of *Callianassa* hemocyanin

Number of chains	$s^o_{20,w}$ (Svedbergs)	$M \times 10^{-5}$	Conditions for stability[a]
1	5[b]	0.72	pH ≥ 8.9, after EDTA treatment, OR: in SDS or 6M GuHCl
6 "monomer"	17.1	4.31	$6.5 \leq$ pH ≤ 8.9, $[M^{2+}]$ 0.01M
12 "dimer"	∿25[b]	8.62[c]	$4.5 \leq$ pH ≤ 5.5, OR: low temp., high concs. of M^{2+}, pH > 7.5
24 "tetramer"	38.8	17.2	pH > 7.5, in high concs. of M^{2+}, room temp.

[a]The pH ranges are approximate, since transitions are not completely sharp. M^{2+} refers to divalent ions (Mg^{2+} and Ca^{2+} have been tried, and appear roughly equivalent).
[b]These values are approximate; extrapolation to C = 0 has not been carried out.
[c]Assumed value. See text

The Polypeptide Chains. If the hemocyanin is subjected to denaturing conditions (SDS or Gu·HCl) or simply "stripped" of divalent ions (Sullivan et al. 1974) and raised to a pH ≥ 8.8, it will dissociate completely into individual polypeptide chains. These chains are apparently homogeneous, or nearly so, insofar as molecular weight is concerned, and each has a mass corresponding to the presence of one O_2 binding site or two Cu atoms (Roxby et al., 1974). The chains are, however, not homogeneous in polyacrylamide gel electrophoresis at pH 8.9, as shown in Figure 1 (Miller, K.I., Eldred, N.W., Arisaka, F., and van Holde, K.E., submitted for publication). At least six components can be clearly resolved in preparations from purified hemocyanin. Of the bands shown in Figure 1, at least five must contain O_2-binding sites, for scanning

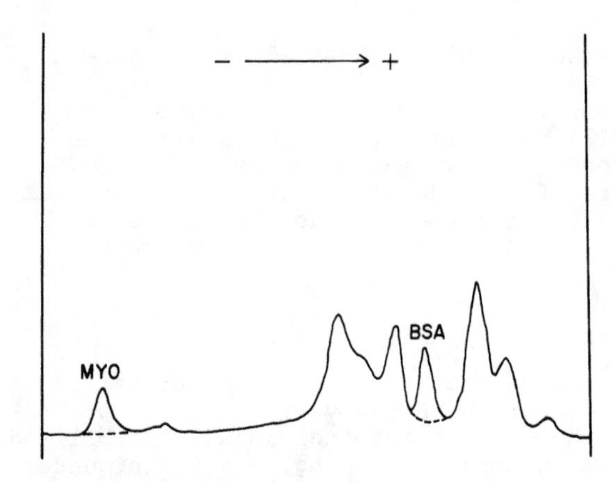

Fig. 1. Heterogeneity of the polypeptide chains of *Callianassa* hemocyanin. Electrophoresis is at pH 8.9, on 5% polyacrylamide. Bovine serum albumin (*BSA*) and myogolbin (*MYO*) have been added as markers. Gels were stained with amido black and scanned at 600 nm

of the gel at 338 nm detects all except the broad band to the extreme left. Whether this component does not contain an O_2 site, or this has been destroyed by the high pH and EDTA conditions is not yet determined. While it is clear that the *Callianassa* hemocyanin is microheterogeneous, the nature of this heterogeneity is not known, except that it must involve charge differences between the chains. The full set of bands is observed in samples taken from individual animals, but proportions of different components vary somewhat between individuals. Note also that there do not appear to be equimolar quantities of the different chains. This leads to the conclusion (see below) that the 17S particles cannot consist of one unique assembly of such chains.

The 17S Monomer. This particle must, by molecular weight measurements, contain six of the polypeptide chains (see Table 1). We refer to it as the monomer because it is the building block from which higher order structures are assembled; it is also the smallest component observed in the hemolymph. In fact, the hemolymph appears to contain two quite different kinds of monomer units. Most (about 85%) are capable of further association; the remaining 15% are incompetent in association, and can be separated by gel filtration (Roxby et al., 1974). All experiments described herein have been carried out with the "competent" fraction.

The variable and nonintegral stoichiometry of the polypeptide chains leads to the proposition that there must be a number of distinguishable 17S species. Polyacrylamide gel electrophoresis of the component hemocyanin, under conditions where the 17S monomers are stable, yields a profile with two major bands and at least three minor bands (N. Eldred, unpublished). The breadth of these bands suggests that the actual composition is far more complex, and further attempts at resolution are in progress.

In electron microscope photographs (van Bruggen and van Holde, unpublished), the *Callianassa* monomers give the usual "square" or "hexagonal" profiles commonly found for 17S arthropod hemocyanins (van Holde and van Bruggen, 1971).

The 25S Dimer. Components of about this sedimentation coefficient are commonly found in arthropod hemocyanins; in fact for many species this represents the highest aggregation state observed (van Holde and van Bruggen, 1971; see Fig. 5). These are invariably found, by molecular weight analysis or electron microscopy, to represent dimers of the 17S monomers. In our earlier experiments with *Callianassa* hemocyanin (Roxby et al., 1974) we found no direct evidence for such a component. The

17S particles appeared to associate directly to a 39S tetramer without detectable intermediates (see below). However, more comprehensive investigation of the association reactions has revealed two conditions under which a dimer appears to be a stable form: (1) In the presence of high concentration of divalent ions at low temperature (4°C) and neutral pH, much of the hemocyanin is, by sedimentation equilibrium measurements, present as dimer (Blair and van Holde, 1976). The association to tetramer, which is the favored reaction at room temperature, appears to be blocked at low temperatures. Under these conditions dimer formation requires the presence of divalent cations. (2) In the pH range between about 4.5 and 5.5, at room temperature, the hemocyanin is almost entirely in 25S dimeric form (see Fig., N. Eldred, unpublished). This association appears to be in some ways distinct from that described in (1) in that it is insensitive to divalent ions. We must consider the pssibility that the dimers found under conditions (1) and (2) may be conformationally distinct.

Structures appearing identical to dimers observed for other arthropod hemocyanins can also be found in electron micrographs of *Callianassa* hemocyanin (van Bruggen and van Holde, unpublished).

The 39S Tetramer. In the hemolymph, most of the hemocyanin is present in a form with $S^{o}_{20,w}$ = 38.8S. In fact, the "competent" hemocyanin is entirely in this form under these conditions. Molecular weight studies (see Table 1) reveal this to be a tetramer of the 17S monomer particles. In the electron microscope, these appear to be tetrahedral structures, quite unlike any other arthropod hemocyanins reported (van Bruggen and van Holde, unpublished).

Tetrameric arthropod hemocyanins have so far been observed, to our knowledge, only in the classes *Arachnida* and *Merostomata* , with the single exception of the crustacean, *Cancer pagurus* (Eriksson-Quensel and Svedberg, 1936). All of those examined to date by electron microscopy were found to be square planar structures (van Holde and van Bruggen, 1971). We had earlier (Roxby et al., 1974) postulated such a structure for the *Callianassa* hemocyanin, but this appears now to be in error. The 39S

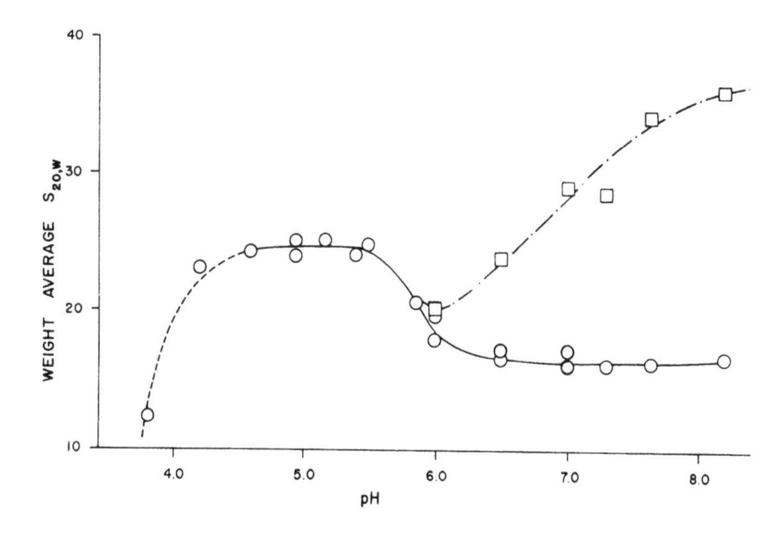

Fig. 2. Weight average sedimentation coefficient as a function of pH. Open circles (-o-): show results in absence of divalent ions; squares (-□-): in 0.5 M[Mg^{2+}]. All data at hemocyanin concentration of approximately 1 mg/ml, in 0.1 ionic strength buffers

tetramer seems to have an absolute requirement for divalent ions for stability. As can be seen in Figure 2, the addition of Mg^{2+} at pH values greater than 6.0 promotes the formation of this complex; it is in this pH range that the hemocyanin also binds divalent ions most strongly (F. Arisaka, unpublished).

Association-Dissociation Equilibria

Having catalogued the aggregation states of *Callianassa* hemocyanin, we turn now to experimental results concerning the equilibria between these states. In our first paper (Roxby et al., 1974) it was reported that the competent hemocyanin component was capable of repeated cycles of association to the 39S form and dissociation to the 17S form by changes in Mg^{2+} concentration. A more severe test of true reversibility, however, is adherance to the mass action law. Figure 3 represents a test of this behavior for the equilibrium under conditions in which

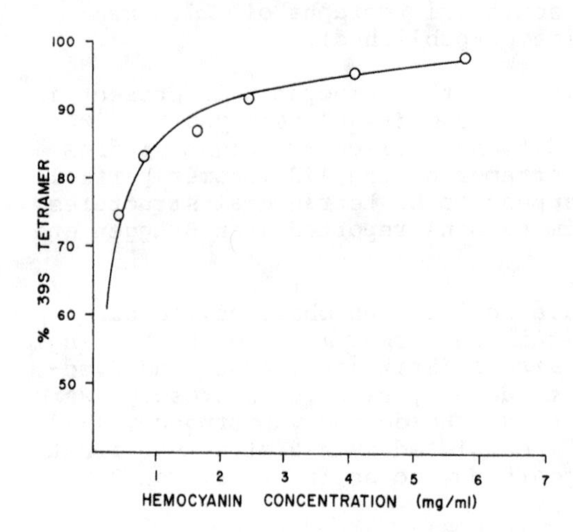

Fig. 3. A test of mass-action law for hemocyanin dissociation. Samples at pH 7.65, 50 mM Mg^{2+}, and 20°C were diluted to concentrations indicated. Sedimentation velocity experiments were carried out, and percentage 39S calculated from the weight average sedimentation coefficient, assuming only monomer and tetramer present. *Solid line:* theoretical curve for a monomer-tetramer equilibrium

only the 17S monomer and 39S tetramer are present in appreciable concentrations (see below). Dilution of the sample leads to partial dissociation, in accordance with the law of mass action.

A detailed analysis of the thermodynamics of this association reaction has been carried out by Blair and van Holde (1976). Using the method of sedimentation equilibrium, we have examined the association-dissociation behavior as a function of both magnesium ion concentration and temperature. It was found that under most conditions the data could be adequately described only if three species, monomer, dimer, and tetramer were postulated. Under no circumstances was the existence of a trimer species required to fit the data. This allowed the description of the equilibrium in terms of two apparent association constants:

$$K_{12}^{obs} = \frac{[\text{dimer}]}{[\text{monomer}]^2} \quad (1)$$

$$K_{14}^{obs} = \frac{[\text{tetramer}]}{[\text{monomer}]^4} \quad (2)$$

The effect of magnesium concentration was then included by postulating reactions:

$$2\ (\text{monomer}) + n\text{Mg}^{2+} \xrightarrow{K_{12}} (\text{dimer}) \quad (3)$$

$$4\ (\text{monomer}) + m\text{Mg}^{2+} \xrightarrow{K_{14}} (\text{tetramer}) \quad (4)$$

where K_{12} and K_{14} are the true equilibrium constants. One then has:

$$\log K_{12}^{obs} = \log K_{12} + n \log [\text{Mg}^{2+}] \quad (5)$$

$$\log K_{14}^{obs} = \log K_{14} + m \log [\text{Mg}^{2+}] \quad (6)$$

The data obtained from a number of sedimentation equilibrium experiments are summarized in Table 1. Linear graphs of log K^{obs} vs. log $[\text{Mg}^{2+}]$

Table 2. Dependence of apparent association constants on $[\text{Mg}^{2+}]$ and temperature[a]

	$[\text{Mg}^{2+}]$ (mol/L)	T ($^{\circ}$C)	K_{12}^{obs} (mol/L)$^{-1}$	K_{14}^{obs} (mol/L)$^{-3}$
A. Mg series				
	0.020	20	$\sim 10^4$	4.4×10^{15}
	0.025	20	2.3×10^5	7.3×10^{16}
	0.030	20	3.9×10^5	1.5×10^{18}
	0.035	20	6.5×10^5	5.5×10^{18}
	0.040	20	1.3×10^6	4.0×10^{19}
B. Temperature series				
	0.050	4	1.3×10^7	4.5×10^{20}
	0.050	10	1.2×10^7	1.0×10^{21}
	0.050	15	1.5×10^7	3.0×10^{21}
	0.050	30	---(b)	4.0×10^{22}

[a] Data taken from Blair and van Holde (1976), with correction of numerical errors (the values as given by Blair and van Holde are 4X too large for K_{14}, 2X too large for K_{12}).
[b] Not enough dimer present to allow calculation of K_{12}.

then allowed evaluation of n and m. The results are $n = 6.6$, $m = 12.7$. A very similar result has been obtained by Arisaka (unpublished) in sedimentation velocity studies of the monomer-tetramer reactions; he finds $m = 11$ from those experiments. Furthermore, a value $n = 5$ was found by Morimoto and Kegeles (1971) for the dimerization of lobster 17S hemocyanin. The similarity in behavior points to a common mechanism in the dimerization process.

The full significance of our results is more evident if the numbers n and m are placed on a per-monomer basis. Then in each case they indicate that about 3 Mg^{2+} are required per monomer. *The same amount (per monomer) is required to form the tetramer as to form the dimer.* In other words, the monomer→dimer step in the association is where Mg^{2+} is required; no additional Mg^{2+} is needed to further associate dimers into tetramers. Comparing this result with that shown in Figure 2, another conclusion can be drawn: the monomer-dimer association requires either divalent ions at pH > 6 or protonation of some groups with a pK_a of about 5.9. It is tempting to suggest that the same groups may be involved in both the protonation and Mg^{2+}-binding reactions. The situation here may be somewhat similar to that observed by Klarman et al. (1972) with the Ca^{2+}-dependent association of the hemocyanin of the mollusc, *Levantina hierosolima*. Further study is in progress.

If the dimer-tetramer transition is independent of divalent ions, upon what does it depend? We have found that this equilibrium is in fact very sensitive to temperature, with high temperatures favoring the tetrameric state. Table 2 also shows the temperature dependence for the equilibrium constants for tetramer formation and dimer formation, deduced from sedimentation equilibrium experiments at different temperatures (Blair and van Holde, 1976). Examination of these data shows another remarkable difference between the two reactions: While the values of K_{12} increase rapidly with increasing T, the values of K_{14} remain almost constant. Thus, the strong temperature dependence lies entirely in the dimer-tetramer step. The data in Table 2B give a linear van't Hoff graph, from which one can estimate the apparent enthalpy change for the formation of one mol of tetramer to be +30 kcal/mol. This positive value suggests that the dimer-tetramer step may be driven by hydrophobic interactions between dimers. Curiously, the dimers formed at low pH (see Fig. 2) do not seem to undergo this reaction. This may mean either that the low pH dimer is different in conformation, or that the protonation of groups at low pH blocks the hydrophobic association sites.

Table 3. Stability conditions for monomer, dimer and tetramer at pH 7.6

		$[Mg^{2+}]$	
		High	Low
Temperature:	High	T	M
	Low	D	M

This combination of different dependencies of the two steps in the reaction on divalent ions and temperature allows one to choose conditions (at pH 7.6) where any one of three components dominates. These are summarized in Table 3.

Discussion

The evidence cited above indicates that in spite of the microheterogeneity exhibited by this hemocyanin, the monomer-tetramer equilibrium is shifted by dilution of the system, and behaves at least semi-quantitatively like a thermodynamically reversible process. How are these observations to be reconciled, since it is difficult to believe that all of the 17S monomers are identical, or will have identical association constants? Indeed, we have shown direct evidence that the 17S particles are themselves heterogeneous. A possible explanation can be seen by examining the response of a simple monomer-dimer equilibrium to changes in concentration. It is easy to show that the fraction of dimer is given by:

$$f_D = \frac{1 + 2\ kC_o \pm \sqrt{1 + 4\ kC_o}}{2\ kC_o} \tag{7}$$

where k is the association constant (on a weight concentration scale) and C_o the total concentration. Clearly f_D is a function of the variable kC_o; it is graphed in Figure 4 on a logarithmic scale. Now suppose we have two components each undergoing an independent monomer-dimer association. In the special case indicated by points A, A', the system will be essentially insensitive to dilution. If the case is not the trivial one in which one C_o is much greater than the other, the k values

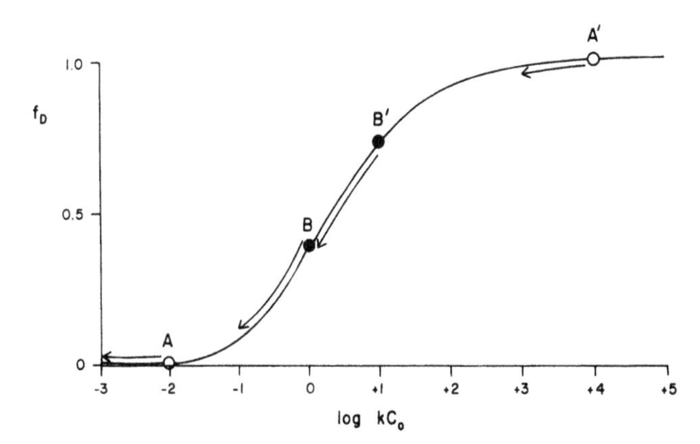

Fig. 4. The behavior of two-component systems capable of monomer-dimer association. *Curve:* general functional relationship between the weight fraction dimer (f_D) and the variable kC_0 (see text). The points *A, A'* represent a system in which the two components have very different kC_0 values. Ten fold dilution (*arrows*) leads to no significant change in the fraction of dimer present, which is determined entirely by the C_0 values. The system *B, B'*, in which the kC_0 values differ by only a factor of ten, is much more sensitive to dilution

must differ by a large factor. This must be the situation observed in some hemocyanin species, where the association constants are highly sensitive to a variable like pH. At a given pH, one component may have a k value many orders of magnitude smaller than the other. One component is essentially all dimer, the other essentially all monomer.

In less drastic cases, microheterogeneity need not lead to insensitivity to dilution. Consider the two-component system represented by points B and B'. Here the kC_0 values differ by a factor of ten. Dilution of the system (represented by arrows) will lead to an observable change in the composition. In such a case, however, very accurate data should show that the dissociation is not occurring according to a simple scheme. Furthermore, any "equilibrium constants" obtained should be suspect, as far as numerical values are concerned.

It may be that we are dealing with a system of this latter type in the study of *Callianassa* hemocyanin. The qualitative conclusion that has been drawn concerning the influence of divalent ions and temperature, for example, should be valid. But numerical results may still represent complicated averages over a host of molecular species. It is of primary importance to carry out such studies with truly homogeneous samples. The preparation of such samples is our primary goal at the moment.

References

Blair, D., van Holde, K.E.: Sedimentation equilibrium studies of a complex association reaction. Biophys. Chem. (1976) (in press)

Carpenter, D., van Holde, K.E.: Amino acid composition, aminoterminal analysis, and subunit structure of *Cancer magister* hemocyanin. Biochemistry 12, 2231-2238 (1973)

DiGiamberardino, L.: Dissociation of *Eriphia* hemocyanin. Arch. Biochem. Biophys. 118, 273-278 (1967)

Eriksson-Quensel, I.B., Svedberg, T.: The molecular weights and pH-stability regions of the hemocyanins. Biol. Bull. 71, 498-547 (1936)

Holde, K.E. van, Bruggen, E.F.J. van: The hemocyanins. In: Subunits in Biological Systems. Timasheff, S.N., Fasman, G.D. (eds.). New York: Marcel Dekker, 1971, pp. 1-55

Klarman, A., Shaklai, N., Daniel, E.: The binding of calcium ion to hemocyanin from *Levantina hierosolima* at physiological pH. Biochem. Biophys. Acta 257, 150-157 (1972)

Konings, W.N., Siezen, R.J., Gruber, M.: Structure and properties of hemocyanins, VI. Association-dissociation behaviour of *Helix pomatia* hemocyanin. Biochem. Biophys. Acta 194, 376-385 (1969)

Lamy, J., Chalons, F., Goyffron, M., Weill, J.: Sur la taille des produits de la dissociation de l'hemocyanine du scorpion *Androctonus australis* (L.). C.R. Acad. Sci. Paris 276D, 419-422 (1973)

Loehr, J.S., Mason, H.S.: Dimorphism of *Cancer magister* hemocyanin subunits. Biochem. Biophys. Res. Commun. 51, 741-745 (1973)

Miller, K.I., van Holde, K.E.: Oxygen binding by *Callianassa californiensis* hemocyanin. Biochemistry 13, 1668-1674 (1974)

Morimoto, K., Kegeles, G.: Subunit interactions of lobster hemocyanin. I. Ultracentrifuge studies. Arch. Biochem. Biophys. 142, 247-257 (1971)

Murray, A.C., Jeffrey, P.D.: Hemocyanin from the Australian freshwater crayfish *Cherax destructor*. Subunit heterogeneity. Biochemistry 13, 3667-3671 (1974)

Roxby, R., Miller, K.I., Blair, D.P., van Holde, K.E.: Subunits and association equilibria of *Callianassa californiensis* hemocyanin. Biochemistry 13, 1662-1668 (1974)

Siezen, R., van Driel, R.: Structure and properties of hemocyanins. VIII. Microheterogeneity of α-hemocyanin of *Helix pomatia*. Biochem. Biophys. Acta 295, 131-139 (1973)

Sullivan, B., Bonaventura, J., Bonaventura, C.: Functional differences in the multiple hemocyanins of the horseshoe crab, *Limulus polyphemus*, L. Proc. Natl. Acad. Sci. 71, 2558-2562 (1974)

Spider Hemocyanins: Recent Advances in the Study of Their Structure and Function

B. LINZEN, D. ANGERSBACH, R. LOEWE, J. MARKL, AND R. SCHMID

Introduction

Spider blood plasma is characterized by its blue color and a high content of protein. Without doubt it is involved in respiratory function, and it is our aim (1) to elucidate this function at the molecular level and (2) to relate the molecular qualities to physiological traits so that, some day in the future, we may arrive at a mechanistic description of respiration in these animals. We shall first describe results of chemical studies and later turn to aspects of oxygen transport in the intact animal. This account should be viewed in conjunction with the report by Dr. Loewe (Loewe et al., Chap. 7, this vol.) on the relation between oxygen binding and quaternary structure.

Results and Discussion

While it was originally believed that almost all of the protein present in spider blood is hemocyanin, recent analyses have revealed a second major blood protein. Polyacrylamide gel electrophoresis (PAGE) resolves the blood proteins of tarantulas (*Dugesiella californica* and *D. helluo*) into two bands, and the blood proteins of *Cupiennius salei* into three bands. Treatment with SDS and mercaptoethanol leads to the same pattern, in each of these species: A predominant band with a molecular weight of about 70,000 daltons, and two minor bands in equal proportion, which correspond to 95,000 and 110,000 daltons. The relation between the original pattern and the SDS-PAGE pattern is shown in Figure 1. Sedimentation studies, spectral analysis, and quantitative copper determinations have shown that the 70,000 dalton band represents the hemocyanin. The native hemocyanin sediments with 37S in the case of the tarantulas, and with 24S and 16S in the case of *Cupiennius*. The 95,000 and 110,000

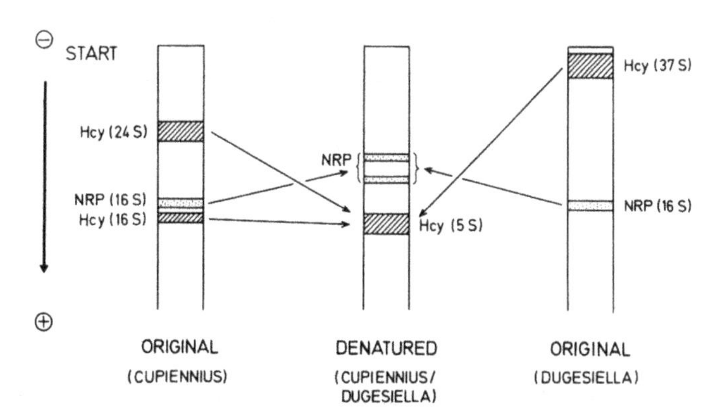

Fig. 1. Separation of native (*right* and *left*) and denatured (*center*) spider hemolymph proteins. Sedimentation coefficients are in brackets. *Arrows*: relations between protein patterns

dalton proteins are constituents of a different macromolecule which also sediments with 16S, and thus accompanies hemocyanin in some experiments, but is devoid of copper. This "non-respiratory protein" is distinguished by a higher IEP, a different amino acid, and carbohydrate composition, only four subunits (an $\alpha_2\beta_2$ structure is assigned to it), and its tetrahedral shape (Markl et al., to be published).

37S hemocyanin of tarantulas, and 24S hemocyanin of *Cupiennius* are routinely isolated by gel filtration (Fig. 2). These preparations appear

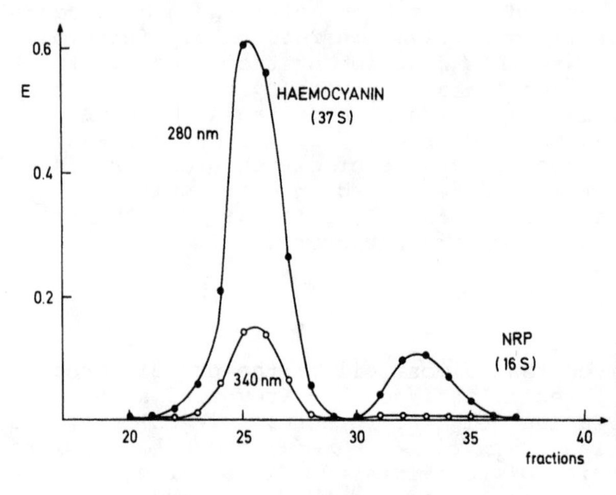

Fig. 2. Separation of *Dugesiella* blood proteins by gel filtration on Biogel A5m. 0.01 M Tris-HCl, pH 7.4, 4°C, column length 110 cm, 10 mg of protein applied

to be homogenous if judged by their movement in the ultracentrifuge, in mixed agarose-polyacrylamide gels, and in conventional SDS-PAGE. By running samples in the system of Kaltschmidt and Wittmann (1970) and Kaltschmidt (1971), a slight separation into three bands was achieved, which will be dealt with below. If we accept a high degree of homo-

Table 1. Properties of hemocyanins isolated from blood of *Dugesiella californica*, *Dugesiella helluo*, and *Cupiennius salei*

	D. californica	*D. helluo*	*Cup. salei*
$s^{o}_{20,w}$ (native hcy's)	36.7 S	36.7 S	23.4 S 15.9 S
$s^{o}_{20,w}$ (subunit)	5.8 S	5.8 S	4.7 S
subunit MW			
(sed. equilibrium)	70,300	-	69,900
(SDS-PAGE)	71,000	74,000	72,000
(copper analysis)	72,500	-	73,500
IEP (pH)	5.0-5.1	5.0	5.2

geneity for the moment we may summarize some of the physical (Table 1) and chemical data. The amino acid composition (Markl et al., to be published) will be discussed in relation to hemocyanin evolution by Dr. Ghiretti-Magaldi (Chap. 35, this vol.). The N-terminus is blocked. The three spider hemocyanins contain ca. 1.4% of neutral carbohydrates;

half of these, 0.7% (about three residues), are represented by glucose, the remainder by mannose, fucose, and arabinose (one residue each). In addition, there are ca. 0.4% of glucosamine (possibly acetylated; two residues). The quantitative data should, of course, be viewed with some reservation.

We have paid particular attention to the problem of the constituent polypeptide chain. Is it possible to break spider hemocyanins to smaller fragments than the functional subunit containing two copper atoms? We have, therefore, denatured the isolated proteins by four different methods - with SDS, urea, guanidinium hydrochloride, and 70% formic acid - and have arrived, in each case, at a molecular weight of ca. 70,000 daltons, the same obtained after alkaline dissociation. We believe that this is the lowest molecular weight obtainable without breaking covalent bonds.

There are, however, some puzzling observations upon cleavage of tarantula hemocyanin by cyanogen bromide. Although the results of these experiments are still of preliminary nature, and their explanation quite speculative, they may lead us to a more clearly defined approach in the future work.

From the methionine content one would expect more than a dozen cyanogen bromide peptides; SDS-PAGE has revealed only seven (Fig. 3). While, on the one hand, it cannot yet be excluded that short cleavage peptides

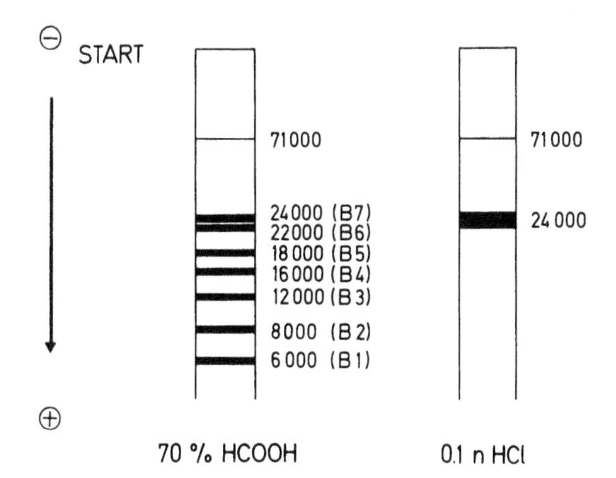

Fig. 3. Separation by SDS-PAGE of cleavage peptides after incubation of *Dugesiella* hemocyanin with 0.5% cyanogen bromide in formic acid (*left*) and hydrochloric acid (*right*). Molecular weights are indicated

have escaped our notice, we have, on the other hand, to account for the peculiar quantitative relations displayed by the seven bands (Table 2). On a molar basis, their amounts should be equal, but as it stands the four larger peptides represent only 40% of the polypeptide chain while the three smaller peptides comprise 60%. Also, the four large peptides appear to represent two pairs, with closely similar molecular weights within each pair. There is one additional observation which we regard as significant. If treated with formic acid alone at room temperature, the hemocyanin is slowly hydrolyzed, yielding four bands after 24 h, viz. at 54,000; 36,000, and 18,000 daltons, plus starting material (71,000 daltons). The half-molecule (36,000 daltons) accounts for 80% of the products.

These results can be reconciled with the conception that spider (and other arthropod) hemocyanins are composed of two equally sized, structurally closely related polypeptide chains in sequence, linked by a

Table 2. Molecular weights and relative quantities of peptides after cyanogen bromide cleavage of *Dugesiella californica* hemocyanin

Peptide	MW	% of 66,000	Peak area	% area	% chain / % area	mol %	mol % (rounded)
B 7/B 6	23,000	34.8	206	32.1	1.08	21.9	20
B 5/B 4	17,000	25.8	173	27.1	0.95	19.2	20
B 3	12,000	18.2	117	18.2	1.00	20.2	20
B 2	8,000	12.1	88	13.8	0.88	17.8	20
B 1	6,000	9.1	56	8.8	1.03	20.9	20
	66,000	100,0	640	100.0	4.94	100.0	100

The peptides were separated by SDS-PAGE, stained, and scanned. The peptides B 7 (24,000 daltons) and B 6 (22,000), and the peptides B 5 (18,000) and B 4 (16,000) were not sufficiently resolved. For calculation they were combined, and intermediate molecular weights (23,000 and 17,000, respectively) assumed.

bond which is particularly sensitive to formic acid. This idea is by no means new, it was advanced ten years ago by Pickett et al. (1966), and was advocated again by Waxman (1975) for lobster hemocyanin. The conception of two constituent polypeptide chains instead of one single giant chain with some 600 amino acids would be especially attractive as it would provide for a certain degree of symmetry required by the active center with its two copper atoms. But this notion may lead us one step further. If the two chains were synthesized independently and, after completion, joined together, then we might expect to find a series of related hemocyanins with the two chains arranged in four different ways, viz. a-a, a-b, b-a, and b-b. There is no need to postulate that these should occur in equal amounts. This could furnish one explanation, among others, for microheterogeneity in hemocyanins (Sullivan et al., 1974). Slight differences in chain length of tarantula hemocyanin were detected by special PAGE techniques (see above); if tarantula hemocyanin is dissociated at pH 9.6, and the products analyzed by PAGE, four bands are observed in the presence of glutathione, and five in its absence (H. Schneider, unpublished experiments). Experiments are under way to verify or to disprove this hypothesis.

Turning now to a consideration of the oxygen binding properties of *D. californica* hemocyanin, we want to point out the marked sensitivity to pH changes, of both oxygen affinity and subunit cooperativity (Fig. 4).

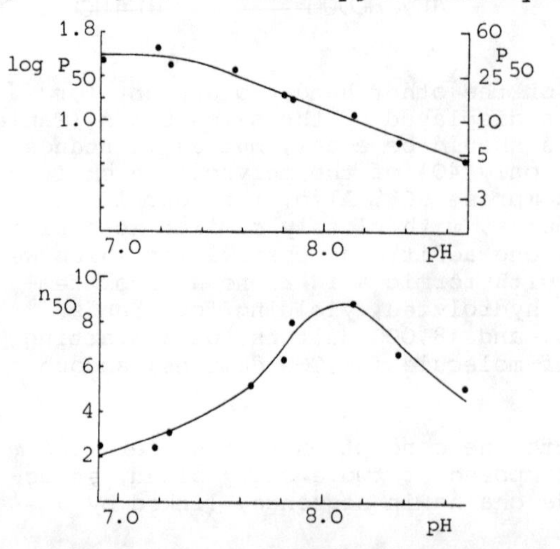

Fig. 4. The change of oxygen affinity (*upper diagram*) and cooperativity (*lower diagram*) with pH, in *Dugesiella californica* 37S hemocyanin. Tris buffers of ionic strength 0.1, 25°C, no Mg^{2+}, no Ca^{2+}

There is a strong positive Bohr effect and a fourfold change in n_{50} within one pH unit. Both effects run independently of each other which indicates that there are different proton accepting groups, each with a particular effect on conformation and ligand binding. Although the binding curves were measured with dialyzed preparations, in the absence of calcium and magnesium, other experiments indicate that in tarantula hemocyanin these divalent cations have relatively little effect on oxygen binding, at least compared to *Cupiennius* hemocyanin. This makes us believe that the binding curves measured in vitro are not too different from the curves in vivo.

The most peculiar feature of Figure 4 is the extremely high Hill coefficient at pH 8 (Loewe et al., 1977) and its steep decrease with falling pH. Beyond doubt, this is a special property built into this molecule to meet the physiological requirements of tarantulas. The hemocyanin of *Cupiennius* exhibits a quite different behavior (Loewe and Linzen, 1975), cooperativity rising in concert with P_{50}. Why then should tarantula hemocyanin be different in this respect? We have tried to find an answer by considering the respiratory physiology of *D. californica*. By introducing micro polarographic electrodes (tip diameter less than 5 micron; Baumgärtl and Lübbers, 1973) into various regions of the spider body we have been able to measure the actual oxygen pressures in "arterial" (postpulmonal) and "venous" (prepulmonal) blood. The animals were held by a clamp attached to their abdomina while the legs were in contact with a styrofoam globe which moved freely in every direction and thus allowed the animals to sit quiet or to "run" at will. P_{O_2} was recorded continuously. Since the delicate electrodes also respond to even slight mechanical displacement they provided us, at the same time, with information on heart frequency and relative stroke intensity.

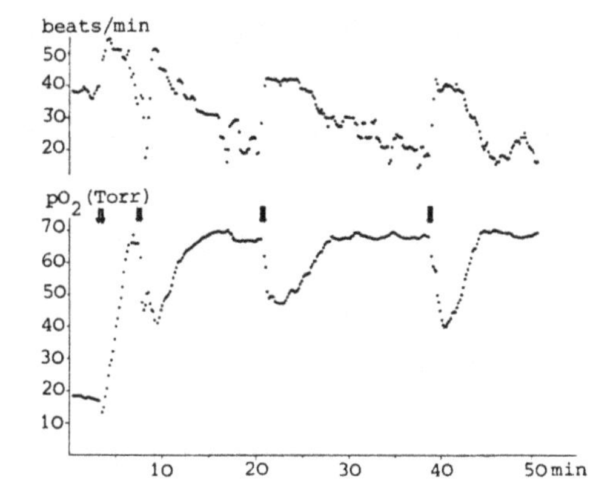

Fig. 5. Recording of arterial (pericardial) oxygen pressure (STPD), and heart frequency in a specimen of *Dugesiella coliformica*. *Double arrows:* periods of stimulation

One such recording (data corrected for barometric pressure and water vapor pressure) is shown in Figure 5. At the beginning the animal was completely relaxed and kept its arterial P_{O_2} at a low (ca. 20 mm Hg) and constant level. Upon stimulation the ventilation was stepped up, and arterial P_{O_2} rose more than three-fold. The high level was again kept very constant, but each burst of activity (running) caused an immediate pressure drop. Minimal P_{O_2}s are near 40 mm Hg in arterial blood but at the base of a leg, the venous pressure may momentarily drop to near-zero. This is a typical pattern, and even the numerical values are close to the average of "resting" arterial oxygen pressure of 28 mm

Hg, and "alert" P_{O_2} of 68 mm Hg. The exact resting blood pH is difficult to measure, it may be somewhere between pH 7.5 and 7.8.

Why are there two levels of arterial oxygen pressure? It is important to realize that a high rate of ventilation implies a high rate of water loss, but tarantulas have to handle water economically. Thus it is readily understood that they should ventilate their book-lungs as little as possible as long as there is a low requirement for oxygen. Under this condition the affinity of the hemocyanin should be high, and sigmoidity pronounced, so that at sites of constant oxygen demand sufficient oxygen could be delivered. When the animal is either active or in a state of alertness, ventilation shifts into "high gear". This is a condition in which also blood pH may be expected to drop, and the oxygen binding curve to be shifted to the right. This alone would improve unloading of the hemocyanin, as long as tissue P_{O_2} is sufficiently low. However, if in addition to the Bohr effect cooperativity decreases, a second important advantage ensues. Hemocyanin unloading begins at higher oxygen pressures. This results in a steeper oxygen gradient for terminal diffusion and thus provides for more rapid delivery of oxygen to the tissues. Since there is no capillary network in spiders, unloading at high pressures would be a most desirable mechanism which could partly compensate for the lack of the former. In fact, at a "low" pH (7.2) tarantula hemocyanin acquires the characteristic of respiratory proteins typical for very active animals.

It is clear that molecular properties and the qualities described by systemic physiology are intimately tied together. In the present example we have considered, beyond the respiratory function proper, the implications of water balance and of histological complexity. Such considerations must be backed by adequate quantitative studies, and it is hoped that some day these data will be available.

Acknowledgments. This work was supported by the Deutsche Forschungsgemeinschaft, by the Stiftung Volkswagenwerk (analytical ultracentrifuge), and the Fonds der Chemischen Industrie. The assistance of Miss M. Holland and of Mrs. H. Geisert is appreciated.

References

Baumgärtl, H., Lübbers, D.W.: Platinum needle electrode for polarographic measurement of oxygen and hydrogen. In: Oxygen Supply. Kessler, M., Bruley, D.F., Clark, L.C., Lübbers, D.W., Silver, I.A., Strauss, J. (eds.). München-Berlin-Wien: Urban & Schwarzenberg, 1973, pp. 130-136

Kaltschmidt, E.: Ribosomal proteins. XIV. Isoelectric points of ribosomal proteins of *E. coli* as determined by two-dimensional polyacrylamide gel electrophoresis. Analyt. Biochem. 43, 25-31 (1971)

Kaltschmidt, E., Wittmann, H.G.: Ribosomal proteins. VII. Two-dimensional polyacrylamide gel electrophoresis for fingerprinting of ribosomal proteins. Analyt. Biochem. 36, 401-412 (1970)

Loewe, R., Linzen, B.: Haemocyanins in spiders, II. Automatic recording of oxygen binding curves, and the effect of Mg^{++} on oxygen affinity, cooperativity, and subunit association of *Cupiennius salei* haemocyanin. J. Comp. Physiol. 98, 147-156 (1975)

Pickett, S.M., Riggs, A.F., Larimer, J.L.: Lobster hemocyanin: properties of the minimum functional subunit and of aggregates. Science 151, 1005-1007 (1966)

Sullican, B., Bonaventura, J., Bonaventura, C.: Functional differences in the multiple hemocyanins of the Horseshoe crab, *Limulus polyphemus* (L.). Proc. Natl. Acad. Sci. 71, 2558-2562 (1974)

Waxman, L.: The structure of arthropod and mollusc hemocyanins. J. Biol. Chem. 250, 3796-3806 (1975)

Scorpion Hemocyanin Subunits: Properties, Dissociation, Association

J. Lamy, J. Lamy, M.-C. Baglin, and J. Weill

Introduction

The scorpion does not represent an important part of the research on the hemocyanin (Hcy) structure, and until recently, only a few electron micrographs (Wibo, 1966; Van Bruggen, 1968), ultracentrifugation (Wibo, 1966), circular dichroism (Witters et al., 1974) and oxygen binding studies (Padmanabhanaidu, 1966) have been reported. It is a dangerous animal, hard to breed, with low blood volume (0.2 to 0.5 ml per adult individual). This does not, however, lessen the choice of this animal in so much as one can miniaturize the study methods. In effect, its blood is at least 95% Hcy, and by an evolutional mystery, the subunits that constitute this molecule have so marked themselves during the ages that they have become easily separable, which does not seem to be the case with other arthropods. It is for these two reasons that we have chosen to study this apparently unsuitable animal.

Material and Methods

Hemolymph (Hl) and Crude Hcy Preparation. Unless otherwise stated, the only species utilized is *Androctonus australis garzonii* (Goyffon and Lamy, 1973). The Hl obtained by intracardiac puncture is centrifuged 10 minutes at 800 g in order to eliminate hematocytes, then crude Hcy is prepared by one of the three following procedures:

1. 3 successive 3-h centrifugations at 127,000 g in a Tris-buffer HCl 0.05 M (in Tris) pH 7.5 and containing $CaCl_2$ 6 mM and $MgCl_2$ 16 mM.

2. 3 successive precipitations at 20°C by 55% saturated ammonium sulfate.

3. Gel filtration on Bio-gel A Agarose 5 m (Bio Rad Laboratories).

The crude Hcy obtained by the three methods presents a ratio OD_{280}/OD_{340} < 4.

Disc Electrophoresis. We have employed the technique developed by Davis (1964). The quantities of protein present in the sample are 200 µl per gel of crude Hcy, 100 µg for the mixture of six subunits and 30 µg for an isolated subunit.

The nomenclature system has been described in other reports (Lamy et al., 1974b). Briefly, for a given species, only those bands of relative mobility superior to 0.30 are taken into account and classed in order of increasing mobility. The indexing letter is the first letter of the sub-species name or the variety where the subunit has first been identified (1_H signifies that the subunit considered has the same mobility as the subunit of *A. australis hector*). The letter is capitalized if the ratio of the concentration of subunit to the entirety of subunit is the same as that in the Hcy where it was initially described. If the ratio is clearly smaller the letter is not capitalized:

e.g.: 6_h signifies the subunit is as mobile as that of subunit 6 *A. australis hector*, but relatively less concentrated.

With this system, the species *A. australis garzonii* is written as:

1_H 2_G 3_H 4_H 5_H 6_H H = hector; G = garzonii

For the experiments in the presence of SDS, the method of Weber and Osborn (1969) has been employed with the following standards: phosphorylase a (MW 94,000), bovine serum albumin (BSA) (MW 66,000), glutamate dehydrogenase (MW 53,000), ovalbumin (MW 46,000), D-aminoacid oxidase (MW 37,000), α-chymotrypsinogen (MW 25,700).

Thin Layer Gel Filtration (TLG). The TLG technique (Pharmacia Fine Chemicals) separates monomers from dimers, hexamers and higher order polymers (34S and 47S).

The advantages are:

1. Small sample quantity (30 µg per subunit).

2. Mild treatment which respects the degree of polymerization to the same extent that column gel filtration does (contrary to disc electrophoresis).

3. Excellent resolution coming essentially from the small sample volume with respect to the length of the development.

4. Speed: 5 h for Sephadex G 150 Superfine and 9 h for Sephadex G 200 Superfine at 20°C placed at 15° angle.

5. Possibility of treating, on the same plate, ten samples containing ions or crystalloids of different concentrations.

6. Duplication technique and coloration with Coomassie Brillant Blue R (Fazekas et al., 1963) permit a storable record.

Two gels have been used: Sephadex G 150 Superfine which excludes those polymers whose sedimentation coefficient is equal to or greater than 16S and Sephadex G 200 Superfine which retards hexamer, but excludes the higher polymers. These two types of gel completely separate subunits 2, 3, 4, 5 and 6 from subunits 1 and X. In all instances, the buffer was 0.05 M Tris-HCl, pH 7.5, with or without divalent cations or urea.

Column Gel Filtration. Four gels have been used in the 0.05 M Tris-HCl buffer, pH 7.5: Sephadex G 100 Superfine, Sephadex G 200, Bio-Gel A Agarose 1.5 m and Bio-gel A Agarose 5 m (Bio Rad Laboratories).

1. Sephadex G 100 Superfine is utilized to purify the dimers (subunits 1 and fraction X) and to eliminate the artificial hexamers of subunits 1, 2, 3, 4, 5 and 6.

2. Sephadex G 200 is employed to separate the residual crude Hcy from those subunits in the dissociation.

3. Bio-gel A 1.5 m efficiently separates crude Hcy (34S) from artificial hexamers, but is inefficient in separating hexamers from dimers and monomers. One uses it in reassociation experiments or in place of Sephadex G 200 in dissociations.

4. Bio-gel A 5 m serves to purify crude Hcy or artificial polymers with sedimentation coefficient superior to 16S which are both retarded.

Conservation. The crude Hcy and subunits prepared as described are conserved at +4°C in the state of suspended protein precipitates in 55% saturated ammonium sulfate. During a storage period of six months, this procedure does not cause changes in the OD_{280}/OD_{340} ratio, nor the dissociation of crude Hcy or artificial polymers. It also does not change the property of crude Hcy and artificial polymers to dissociate in 1 M urea or by pH higher than 9.6. The capability of isolated subunits to give artificial polymers is conserved as well.

Antisera Preparations. Antisera specific for isolated subunits and a mixture of non-fractioned subunits were prepared as follows: four injections of 1 mg of subunit added to Freund's adjuvant were given to "Giant Flemish" rabbits at one week intervals; at the end of four weeks, after verification of the antibody level, the animals were sacrificed and the serum combined with sodium azide (0.2 g per ml) was stored at -30°C.

Results

Isolation of the Six Subunits According to Lamy et al. (1976a). The fractioning plan is given in Figure 1. For the dissociation by 1 M urea, the crude Hcy is extensively dialyzed against 0.05 M Tris-HCl buffer, pH 7.5,

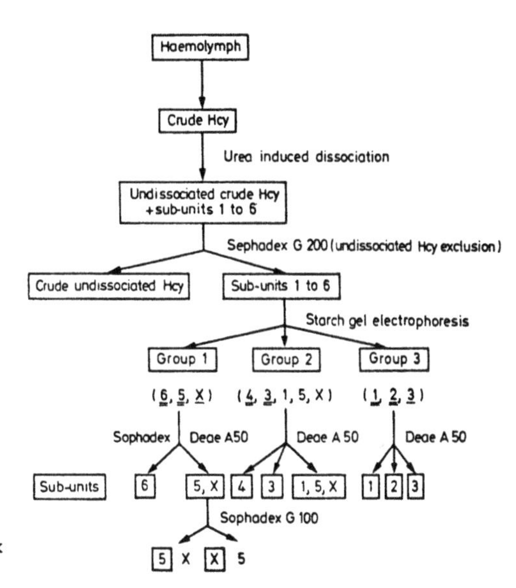

Fig. 1. Isolation diagram of the Hcy six subunits of *A. australis garzonii*

then the urea concentration is adjusted to 1 M. After two days, the non-dissociated molecules (less than 10%) are eliminated by gel filtration using Sephadex G 200. A non-fractioned mixture of six subunits is therefore obtained.

A preparative starch block electrophoresis (Müller and Kunkel, 1956) (30 h of migration in a buffer of 0.15 M veronal-NaOH pH 8.6, then elution by NaCl 9 per mille) gives a group fractioning, which allows one to deposit the already purified samples on the DEAE Sephadex A 50, where the samples are limited to one or two principal fractions (underlined by a double line in Fig. 1).

Each group then undergoes a DEAE Sephadex A 50 column fractioning. For all three groups 0.1 M sodium phosphate buffer pH 6.8 is employed to carry out the absorption, but the elutions are different. In the first group, the pH is progressively lowered to 5.7. This permits the purification of subunit 5 in equilibrium with its X dimer, then the conductivity, initially 9.10^{-1} Ω^{-1}. cm^{-1} is raised to 17.10^{-1} Ω^{-1}. cm^{-1} by the addition of NaCl, which in turn causes the elution of subunit 6. The subunit 5 and X can then be separated by Sephadex G 100 gel filtration, but rapidly, there is an interconversion of the two forms. In the second group, the subunit 3 passes without being absorbed, then the pH is lowered to 5.7, the subunit 1, 5 and X, which exist only as traces, are eluted around pH 6, and a pure subunit 4 is obtained at pH 5.7. In the third group, the subunit 3 (contaminant) passes without being retained, then with the pH being lowered to 5.7, the subunit 1 is eluted; finally the initial conductivity of 9.10^{-1} Ω^{-1}. cm^{-1} is raised to 17.10^{-1} Ω^{-1}. cm^{-1}; the ionic force gradient then liberates subunit 2.

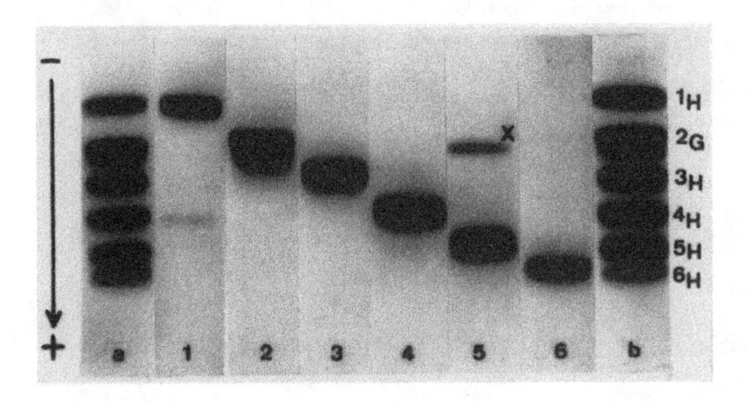

Fig. 2. Polyacrylamide gel electrophoresis of the six isolated subunits. *a*: subunits 1 to 6 obtained from crude Hcy dissociation, *1* to *6*: isolated subunits 1 to 6, *b*: mixture of isolated subunits

Subunit Properties. (a) *Electrophoresis.* Figure 2 shows the polyacrylamide gel electrophoresis of the six subunits prior to their isolation, isolated, and then recombined in the initial proportions. Note that except for the subunit 5_H, all the fractions are pure. The impurity associated with subunit 5_H has been given the name "X fraction" and subsequently identified as a 5_H dimer.

(b) *Sedimentation Coefficient.* The Ultracentrifugation Station of the National Scientific Research Center (C.N.R.S.) performed the analytic ultracentrifugation of the isolated subunits according to the following experimental conditions: 75 mM veronal-NaOH buffer, pH 8.3; protein concentration 3.8 mg/ml; temperature 20°C. The results are presented in Table 1. It is shown that subunit 1 has a sedimentation coefficient in the vicinity of 7S, clearly greater than that of the five other subunits.

Among the five other subunits, two present a particularity: subunit 2 (5.52S) for which an asymetric peak is found in ultracentrifugation presents a trail in disc electrophoresis (Fig. 2); the significance of this phenomenon is not clear and subunit 5 is heterogeneous and is composed of 82% of 4.74S and 18% of 6.60S. Other arguments presented elsewhere show that the 6.60S fraction is the "X fraction".

Table 1. Sedimentation coefficient (in Svedberg units) of isolated sub-units

Svedberg unit	1	2	3	4	5	6
Monomer	-	5.52	4.53	4.46	4.74	4.73
Dimer	6.81	-	-	-	6.60	-

SDS Electrophoresis. The results are given in Figure 3. It is seen that in each case, the molecular weight of the major fraction is within 65,000 and 71,000.

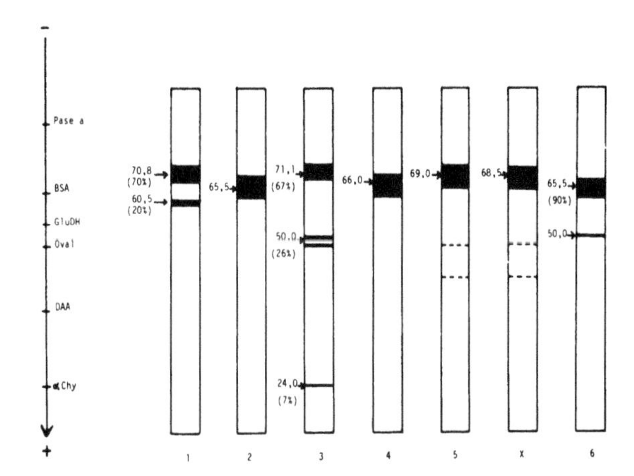

Fig. 3. SDS disc electrophoresis of Hcy isolated subunits of *A. Australis garzonii* (MW × 10^{-3})

Subunit 5 and the X fraction present exactly the same diagram, in so much as concerns the major fraction as well as the impurities, which is an argument in favor of the dimeric nature of the "X fraction".

Subunit 1 is certainly equally of a dimeric nature, the value of 70,800 obtained for the major fraction (≃ 70%) being incompatible with the sedimentation constant (6.8S) and the gel filtration. The nature (MW 60,500) and the proportion (≃ 20%) of the minor fraction does not allow one to suppose, for the moment, that subunit 1 is the result of the union of two among the five other subunits.

The significance of the heterogeneity of subunit 3 is not clear.

Gel Filtration and TLG. In TLG on Sephadex G 150 Superfine, subunit 2, 3, 4 and 6 migrate as one spot slightly slower than that of the BSA monomer, which moves similarly on a column of Bio-gel A 1.5 m, Sephadex G 200, or Sephadex G 100.

Subunit 1, less retarded under the same conditions, is totally separated from the five others in TLG, and migrates slightly behind the BSA dimer. On a column, the separation is only possible when the ratio of concentrations of subunit 1 to the other subunits is not too highly unfavorable.

Subunit 5, always accompanied by the "X fraction", gives two spots in TLG, one at the level of subunits 2, 3, 4, 6 and the other at the level of subunit 1. The second spot, recuperated by scraping and submitted to a disc electrophoresis is found to consist mostly of "fraction X", but also of subunit 5; the proportion is inverted if one considers the spot which migrates to the level of subunits 2, 3, 4, 6.

Immunoelectrophoresis. A preliminary study done with the serum anti-subunits 1 to 6 has allowed us to show that the subunit 5 and the X fraction present a complete antigenic identity (total fusion of the subunit 5 precipitation arc with that of the X fraction), and that subunits 2, 4, and 5 are antigenically pure and can only present a partial antigenic similarity amongst themselves (intersecting arcs). On the other hand, if some of the six subunits present common determinants, there is not a total identity between any two of them. Work is continuing to specify and complete these results.

Genetics. The relatively large number of different subunits found in the scorpion seems to be a contradiction of that which is reported in other arthropods, such that one might wonder if this is not an instance of artifacts. While the probability of such an occurence is slight, taking into consideration the homogeneity and the convergence of the results, the demonstration of mutations carried on subunits 2 and 6 permits the elimination of this hypothesis, at least for the subunits concerned.

In 1970, Goyffon et al. proposed to use the Hcy dissociation products proteinogram as an element of scorpion classification. In effect, on the one hand, it was noted that in numerous cases (Lamy et al., 1971) differences allowed the identification of the animal up to the species (Goyffon et al., 1970, 1973), or even the sub-species and, on the other hand, the Hcy subunits proteinogram is remarkably consistent from one individual to another, and throughout the animal's life, and it permits a simpler identification than the morphological characteristics habitually utilized for classification.

This method has enabled us to isolate and define a new scorpion subspecies, *A. australis garzonii* (Goyffon and Lamy, 1973) for which the subunit 2 mobility is less in disc electrophoresis than that of *A. australis hector*. This work was confirmed with the capture of a female in Tunisia for which the Hcy subunit proteinogram was: 1_H 2_g 2_h 3_H 4_H 5_H 6_H. This female, fertilized in the wild, bore 68 offspring of which 30 were of the type pure *garzonii* (1_H 2_G 3_H 4_H 5_H 6_H) and 36 were identical to the mother's type (Lamy et al., 1974a). This proprtion was that which one could expect from a natural crossing with type *garzonii* male, with the hypothesis of two alleles G and H for subunit 2 and autosomal transmission.

Continuing, we have successfully raised and crossed in captivity the offspring from a type *garzonii* female (captured in Tunisia) and shown that the subunit mobility is constant at least for two generations (Lamy et al., 1975). More recently (Lamy et al., 1974b) a captured female, double heterozygous for subunit 2 and 6, presenting the following type: 1_H 2_g 2_h 3_H 4_H 5_H 6_a 6_h (A for *Africanus*) has given birth to 28 offspring of which 14 are the same type as herself, 8 *garzonii*, and 6 *africanus* (1_H 2_H 3_H 4_H 5_H 6_A). The fact that only the double recombinants had been obtained shows that subunits 2 and 6 are transmitted by an autosomal mode and that the two genes are situated on the same chromosome. These experiences have shown equally that the three sub-species *garzonii*, *hector* and *africanus* were interfertilizable, which does not show the traditional system based on morphology. As a result, a large number of captured animals have been examined and other mutants discovered, but the mutations are always carried on subunits 2 and 6. To this day no mutant on subunits 1, 3, 4, and 5 have been discovered.

Dissociation. Three methods have been utilized: (1) Hemolymph freezing (3 freezings and thawings at 10-min intervals) (Lamy et al., 1970). (2) Extensive dialysis against the buffer 0.05 M Tris-HCl pH 7.5 in

Fig. 4. Influence of ammonium sulphate on Hcy dissociation provoked by urea 1 M. Direction of the TLG development indicated by *arrow*

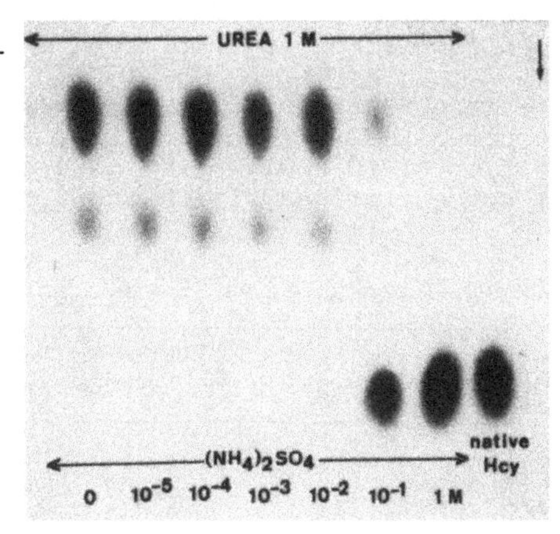

order to eliminate the cations, then urea addition just until a 1 M concentration is achieved. (3) Extensive dialysis against a carbonate - bicarbonate buffer of an ionic force of 0.1 at a pH $\geq$ 9.5.

The dissociation products analysis shows in the three cases six subunits in equal quantity by disc electrophoresis and two spots in TLG (Fig. 4) of which there is a large spot composed of the monomeric subunits, 2, 3, 4, 5 and 6, and a small spot containing essentially subunit 1 (dimer). The X fraction which also migrates in this spot is present in small quantities under these conditions.

By analytical ultracentrifugation, a fraction of 7S and a heterogeneous peak around 4.8S is found; this 4.8S peak is six times more abundant than the 7S peak (Lamy et al., 1973).

The residual Hcy proportion varies with the dissociation method. By freezing (without previously eliminating the ions), it is approximately 30% and can be reduced by increasing the number of freezings and thawings. By urea 1 M (with preliminary dialysis), the dissociation is slow (40% in 4 h) and only becomes subtotal (> 90%) after two days of dialysis. Alkaline dissociation with prior dialysis gives identical results.

Effect of Cations on Dissociation. Preliminary gel filtration experiments using a Bio-Gel A 5 m column (Lamy et al., 1976b) have shown that dissociation induced by 1 M urea or by alkaline pH is only possible after complete elimination of divalent ions by either dialysis or EDTA. As a consequence, the following protocol was adapted for the dissociation studies.

First the crude Hcy is extensively dialyzed against the buffer Tris-HCl buffer, pH 7.5 containing the salt to be studied, then a 7 M urea solution is added to the counter dialysis liquid in such a manner as to bring the urea concentration to 1 M. After two days, TLG in Sephadex G 150 Superfine is carried out as indicated. For each salt, the following concentrations have been studied: 10^{-5}, 10^{-4}, 10^{-3}, 10^{-2}, 10^{-1}, 1 and 2 M.

Figure 4 presents an example of the results obtained with ammonium sulphate. The higher spot corresponds to subunits 2 + 3 + 4 + 5 + 6, the intermediate spot contains subunit 1 (and eventually X), and the lower spot excluded corresponds to the residual crude Hcy. It is seen that at a concentration of 0.1 M the protection is important and that it is total at 1 M. For all concentrations less than or equal to 10^{-2} M, dissociation is complete.

The results obtained for other salts are presented in Table 2. It should be noted that all the divalent ions give total protection at a concentration of 10^{-2} M. In the case of $MgCl_2$, this protection disappears at concentrations of 1 M and 2 M.

Table 2. Influence of various cations upon 1 M urea-induced dissociation

	Concentrations M						
Salt	10^{-5}	10^{-4}	10^{-3}	10^{-2}	10^{-1}	1	2
Li_2SO_4	O	O	O	+	+++	++++	a
NaCl	O	O	O	+	++	+	+
KCl	O	O	O	O	++	++	++
$(NH_4)_2SO_4$	O	O	O	O	+++	++++	a
$MgCl_2$	O	++	+++	++++	++++	O	O
$CaCl_2$	O	++	+++	++++	++++	a	a
$MnSO_4$	O	++	++++	++++	++++	a	a

a: unassayed concentration.
++++: total protection (dissociation = O).
+++: 3/4 protection.
++: 1/2 protection.
+: beginning of protection (threshold of apparition of upper spot).

Among the monovalent cations, only lithium and ammonium give complete protection and the threshold of protection is clearly higher (100 times greater) than that of divalent cations. The case of sodium is exceptional in that a maximum is observed at 0.1 M.

Association. (a) *From Non-Isolated Subunits 1 to 6.* Calcium (6 mM) and magnesium (16 mM) are added to the non-fractioned mixture of the six subunits at physiological concentrations for 96 hours.

Two types of analysis are then performed:

1. Gel filtration on Bio-gel A 1.5 m in the presence of calcium and magnesium and analysis of the obtained peaks by disc electrophoresis and analytic ultracentrifugation.

2. TLG followed by a disc electrophoresis of the gel gleaned at the level of the spots (the polymeric fraction spots are treated by 1 M urea in the absence of calcium and magnesium before electrophoresis).

Identification by Analytical Ultracentrifugation of the Bio-gel A 1.5 m Column Fractioning Products. The excluded peak representing 41% of the recuperated proteins, contained two principal polymers with sedimentation coefficients of 34.3S (77.8%) and 46.8S (16.3%).

The two slower and poorly separated peaks were recombined (retarded fraction) and then submitted to ultracentrifugation, seven days later. The sedimentation coefficients of the four protein species present were: 4.98S (13%); 7.6S (19.8%); 15.12 (53.7%); 33.95S (13.5%), total protein concentration = 4.7 mg/ml.

These results indicate that the excluded fraction of S = 46.8 does not physiologically exist in the scorpion, while the "34S" fraction represents the native form of Hcy, and the presence of 13.5% of a "33.95S" polymer in the slower peaks, while this species should be found to be excluded, is explicable by the seven days delay between the Bio-gel fractioning and the ultracentrifugation. It is probable that reassociation occurred during this period.

The discrepancy between the ratio of the light fractions R = 5S/7S in the slower fraction (R = 0.65) and in the crude Hcy (R = 5) is explained by the presence in the 7S fraction of a large quantity of "X fraction".

Subunit Composition of the Different Species Present in the Medium after Reassociation. The TLG experiments on Sephadex G 200 Superfine shows four spots after reassociation, of which the first contains the subunit monomers, the second the dimers, the third less retarded than the dimers corresponds to the 16S hexamers, and the fourth is found at the exclusion. The subunit composition of these four spots obtained by densitometry of the disc electropherograms, is shown in Table 3. It is seen that

Table 3. Subunit composition of TLG spots after Ca and Mg-induced reassociation of an unfractionated mixture of the six sub-units

	Sub-unit or fraction as percentage of total of each spot						
Spot	1	2	3	4	5	X	6
Monomer (≃ 5S)			35.6	9	25		31
Dimer (≃ 7S)	38					46	
Intermediary (16S)		28	7	28	11		25
Exclusion (34S + 47S)	13	20	12	20	18		17

the excluded polymers contain all subunits present in the crude Hcy, but in different proportions, subunits 1 and 3 being in lesser proportion. The hexamers contain subunits 2, 4 and 6 in equivalent proportions, subunits 3 and 5 in reduced proportion, and an absence of subunit 1. The dimer spot evidently contains subunit 1, but also the X fraction in a high proportion; it is possible that the tranformation of subunit into X ends in eliminating a non-negligible amount of subunit 5 from the circuit. The monomeric spot contains above all subunits 3, 6 and 5, a little of subunits 4, and no subunit 2.

These results are complex to interpret; however, it is important that the addition of calcium and magnesium allows one to obtain one or many excluded polymers in Sephadex G 200 which contain all the subunits present in crude Hcy. The fact that the proportion of the six subunits is different in the crude Hcy and in the polymers is perhaps due to the heterogeneity of the fraction (34S + 47S). The absence of subunit 1 in the 16S hexamer fraction is equally remarkable.

(b) *From Isolated Subunits (Preliminary Study).* The following results were obtained:

- The polymer obtained from a mixture of the six isolated subunits, excluded in TLG on Sephadex G 200 Superfine, contains the six subunits in proportions not unlike that of the mixture 34S + 47S of Table 3.

Subunit 1 is absolutely necessary to obtain a polymer greater than 16S (hexamer).

Subunit 4 is the only one capable of polymerizing in the absence of other subunits provoked by divalent cations, to form an artificial hexamer (sedimentation coefficient 14.9S).

- The equimolar binary mixture of subunit 4 and 1 furnishes a hexamer exclusively constituted of subunit 4. All other equimolar binary mixtures based on subunit 4 (4 + 2, 4 + 3, etc.) lead to a hexamer of binary composition, but in variable proportions (probable coexistence of different hexamers).

- No binary mixture based on subunit 1, polymerizes into a 16S compound, except for the combination 1 + 4 which, it appears, does not incorporate subunit 1. Among the other binary mixtures tested, only the combination 2 + 6 has proved capable of polymerizing into a 16S compound.

- All the higher order mixtures (tertiary, quaternary, etc.) based on subunit 4 lead to one or more hexamers which contain all the subunits present in the reaction medium except subunit 1.

- The equimolar mixture of subunits 1, 2, 3, 5, 6 leads to an almost complete absence of any 16S compound and a total absence of higher order polymers.

Discussion

On the Monomers. The heterogeneity of monomers (4.5S to 5.5S), the molecular weight of which is commonly admitted to be around 75,000 daltons (van Holde and van Bruggen, 1971), is quite remarkable in the scorpion *A. australis*. The presence of 5 different monomers, easily identifiable by their charge, considerably simplifies the task of trying to comprehend the architecture of the native molecule.

The particular ability of subunit 4 to polymerize in one probably hexameric 16S compound is remarkable, as is its ability to combine with the four other subunits to give binary, tertiary, etc. 16S polymers. These special properties probably assign it a "key stone" role in the quaternary structure of Hcy.

Subunits 2 and 6 which mutate easily seem to play a less essential role.

Subunit 5 by its power of dimerization presents a special interest because it can permit the union of the hexamers or dodecamers.

The heterogeneity of these monomers has recently been confirmed by Sugita and Sekiguchi (1975).

On the Subunit 1 and the X Fraction. The dissociation products of scorpion Hcy contain a molecule, the nature of which is indisputably dimeric, which has been confirmed also by Sugita and Sekiguchi (1975), that we have, possibly abusively, named subunit 1. The fact that subunit 1 resists all mild dissociation procedures leads one to believe that the native Hcy molecule exist as a dimer and not a monomer. To dissociate it we must resort to stronger procedures such as SDS or urea 4 M.

The presence of this dimer poses a serious question about the size of the smallest molecule which contains all the subunit. The fact that the molar concentration of subunit 1 is equal to half that of the other subunits forces us to admit that we are dealing with a *dodecamer* whose sedimentation coefficient will be about 24S and whose structure will be: $1, 2_2, 3_2, 4_2, 5_2, 6_2$.

We believe that subunit 1 is probably one of two linking points existing between the hexamers, the other possibly being assured by the X fraction, dimer of subunit 5. In this latter case, however, the link would be looser, subunit 5 and the X fraction being easily interconvertible.

We have no real argument that allows us to suppose that subunit 1 is a dimer of one or two of the five other subunits.

On the Artificial "Hexamers". There has never been any "16S hexamer" among the dissociation products obtained by any of the processes studied after extensive dialysis of the scorpion Hcy.

If one admits that subunit 1 plays a linking role in the dodecamer, it is clear that one cannot obtain a single hexamer type, but only a combination of a heptamer and a pentamer. If, besides its linking role, one attributes a role in the cohesion of the molecule to subunit 1, one conceives that the dodecamer dissociation would make the molecule explode directly into the monomer and dimer.

As for the "hexamers" resulting from the reassociation, their degree of polymerization implies that they do not contain any subunit 1.

Nevertheless, when the six subunits are present at the same concentration, one realizes that the ratio of the concentrations is favorable for the formation of a "16S hexamer" rather than a dodecamer. In effect, the 5 monomers can be incorporated in a large number of possible "hexamers", which constitute a kind cul-de-sac, while subunit 1, which can occupy only one correct position in the dodecamer, must be incorporated at the precise moment of the polymerization.

Another point deserves attention: to our knowledge, the presence of "hexamers" in the dissociation products has never been reported in arthropods, simultaneously with that of a dimer (7S). This is easily comprehensible if one considers that the subunit 1, which exists in the scorpion in a dimeric form, can exist in a monomeric form in other arthropods; the dissociation of the eicosatetramer (34S) can thus end up 4 stable "16S hexamers".

The Eicosatetramers. The electromicroscopic literature shows that the 34S compound in scorpions possesses roughly, a square structure, formed by the juxtaposition of two rectangles (24S) separated by a transverse interval crossed by one or two bridges. This notion of double symetry is in agreement with the hypotheses presented above, because it necessitates two different types of linkage between the hexamers.

A preliminary ultramicroscopic study of the products of artificial reassociation (34S + 47S) isolated on a column of Bio-Gel A 1.5 m shows that one part of the molecules was identical to those of crude Hcy and the other larger part had the square form composed of 4 elementary squares, the study of which is currently ongoing.

References

Bruggen, E.F.J. van: Electron microscopy of haemocyanins. In: Physiology and Biochemistry of Haemocyanins. Ghiretti, F. (ed.). New York: Academic Press, 1968, pp. 37-48

Davis, B.: Disc electrophoresis. II. Method and application to human serum proteins. Ann. N.Y. Acad. Sci. 121, 404-427 (1964)

Fazekas de St Groth, S., Webster, R.G., Datyner, A.: Two new staining procedures for quantitative estimation of proteins on electrophoretic strips. Biochem. Biophys. Acta 71, 377-391 (1963)

Goyffon, M., Lamy, J., Vachon, M.: Identification de trois especes de Scorpions du genre *Androctonus* a l'aide du proteinogramme de leur hémolymphe en gel de polyacrlyamide. C. R. Acad. Sci. Paris 270D, 3315-3317 (1970)

Goyffon, M., Stockmann, R., Lamy, J.: Valeur taxonomique de l'electrophorèse en disques des proteines de l'hémolymphe chez le Scorpion: étude du genre *Buthotus* (Buthidae). C. R. Acad. Sci. Paris 277D, 61-63 (1973)

Goyffon, M., Lamy, J.: Une nouvelle sous-espèce d'*Androctonus australis* L. (Scorpions, Buthidae): *Androctonus australis garzonii N. ssp.* Caracteristiques morphologiques, écologiques et biochimiques. Bull. Soc. Zool. France 98(1), 137-144 (1973)

Holde, K.E. van, Bruggen, E.F.J. van: The hemocyanins. In: Subunits in Biological Systems, Part A. Biological Macromolecules Series. Timasheff, S.N., Fasman, G.D. (eds.). New York: Marcel Dekker Inc., 1971, Vol. V, pp. 1-53

Lamy, J., Richard, M., Goyffon, M.: Sur les modifications des electrophorégrammes en gel de polyacrylamide des proteines de l'hémolymphe des Scorpions *Androctonus australis* (L.) et *Androctonus mauretanicus* (Pocock), provoquees par la congelation. C. R. Acad. Sci. Paris 270D, 1627-1630 (1970)

Lamy, J., Goyffon, M., Sasse, M., Vachon, M.: L'electrophorèse des proteines de l' hémolymphe en gel de polyacrylamide, premier critère biochimique pour l'identification des Scorpions. Biochimie 53, 249-251 (1971)

Lamy, J., Chalons, F., Goyffon, M., Weill, J.: Sur la taille des produits de la dissociation de l'hémocyanine du Scorpion *Androctonus australis* (L.). C. R. Acad. Sci. Paris 276D, 419-422 (1973)

Lamy, J., Le Pape, G., Goyffon, M., Weill, J.: Sur l'obtention de Scorpions de la sous-espece *Andoctonus australis garzonii* (Goyffon et Lamy) a partir d'une femelle hybride sérologique. Confirmation de la validité des critères de détermination utilisés. C. R. Acad. Sci. Paris 278D, 129-132 (1974a)

Lamy, J., Le Pape, G., Weill, J.: Heterogeneité de l'espèce *Androctonus australis* (L.). (Scorpions, Buthidae). Création d'une nouvelle sous-espèce *A. australis africanus* N. ssp. sur la base de critères biochimiques et génétiques. C. R. Acad. Sci. Paris 278D, 3223-3226 (1974b)

Lamy, J., Le Pape, G., Weill, J.: Sur le mode de transmission genetique des fractions legeres de l'hemocyanine du Scorpion *Androctonus australis garzonii* (Goyffon et Lamy) sur deux generations. C. R. Acad. Sci. Paris 280D, 347-350 (1975)

Lamy, J. Pensec, J.F., Weill, J.: Methode d'isolement de six sous-unites de l'hemocyanine du Scorpion *Androctonus australis garzonii*. C. R. Acad. Sci. Paris D in press (1976a)

Lamy, J., Leclerc, M., Lamy, J., Weill, J.: Effet des ions divalents sur la dissociation de l'hémocyanine du Scorpion *Androctonus australis garzonii*, provoquée par l'urée 1 M. (to be published)(1976b)

Müller-Eberhard, H.J., Kunkel, H.G.: The carbohydrates of γ-globulines and myeloma proteins. J. Exp. Med. 104, 253-259 (1956)

Padmanabhanaidu, B.: Physiological properties of the blood and haemocyanin of the scorpion *Heterometrus pulvipes*. Comp. Biochem. Physiol. 17, 167-181 (1966)

Sugita, H., Sekiguchi, K.: Heterogeneity of the minimum functional unit of hemocyanins from the spider (*Argiope bruennichii*), the scorpion (*Heterometrus* sp.), and the horseshoe crab (*Tachypleus tridentatus*). J. Biochem. (Tokyo) 78, 713-718 (1975)

Weber, K., Osborn, M.: The reliability of molecular weight determinations by dodecyl sulfate-polyacrylamide gel electrophoresis. J. Biol. Chem. 244, No. 16, 4406-4412 (1969)

Wibo, M.: Recherches sur les hémocyanines des arthropodes: Constantes de sédimentation et aspects morphologiques. Univ. Louvain, Lab. Chim. Physiol. Mémoire présenté au concours des bourses de voyage (1966)

Witters, R., Goyffon, M., Lontie, R.: Étude de l'hémocyanine de Scorpion par dichroisme circulaire. C. R. Acad. Sci. Paris 278D, 1277-1280 (1974)

Subunit Association and Oxygen-Binding Properties in Spider Hemocyanins

R. LOEWE, R. SCHMID, AND B. LINZEN

Introduction

It is well known that respiratory pigments which are not confined to cells, but occur dissolved in the hemolymph, are always present as high molecular-weight aggregates. The physiological reasons for this may be the necessity (1) to keep the relative viscosity of the hemolymph low and (2) to keep the colloid osmotic pressure within physiological limits, while at the same time a high oxygen capacity is desirable.

The basic building stone of arthropod hemocyanins is a 16S oligomeric protein composed of six polypeptide chains, which provides for a high degree of allosteric interaction and functional flexibility. While in some animals it is the only state of aggregation, there are other species in which the dimer (24S) or the tetramer (37S) of this building stone are observed. We were interested to learn whether these higher orders of association would affect the affinity towards oxygen and the shape of the binding curve. To this end, we studied the hemocyanin of the tarantula, *Dugesiella californica* which occurs as tetramer, and the hemocyanin of the lycosid spider *Cupiennius salei* which is present as 24S and 16S species.

Results and Discussion

In this study, there were two points which required particular consideration: (1) Under the experimental conditions examined by us, tarantula hemocyanin does not dissociate into 24S dimers and 16S monomers but directly to the 6S subunit. However, if this subunit is dialyzed against neutral (pH 7.2 or 7.5) buffers, containing 1 mM Mg^{2+}, reassociation ensues. Such reassociation mixtures yielded, upon ultracentrifugal analysis, both 16S and 37S particles. Fortunately, the reassociation process is very slow, so that the components of the mixture can be separated by gel chromatography (Fig. 1). (2) While *Cupiennius* provides us with native 24S and 16S hemocyanin, the 16S hemocyanin is contaminated with a second, nonrespiratory protein. In addition, we could not be sure from the beginning whether 16S and 24S particles were really made up of identical subunits. Therefore, we again first dissociated the hemocyanin and then looked for conditions by which either 16S or 24S particles could be reassembled. While reassembly of the 16S particle is not difficult, the 24S aggregate was only obtained in the presence of 40 mM Ca^{2+}.

Native and reconstituted hemocyanins were compared by polyacrylamide gel electrophoresis (PAGE), electron microscopy, and sedimentation analysis. The electrophoretic mobilities in PAGE and the sedimentation coefficients were identical for both *Dugesiella* and *Cupiennius* hemocyanins. The appearance of the hemocyanins in negatively stained EM preparations is shown in Figures 2 and 3. No difference in the shape and dimensions of the reassembled molecules is seen.

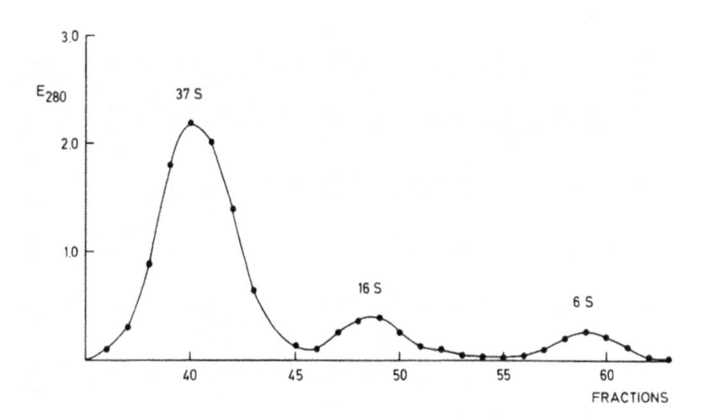

Fig. 1. Separation of reconstituted *Dugesiella* hemocyanin components by chromatography on Biogel A 5 m, 200-400 mesh; protein concentration 2 mg/ml, Tris/HCl buffer pH 7.5, 1 mM Mg^{2+}

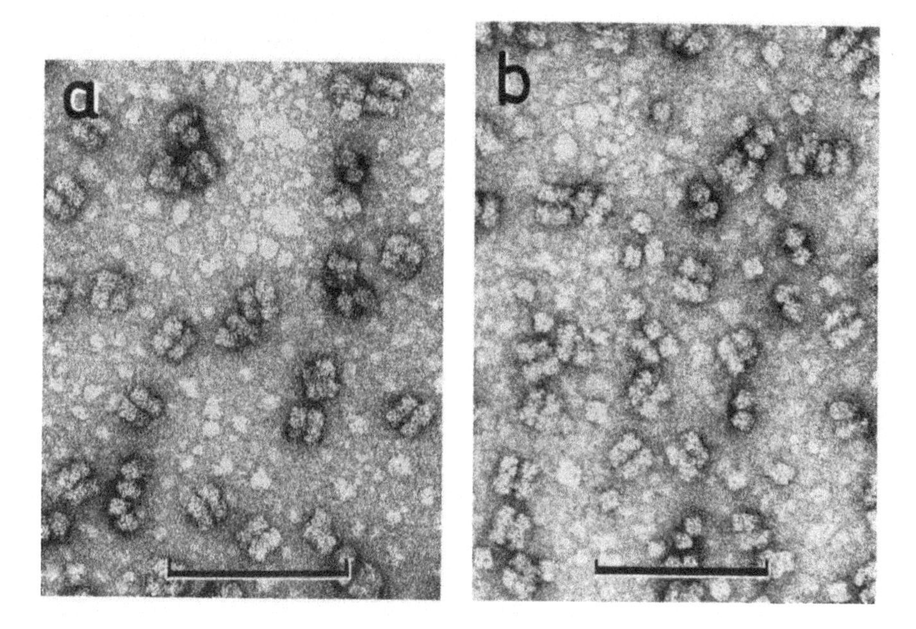

Fig. 2a and b. Electron micrographs of negatively stained *Dugesiella* hemocyanin. (a) native, (b) reconstituted hemocyanin. Tetramers measure 190 × 200 Å, monomers about 90 Å in diameter (*hexagons*) and 90 × 80 Å (*rectangles*) (× 240,000). Bar: 1000 Å

The oxygen-binding curves of native and reconstituted hemocyanins were measured by the continuous polarographic method described previously (Loewe and Linzen, 1975). The data are summarized in Tables 1 and 2. While on the one hand, the binding properties are not identical between native and reconstituted hemocyanin particles, these differences are not dramatic, they may be caused by a partial loss of copper. Also, they do not detract from our main conclusions.

In *Dugesiella* the association of 6S subunits to 16S molecules does not alter the high oxygen affinity (P_{50} = 6 mm Hg); some cooperativity, however, appeared (n_{50} = 1.4). Further association to 37S results in

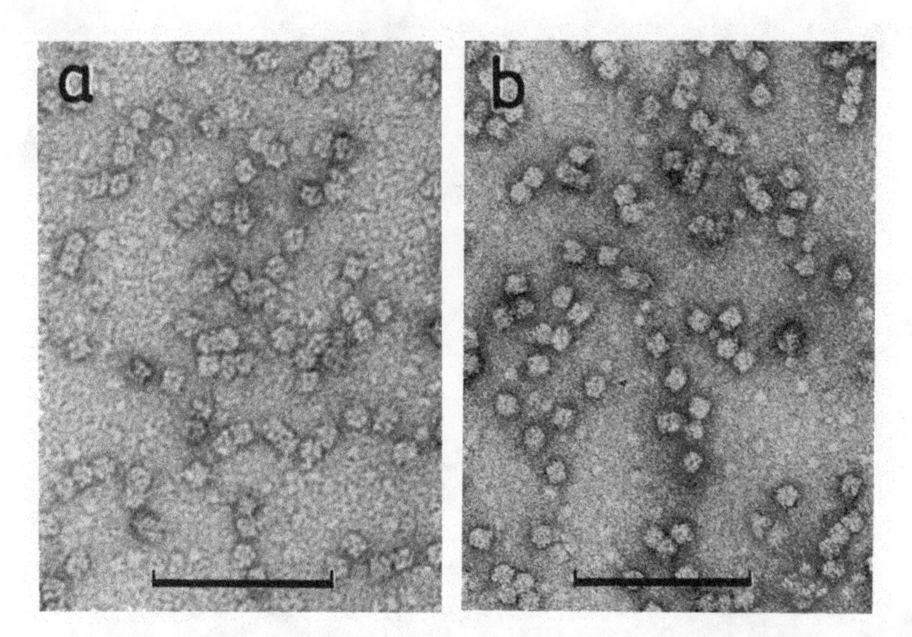

Fig. 3a and b. Electron micrographs of negatively stained *Cupiennius* hemocyanin. (a) native (× 247,500), (b) reconstituted (× 240,000). Dimers measure 90 × 190 Å, monomers about 90 Å in diameter (*hexagons*), or 90 × 80 Å (*rectangles*). Bar: 1000 Å

Table 1. Oxygen binding data of *Dugesiella* hemocyanin measured in 0.1 M Tris/HCl buffer in the presence of 1 mM Mg^{2+} at 25°

	Native		Reconstituted	
	P_{50}	n_{50}	P_{50}	n_{50}
37S	28.1 ± 0.6	2.4 ± 0.2	21.4 ± 0.7	1.6 ± 0.1
16S	-	-	6.1	1.4
6S[a]	5.0 ± 1.0	1.0	-	-

N = 3, where S.D. is indicated.

[a]in $NaHCO_3/Na_2CO_3$ buffer pH 9.6, 0.1 ionic strength.

Table 2. Oxygen-binding data of *Cupiennius* hemocyanin measured in 0.1 M Tris/HCl buffer pH 7.5-7.2 in the presence of 40 mM Ca^{2+} at 25°

	Native[a]		Reconstituted[b]	
	P_{50}	n_{50}	P_{50}	n_{50}
24S	59.0 ± 2.0	4.6 ± 0.1	48.2	3.7
16S	40.0 ± 4.0	3.3 ± 0.1	38.0	3.0
16S[c]	-	-	35.6	3.1
5S[d]	2.0	1.0	-	-

N = 3, where S.D. is indicated.
[a]in 0.1 M Tris/HCl buffer pH 7.5.
[b]in 0.1 M Tris/HCl buffer pH 7.2.
[c]reconstituted from 5S, which have been obtained from 24S by dissociation.
[d]in $NaHCO_3/Na_2CO_3$ buffer pH 9.6, 0.1 ionic strength; measured at 20°.

a remarkable decrease of affinity (P_{50} increases from 6 to about 21 mm Hg) and in significant cooperativity. It may be concluded that tetramerization of 16S monomers has a greater effect on oxygen binding than formation of the 16S molecule.

In *Cupiennius* hemocyanin, in contrast, subunit association to the 16S species results in a strong decrease of affinity and simultaneously in a significant increase of cooperativity. This is clearly seen both in the native and reassembled 16S hemocyanin. In addition, there is no appreciable difference, whether hemocyanin reassembly starts with 5S material obtained from 24S hemocyanin or with 5S material obtained from 16S hemocyanin. Cooperativity is remarkably high. Formation of the 24S dimer results in a further lowering of the affinity and further increase in cooperativity.

It is concluded from these data that allosteric interaction is not limited by the confines of the 16S particle but extends through the integral dimer, or tetramer. Association of the "physiological subunit" thus has strong implications for the functional properties of hemocyanin and, as may be expected, for the modulation of these functions.

The importance of this higher order of association can also be appreciated from the extremely high Hill coefficients measured in spider hemocyanins. In *Dugesiella*, the n_{50} rises above nine at a pH of 8.2 which is somewhat above the physiological pH. In order to ascertain this value, oxygen-binding curves were measured both in the direction of oxygen unloading (our routine procedure) and of oxygen loading where the response time of the Clark electrode is critical (a slow response would alter the shape of the curve and lower the n_{50}). We used electrodes covered only with a 6 μ Teflon membrane. The comparison of the two curves reveals only a very small difference in P_{50} and n_{50} (Fig. 4a and b). Although measuring the downward curve took 30 min, and

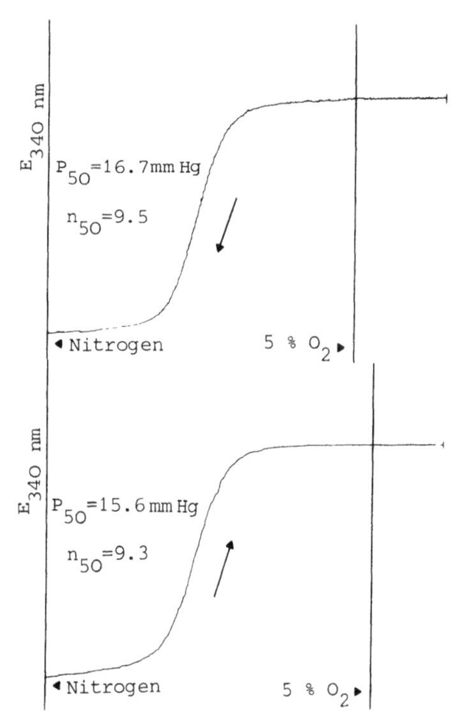

Fig. 4. Original recording of an oxygen-binding curve of *Dugesiella* 37S hemocyanin (25°, Tris/HCl pH 8.2, 0.1 ionic strength). *Above*: "deoxygenation curve" (30 min); *below*: "oxygenation curve" (3.5 min)

measuring the upward curve only 3.5 min, the shape was identical, for all practical purposes. The samples were analyzed in the ultracentrifuge after the experiment to verify that no dissociation had occurred.

Since n represents the minimal number of interacting subunits, it may be concluded that the 16S particle is not the "allosteric unit" in spider hemocyanins, but that homotropic interaction is affected by all higher states of association.

Acknowledgments. We should like to thank Prof. E.F.J. van Bruggen, Miss Wilma Shutter, and Mr. Jan van Breemen, for hospitality and help with the electron micrographs. The work was supported by the Stiftung Volkswagenwerk which provided the analytical ultracentrifuge, the Deutsche Forschungsgemeinschaft (Li 107/18+19), and the European Molecular Biology Organisation which provided a short-term fellowship to R.L. The assistance of Mrs. H. Geisett is gratefully appreciated.

Reference

Loewe, R., Linzen, B.: Haemocyanin in spiders. II. Automatic recording of oxygen binding curves, and the effect of Mg^{++} on oxygen affinity, cooperativity, and subunit association of *Cupiennius salei* haemocyanin. J. Comp. Physiol. 98, 147-156 (1975)

Heterogeneous Subunits of the Hemocyanins from *Jasus edwardsii* and *Ovalipes catharus*

H. A. ROBINSON AND H. D. ELLERTON

Summary

The hemocyanins of the spiny lobster, *Jasus edwardii* and the swimming crab, *Ovalipes catharus* have been studied at various pH values using gel filtration, sedimentation velocity and polyacrylamide gel electrophoresis (PAGE) techniques, in the presence or absence of added Ca^{2+}, and analyzed for the relative proportions of components. Electrophoresis experiments have shown that the single bands isolated by gel filtration display multiple bands by electrophoresis, reflecting subunit heterogeneity of the monomer species.

The subunits of the hemocyanins were analyzed at pH 7.8, 8.8 and 10.6 by the PAGE technique and varying gel concentrations to obtain Ferguson plots. At pH 7.8, the hemocyanin which sediments as a 25S particle is seen as three bands by electrophoresis, and these are charge isomers of a size isomer. The 16S particle shows two bands by electrophoresis, which are also charge isomers of a size isomer. Similar bands were observed at pH 8.8. At pH 10.6, the 5S hemocyanin from both species produced three bands on electrophoresis. Molecular weights obtained for these 5S bands were, for *J. edwardsii*, 84,000 (72%), 96,000 (10%) and 116,000 (18%), and for *O. catharus* , 86,00 (89%), 92,000 (5%) and 115,000 (6%).

In electrophoresis experiments performed in sodium dodecyl sulfate (SDS) with 2-mercaptoethanol or dithiothreitol (DTT), molecular weights obtained with whole purified sera were, for *J. edwardsii*, 64,000 (12%), 98,000 (13%) and 110,000 (75%), and for *O. catharus*, 69,000 (9%), 101,000 (27%) and 112,000 (64%).

Preliminary experiments on the reassociation of the hemocyanins indicate that it is possible to get association to 16S and 25S particles at pH 7.0 after prior dissociation at pH 8.8. For samples which had been dissociated at pH 10.6, some reassociation to 25S particles occurred at pH 7.0 for *O. catharus* hemocyanin, but the 16S hexamer was the largest particle observed under these conditions for *J. edwardsii*.

A shift is reported in the UV absorption spectrum of the hemocyanins under conditions where all divalent metal cations are removed.

Introduction

Crustacean hemocyanin has been shown to exhibit components with sedimentation coefficients of 5S, 16S, 25S, 34S, 39S and 60S (Van Holde and Van Bruggen, 1971; Roxby et al., 1974). On increasing the pH from 7.0 to 10.6 in the absence of calcium or magnesium, successive dissociation occurs until at a pH greater than 10, the 5S subunit is predominant. Progress so far has shown that the main 16S and 25S components correspond to molecular weights of 430,000 to 490,000 and 940,000 to 950,000 respectively while the 5S subunit has been shown to have a molecular weight range of 65,000-90,000 depending upon the species

(Di Giamberardino, 1967; Moore et al., 1968; Ellerton et al., 1970; Murray and Jeffrey, 1974; Ellerton et al., 1976). In addition, smaller units of 25,000 daltons (Salvato et al., 1972) and 35,000 daltons (Pickett et al., 1966) have been reported.

For some time, the nature of the smallest functional subunit of hemocyanin has been under question. Results to date indicate that the hexamers (16S) and dodecamers (25S) are formed from the combination of heterogeneous monomers. The 5S component would appear from association-dissociation and oxygen studies to correspond in size to a "protomer" bearing one oxygen-binding site while other evidence indicates that it is comprised of several components close in molecular weight. For example, polyacrylamide gel electrophoresis studies of Loehr and Mason (1973) and Carpenter and Van Holde (1973) on *Cancer magister* hemocyanin have indicated the presence of two constituents derived from the 5S subunit. This observation is similar to recent findings of Murray and Jeffrey (1974) for gel electrophoresis studies with *Cherax destructor;* they were able to isolate, in the presence of SDS, three different monomers in the ratio of about 7:2:1. Sedimentation equilibrium studies revealed the main monomer to have a molecular weight of 74,700 while estimates on the other two monomers gave values near 75,000. When treated with SDS and DTT, the 17S component obtained by gel filtration at pH 10 was found to contain two different monomers. Ferguson plots from samples electrophoresed at pH 10 yielded a dimer and two monomers.

It is essential to derive the possible effects of these subunit heterogeneities on the reassociation process and the oxygen transport system. To this end, initial studies into the number and size of hemocyanin subunits were undertaken. Parallel experiments utilizing gel filtration, sedimentation velocity and polyacrylamide gel electrophoresis (PAGE) were made on the hemocyanin components from two species. Suitable material for study was obtained from a swimming crab *Ovalipes catharus* located in shallow ocean inlets in the North Island of New Zealand, and from the spiny lobster *J. edwardsii,* which is indigenous to the deeper waters of Cook Strait and other coastal regions of New Zealand.

Experimental

Preparation of Hemocyanins. Live *Ovalipes* crabs were bled by leg puncture whereas the hemolymph from *Jasus* was extracted by ventral tail perforation. Hemolymph extracted from *Ovalipes* (10-20 ml per animal) had a physiological pH of 7.4 and contained 30-50 mg/ml of hemocyanin, whilst the hemolymph obtained from *Jasus* (30-60 ml per animal) had a physiological pH of 6.9 and contained 50-70 mg/ml hemocyanin. The collected blood was allowed to clot (about 2 h) and the resultant gel extracted with several volumes of 0.1 M Tris - 0.01 M $CaCl_2$ (pH 7.0). The decant from each extraction was centrifuged for 15 min at 15,000 r.p.m. (26,000 g) at $4^{\circ}C$ to remove particulate matter; this was followed by filtration of the pooled supernatants through a 0.45 μm Millipore filter. Freed of bacteria, the blood was centrifuged on a Beckman Model L preparative ultracentrifuge for 18 h at 30,000 r.p.m. (90,700 g). The clear supernatant was discarded and the blue pellet washed and resuspended in 0.1 M Tris - 0.01 M $CaCl_2$ (pH 7.0). After a second spin for 18 h at 30,000 r.p.m. and subsequent resuspension, *Jasus* and *Ovalipes* hemocyanins exhibited $O.D._{278}/O.D._{338}$ ratios of 4.0 and 4.74 respectively. The final hemocyanin solution was flushed with carbon monoxide until disappearance of all blue color, ampuled and stored at $4^{\circ}C$. Hemocyanin solutions can be preserved sealed in this manner for a period of one year and possibly longer without significant change in oxygen-binding capacity (De Ley and Lontie, 1970). Samples were removed from ampules prior to use, millipored and re-equilibrated with nitrogen-free oxygen or filtered air.

Aliquots of hemocyanin were dialyzed for two to three days with several changes against 0.1 M Tris-HCl - 0.01 M $CaCl_2$ at pH 7.0, 0.1 M Tris - HCl at pH 8.8, and 0.05 M glycine-NaOH - 0.002 M EDTA at pH 10.6, using standard recipes (Long, 1961). Column chromatography of the material dialyzed at pH 7.0 and pH 8.8 was performed on Biogel Agarose 5M whereas samples dialyzed at pH 10.6 were fractionated on Sephadex G-200. "Stripped" samples of protein (Sullivan et al., 1974) were prepared by exhaustive dialysis at pHs 7.0, 8.9 and 10.6 in buffers containing 0.01 M EDTA.

Concentrations of hemocyanin solutions were determined from dry weights of samples dialyzed at pH 7.0, 8.8 and 10.6 in the above buffers. Volumes of 5 ml of both protein and dialysate were lyophilized under vacuum at $0^{o}C$ then heated to $110^{o}C$ to constant weight. Dialyzed protein solutions were diluted in their respective buffers and their absorptions measured at 280 nm on a Pye Unicam or Beckman DB spectrophotometer.

Ultracentrifugation. All experiments were carried out at $20^{o}C$ on a Spinco Model E analytical ultracentrifuge equipped with RTIC unit. Single-sector 12 mm aluminium centerpieces ($4^{o}C$) were used with an AN-D rotor for a protein concentration range of 1-6 mg/ml. In general, speeds used for the 25S, 16S and 5S particles were 40,000, 48,000 and 56,000 r.p.m. respectively. The Schlieren patterns obtained were photographed on Kodal Metallographic plates and measured on a Nikon two-way comparator. Sedimentation coefficients were calculated with a computer program incorporating a least-squares analysis; values obtained were corrected to the viscosity of water at $20^{o}C$. The relative viscosities of the solutions used were measured in a Ubeloholde viscometer with a flow time of 325 s for distilled water at $25.00^{o}C \pm 0.01^{o}C$.

Polyacrylamide Gel Electrophoresis. Introductory studies were conducted using the continuous buffer system of Weber et al. (1972), but for the majority of experiments, the discontinuous system of Loehr and Mason (1973) was employed. Gels were cast at the temperature of the run in glass tubes with dimensions of 8.5 cm by 0.7 mm for proteins in SDS, and 12-22 cm in length for studies on native hemocyanins.

In general, 4-12% gels were polymerized with an acrylamide to N,N'-methylenebisacrylamide ratio of 37:1 in the presence of the gel buffer (Table 1), 0.1% N,N,N',N'-tetramethylenediamine, and 0.1% ammonium persulfate. Additional gels were prepared for the discontinuous system in a 40:1 ratio with 0.05% ammonium persulfate. Buffering systems are outlined in Table 1.

Native proteins, "stripped" proteins, and pooled samples from gel elution profiles were all prepared at pH 7.0, 8.9 and 10.6. All protein solutions were prepared for electrophoresis by dissolving them in the appropriate gel buffer, containing in addition 20% sucrose and 0.07% bromophenol blue. Denaturation of protein with SDS and 2-mercaptoethanol or DTT was carried out in the buffer indicated (Table 1) at $100^{o}C$ for 2 Min. Both denatured and undenatured samples were loaded on to gels in the order of 1-20 μg per gel (1-100 μl vol). In the case of runs with DTT, the gels were incubated at $4^{o}C$ for several hours in a 0.06% solution of this reagent in 0.035 M Tris-sulfate (pH 8.8) containing 10^{-4} M EDTA and 0.2% SDS.

In experiments with SDS, both hemocyanins were checked for possible proteolytic cleavage of the molecule during incubation by reacting 0.9 ml of protein at $20^{o}C$ for 2 h with 0.1 ml of 0.0125 M phenylmethylsulfonyl fluoride (solution in 75% isopropanol), and then subjecting them to electrophoresis as described. No significant change in the bands obtained was observed.

Table 1. Conditions for polyacrylamide gel electrophoresis

Buffer system Gel	Tank	Acrylamide concentration (%, w/v)	Current m A/gel	Gel length (cm)	Approximate time at current (h)
0.035 M Tris-sulfate (a) pH 7.8 (b) pH 8.8	0.03 M Tris-acetate (a) pH 7.3 (b) pH 8.3	3 to 4	4	14	(a) 3-4 (b) 2-3
		5 to 7	4	18 & 22	(a) 4-5 (b) 3-4
Fairbanks et al. (1971); Loehr and Mason (1973).					
0.035 M Tris-sulfate, 10^{-4} M EDTA, 0.2% SDS pH 8.8	0.03 M Tris-acetate, 10^{-3} M EDTA, 0.02% SDS pH 8.3	4	2	(a) 8.5	2
			7	(b) 12.5	3.5
		6	4	(a) 7.0	1.5
			8	(b) 12.5	4
		10	4	7.0	2
Fairbanks et al. (1971); Loehr and Mason (1973).					
0.05 M glycine-0.002 M EDTA pH 10.6		4 to 12	7	(a) 8.5 (b) 12.5	2-2.75 3-3.75
Murray and Jeffrey (1974)					
0.06 M NaH_2PO_4 H_2O-0.144 M Na_2 $HPO_4 \cdot 7H_2O$, 0.2% SDS pH 7.6 (1:8 dilution used)		5	8	8.5	5
		10	8	(a) 8.5	7
				(b) 12.5	8.5
Weber et al. (1972)					

Electrophoresis was performed on a Shandon apparatus at half the gel amperage indicated in Table 1 for 30 min and then at the full amperage for the duration of the run. Discontinuous systems were run at 4°C while continuous systems were operated at 20°C.

Following electrophoresis, gels were stained according to the method of Fairbanks et al. (1971). To remove SDS, gels were soaked in 25% isopropanol including 10% acetic acid for 6 h and then stained overnight in a solution containing 25% isopropanol, 10% acetic acid and 0.05% Coomassie Brilliant Blue R. This was succeeded by treatment for 6 h in a solution containing 10% isopropanol, 10% acetic acid and 0.005% Coomassie Brilliant Blue R. Destaining was done for 24 h in 10% acetic acid on a shaker bath, and then the gels were equilibrated in 7.5% acetic acid before being measured, and photographed with transmitted light. Selected gels were scanned at 595 nm on a Chromoscan (Joyce, Loebl and Co.). A second staining technique was also employed using 0.04% (w/v) Coomassie Brilliant Blue G in 3.5% (w/v) perchloric acid (Reisner et al., 1975) with destaining in 3.5% perchloric acid. The value of the method appears to be the speed with which the bands appear, with a minimum of background staining. However, it did not prove to be as sensitive for the staining of SDS gels, and in addition there is a less of color intensity in the bands upon standing in 3.5% perchloric acid during destaining.

In some selected gels, the bands were cut out and the protein-dye complex was extracted with 25% pyridine. After shaking for 48 h, the solutions were measured at 605 nm (Fenner et al., 1975). Results for the proportions of components in a gel pattern using this method compared favorably with those obtained from the Chromoscan.

With SDS-polyacrylamide gel runs, the following protein standards were used concurrently to estimate molecular weights (Shapiro et al., 1967): cytochrome c, myoglobin, carboxypeptidase a, trypsin, chymotrypsinogen, pepsin, catalase, phosphorylase a, ovalbumin, bovine serum albumin, and β-galactosidase.

Native hemocyanins dialyzed for three days at pH 10.6 in 0.05 M glycine NaOH with 0.002 M EDTA, together with six protein standards, were electrophoresed on 4% to 12% gels at pH 10.6. Similar runs were carried out at pH 7.0 in native materials using gels of 3 to 7%. A plot of log R_F against T was made according to the Ferguson (1964) relationship

$$\log R_F = \log (R_F)_o + K_R T$$

where R_F is the mobility of the protein relative to that of bromophenol blue, T is the total gel concentration, K_R is the retardation coefficient, and $(R_F)_o$ is R_F value at a gel concentration of zero. Values of K_R were obtained for both hemocyanins and standards, and molecular weights of the hemocyanins were obtained by interpolation from a plot of K_R against molecular weight for the standards.

Analysis for Metal Ions. The copper content of hemocyanin solutions of *Jasus* and *Ovalipes* was determined on a Perkin Elmer 306 or Varian Techtron Model 70 flame atomic absorption spectrophotometer. The hemocyanins had been purified by ultracentrifugation, followed by gel chromatography on Biogel A-1.5 M. In addition, analysis for Na, K, Ca, Mg, Fe, Ag, Sr, Cd, Co, Ba, Zn, Mn, Ni, Au, Pt were performed on serum ultrafiltrates prepared by the low pressure ultrafiltration method of Everall and Wright (1958). Corrections for molecular absorption were made using a deuterium lamp continuum source. Any partial ionization effects were overcome by addition of approximately 1500 ppm $SrCl_2$ or $CsCl_2$ to the samples. All standards employed buffering conditions identical to those of the protein samples. In some cases, 10 mM EDTA was added to the samples to chelate the metals and thereby improve their detection level.

The metal analyses are presented in Table 2. The copper content was found to be 0.165% for *Jasus* and 0.175% for *Ovalipes*, corresponding to subunit molecular weights of 77,000 and 72,600 respectively, assuming there are two copper atoms per monomer. The value for *Jasus* compares with a molecular weight of 81,000 for the monomer from sedimentation equilibrium studies (Ellerton et al., 1976).

The presence of relatively high concentrations of Ca^{2+} and Mg^{2+} in the ultrafiltrate led us to include divalent cations (usually 0.035 M or 0.01 M Ca^{2+}) in buffer systems where undissociated protein was being studied.

Reassociation Studies. In order to assess the ability of the two hemocyanins to reassociate, and the influence of the divalent cation Ca^{2+} on this reassociation, we dialyzed samples of our stock proteins from pH 7.0 to either pH 8.8 or 10.6. The samples at pH 8.8 were then dia-

Table 2. Analyses for metals in the serum ultrafiltrate of *J. edwardsii* and *O. Catharus* hemocyanins from atomic absorption spectroscopy

Element	Concentration mol/l	
	Jasus	*Ovalipes*
K	0.007	0.007
Na	0.481	0.311
Ag	0.0	0.0
Sr	0.6×10^{-4}	0.4×10^{-4}
Ba	0.0	0.0
Zn	0.0	0.0
Fe	0.0	0.0
Ca	0.035	0.012
Mg	0.014	0.009
Mn	0.44×10^{-5}	0.5×10^{-6}
Cu	0.0	0.0

lyzed back to pH 7.0, whereas samples at pH 10.6 were dialyzed either to pH 8.8 or 7.0. Similar experiments were made both with and without Ca^{2+} in the buffer when the pH was lowered. The dialyzed samples at pH 7.0 and 8.8 were then analyzed by gel electrophoresis at pH 7.8 and 8.8 respectively.

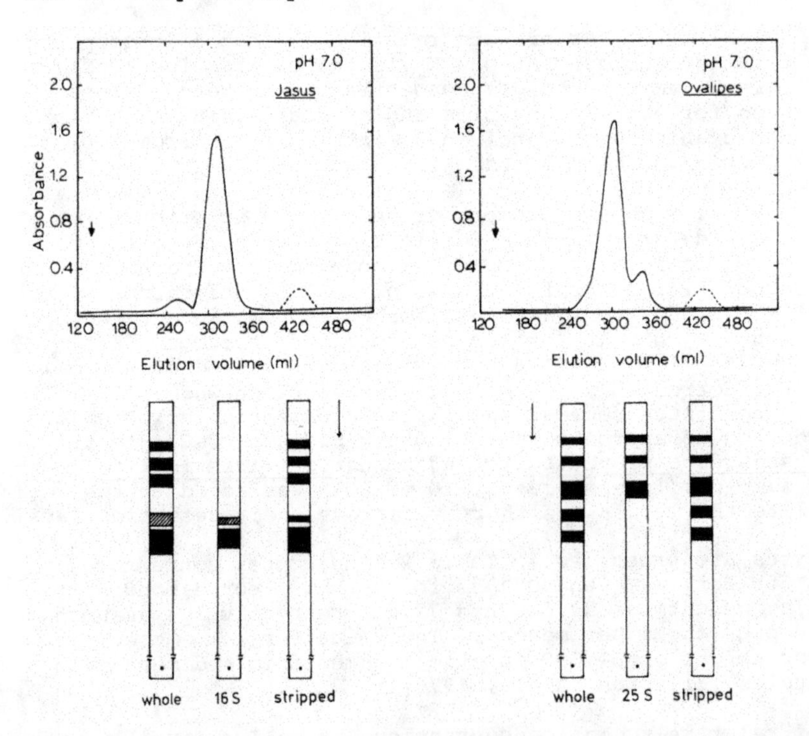

Fig. 1. Gel chromatography (*upper*) and gel electrophoresis (*lower*) of *J. edwardsii* and *O. catharus* hemocyanins. Fractionation was performed at pH 7.0 and 4°C on Biogel A-5M on a 2.5 cm × 84 cm column with a flow rate of 25 ml. *Arrow*: void volume, as determined with blue dextran. Absorbance was measured at 280 nm, except for *dashed lines*, which represent bands with an absorbance at 256 nm (no 280 nm absorbance), which eluted after the hemocyanin bands. Gel electrophoreses were run at pH 7.8; "*16S*" and "*25S*" bands: electrophoresis of samples from the main peaks obtained by column chromatography of *Jasus* and *Ovalipes* respectively

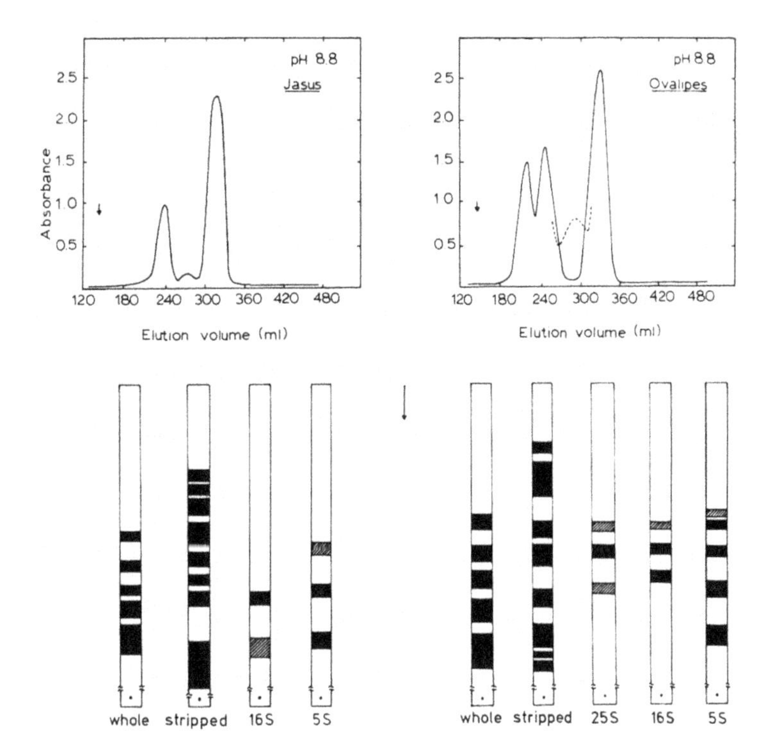

Fig. 2. Gel chromatography (*upper*) and gel electrophoresis (*lower*) of hemocyanins at pH 8.8. Other details are the same as in Figure 1, except that the *dashed line* for *Ovalipes* represents the absorbance at 280 nm of a hemocyanin species of intermediate size that appeared in some elutions

Results and Discussion

The elution profiles from gel chromatography, and the electrophoretic patterns, at the three pH values studied, are presented in Figures 1-3. The relative proportions of components present were obtained from integration of the elution profiles and from the sedimentation velocity experiments, and are summarized in Table 3.

Native Hemocyanins at pH 7.0–7.8. There is general agreement between techniques on the relative proportions of 25S and 16S components present. Approximately 80% of *Ovalipes* hemocyanin is observed in the 25S form, whereas *Jasus*, in the presence of Ca^{2+}, contains only 6% in the 25S form. Earlier studies on *Jasus* (Ellerton et al., 1976) had been made in the absence of divalent metal cations, and under those conditions no 25S particle was observed. In addition, we dialyzed the two hemocyanins in concentrations of $CaCl_2$ ranging from 0 to 100 mM Ca^{2+}, and no change in the basic electrophoretic pattern was observed. Also, samples of hemocyanins that had been "stripped" with EDTA at pH 7 showed no changes in pattern on electrophoresis, in contrast with similar treatment of the protein at pH 8.9, where stripping produced additional bands of reduced mobility not seen in untreated pH 8.9 materials.

The electrophoretic bands from samples separated as "25S" and "16S" fractions by elution on the Biogel agarose 5M column were compared with bands obtained from the unfractionated hemocyanin to enable us

Table 3. Analyses of components of *Jasus edwardsii* and *Ovalipes catharus* hemocyanins by various physical techniques

1. At pH 7.0-7.8

Species	Technique	pH	Ca + present	% Analyses of components[a] 25S	16S
Jasus	Gel filt.	7.0	0.01 M	4	96
	Sed.	7.0	0.01 M	6	94
	Sed.	7.0	0	-	100
	P.A.G.E.	7.8	0.01 M	5.6 (0.3+0.4+4.9)	94.4 (4.5+89.9)
Ovalipes	Gel filt.	7.0	0.01 M	84	16
	Sed.	7.0	0.01 M	78	22
	P.A.G.E.	7.8	0	85 (5+2+78)	15 (9+6)
			0.01 M	82 (2+3+77)	18 (9+9)
			0.01 M	(3+4+75)[b]	(10+8)[b]
			0.035 M	84 (1+4-79)	16 (8+8)

2. At pH 8.8, no Ca^{2+}

Species	Technique	%Analyses of components 25S	16S	Int.	5S
Jasus	Gel filt.				
Jasus	Gel filt.	0	21	11	68
	Sed.	0	52	0	48
	P.A.G.E.	0	40	0	60
Ovalipes	Gel filt.	25	35	0	40
	Gel filt.	31	33	17[c]	19
	Sed.	21	29	0	50
	P.A.G.E.	32	12	0	56

3. At pH 10.5, no Ca^{2+}

Species	Technique	% Analysis of Components[a] Peak 1	Peak 2	5S
Jasus	Gel filt.			
	1. Glycine-EDTA	35	11	54
	2. 0.04 M $NaHCO_2$-NaOH	61	-	39
	P.A.G.E. (Glycine-EDTA)	-	-	(72+10+18)
Ovalipes	Gel filt. 1. Glycine-EDTA	18	12	70
	2. 0.04 M $NaHCO_2$-NaOH	57	9	34
	P.A.G.E. (Glycine-EDTA)	-	-	(89+5+6)

[a] Figures in brackets represent relative percentages in each gel band within a sedimenting component.
[b] Separate analysis obtained by extraction of bands with 25% pyridine (Fenner et al., 1975).
[c] Appearance of an intermediate component was not consistent.

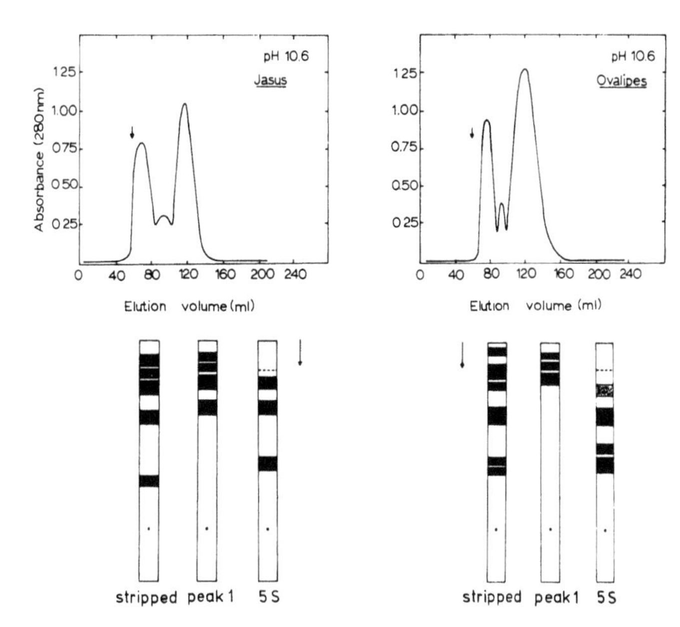

Fig. 3. Gel chromatography (*upper*) and gel electrophoresis (*lower*) of hemocyanins at pH 10.6. Fractionation was performed at pH 10.6 and 4°C on Sephadex G-200, on a 2.5 cm × 36 cm column, with a flow rate of 9 ml/h

to assign the origin of the five electrophoretic bands observed with both *Jasus* and *Ovalipes*. The upper three bands were asigned to the 25S particle, and the lower two bands to the 16S particle.

Staining of the electrophoretic bands of both hemocyanin species for carbohydrate according to the method of Kapitany and Zebrowski (1973) indicated that all of the bands were glycoproteins.

At pH 7.0, in the presence of 0.01 M Ca^{2+}, the ratios of OD_{278}/OD_{338} for *Jasus* and *Ovalipes* were 4.0 and 4.74 respectively after preparative ultracentrifugation. Separation of 25S and 16S components by gel chromatography gave ratios of 4.8 and 4.6 respectively for each particle for *Jasus* and 4.9 and 14.4 for *Ovalipes*. It is interesting to note the lower oxygen affinity of the 16S *Ovalipes* fraction, which may have a bearing on the observed heterogeneity of 25S and 16S electrophoretic bands.

In order to study the apparent heterogeneities of the 25S and 16S fractions from gel filtration, Ferguson plots were made from the data on 3-7% gels. The 16S bands for both *Jasus* and *Ovalipes* displayed parallel lines (identical K_R values) typical of the charge isomer phenomenon described by Hedrick and Smith (1968). Likewise, the 25S bands displayed a second set of parallel lines, indicating that they too are charge isomers. The K_R values of these bands are in the order of twice those obtained for the 16S particle, indicating that the 25S particle is a dimer of the smaller particle. The scarcity of suitable protein markers in this molecular weight range makes computation of molecular weights from K_R values difficult, but it has already been shown from sedimentation equilibrium that the molecular weight of the 16S particle of *J. edwardsii* hemocyanin is 490,000 (Ellerton et al., 1976). The K_R values of the 16S components of *Jasus* and *Ovalipes* are close, indicating that the molecular weight of the 16S particles of both species are approximately 490,000. Similar sedimentation studies on *O. catharus* are currently in hand and will be presented in a future paper.

Native Hemocyanins at pH 8.8. At pH 8.8 differences in the percentage analysis of 25S, 16S and 5S components were observed by comparison of the data from gel filtration, sedimentation and gel electrophoresis. It is known from sedimentation velocity studies of *Jasus* hemocyanin at various pH (Ellerton et al., 1976) that the protein is in a state of transition to the 5S dissociated form at this pH so it seems reasonable to ascribe differences in the analysis to time- and temperature-dependent changes.

Gel electrophoresis on 14 and 18 cm length gels gave five bands for both species of native hemocyanin. On the other hand, samples that had been dialyzed with EDTA present in the Tris-glycine buffer showed eight bands on electrophoresis; some of the bands moved with lower mobilities than was observed for the five bands in the absence of EDTA. The five bands obtained with the native hemocyanins show close similarities to the hemocyanin bands reported by Maguire and Fielder (1975) on several species of portunid crabs.

We also performed gel electrophoresis on samples isolated by narrow pooling from the 25S, 16S and 5S peaks from gel chromatography. The 5S peak of *Ovalipes* was most evident in the fifth band, together with small amounts of the other four bands. 5S protein from *Jasus* electrophoresis was found mainly in the fifth band, with smaller amounts in two other bands. The 25S and 16S peaks of *Ovalipes* electrophorese as mixtures in varying proportions in the three upper bands. The 16S component of *Jasus* shows traces in the fifth band, and a larger band of lower mobility in a position corresponding to the third or fourth band of whole protein. The results indicate that the systems are in transition between associated and dissociated states. Busselen (1970), studying *Carcinus maenas* hemocyanin at pH 9 found that components separated by electrophoresis showed multiple bands when they were re-electrophoresed after separation. It was concluded that re-equilibration was taking place between components in that system.

Native Hemocyanins at pH 10.6. Gel electrophoreses on samples from the first and 5S peaks of fractions separated on a Sephadex G-200 column were compared with bands produced by an unfractionated sample. The 5S component of both hemocyanins produced the three bands of greatest mobility Sometimes the 5S peak also yielded minor traces of bands representing higher association states of the protein.

Ferguson plots were made with 4-12% gels, 12 mm in length. A plot of R_F against gel concentration for the three main bands from the 5S peak of both hemocyanins is presented in Figure 4. The retardation coefficients, K_R, of the protein standards are plotted against molecular weight in Figure 5, and the molecular weights of the hemocyanin bands for the two species were deduced by interpolation. The results are tabulated in Table 4. The close to parallel nature of the Ferguson plots for M_1 and M_2 suggest that these particles are similar in size but different in charge, within the error of the technique. The third component has a much higher molecular weight than M_1 and M_2, and it appears to be a size isomer.

In recent studies with spider, scorpion and horseshoe crab hemocyanins (Sugita and Sekiguchi, 1975) multiple monomer bands were observed upon electrophoresis. Ferguson plots suggested that the bands in each species were charge isomers, as with M_1 and M_2 from our work. Further, the data of Murray and Jeffrey (1974), show a striking similarity with

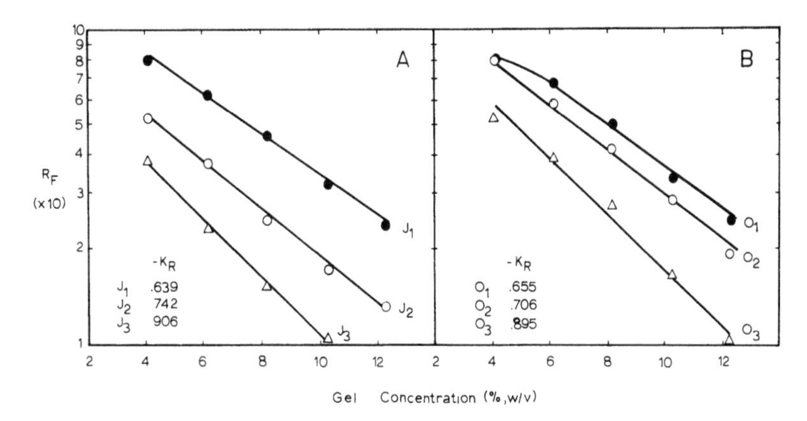

Fig. 4A and B. Determination of values of the retardation coefficient, K_R, for 5S hemocyanin from (A) *J. edwardsii*, and (B) *O. catharus*, from a plot of R_F against gel concentration. Dialyzed native hemocyanins were electrophoresed on polyacrylamide gel at pH 10.6

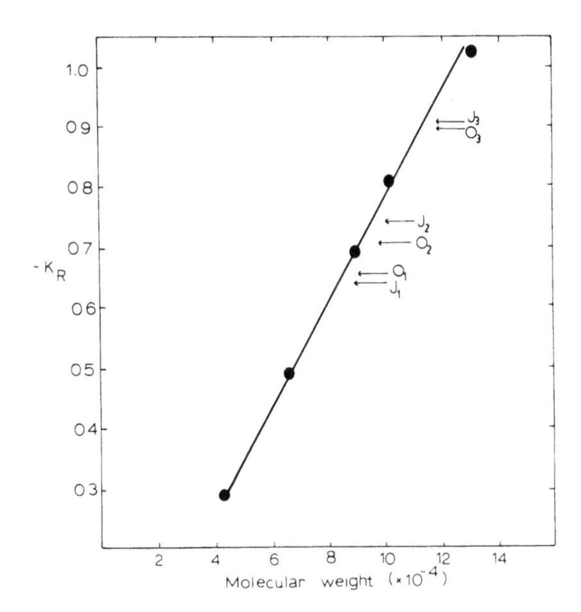

Fig. 5. Determination of the molecular weights of 5S hemocyanins from polyacrylamide gel electrophoresis at pH 10.6. K_R was plotted against molecular weight for standard proteins. J_1, J_2 and J_3 for *Jasus*, and O_1, O_2 and O_3 for *Ovalipes* correspond to monomers M_1, M_2 and M_3 respectively in Table 4

the results obtained from our electrophoreses; *Cherax destructor* shows two bands M_1 and M_2 which may be charge isomers, and a third band M_3, of higher molecular weight. Murray and Jeffrey believe that M_3 is a dimer, and in SDS they obtain a monomer band M_3 intermediate in molecular weight between M_1 and M_2.

The molecular weights of M_1, for both species are very close (M_1 = 84,000 for *Jasus* and 86,000 for *Ovalipes*), as also are the molecular weights of M_2 and M_3 in each case. However, the ratios M_1:M_2: M_3 are

Table 4. Molecular weights of 5S hemocyanins at pH 10.6 by polyacrylamide gel electrophoresis

Component	*J. edwardsii*		*O. catharus*		*C. destructor*[a]	
	MW	% Component	MW	% Component	MW	% Component
M_1	84,000	72	86,000	89	70,000	42
M_2	96,000	10	92,000	5	72,000	12
M_3	116,000	18	115,000	6	132,000	10

[a]Data of Murray and Jeffrey (1974) at pH 10.0. The percentages shown in their data are based on the total protein bands observed.

quite different; the proportion of M_1 is much greater than the proportions of M_2 or M_3. In other studies, Loehr and Mason (1973), investigating *Cancer magister* hemocyanin, obtained monomer subunits of 76,000 and 84,000 in approximately a 1:1 ratio, whereas for *C. destructor*, Murray and Jeffrey (1974) reported a ratio of M_1:M_2 of 3:1. From our work, we have obtained a ratio of 7:1 for *Jasus* and 17:1 for *Ovalipes* for M_1: M_2. It is tempting to speculate on the significance of these differences, particularly on the possible influence upon association of M_1, M_2 and M_3 to the 16S and 25S particles. We are presently investigating the separation of M_1, M_2 and M_3 by ion exchange chromatography with a view to gaining a clearer understanding of the subunit structure of these larger particles.

Gel Electrophoresis in SDS. Electrophoresis experiments in the presence of SDS and dithiothreitol or 2-mercaptoethanel were performed on both species of hemocyanins. Molecular weights of the resulting protein bands were interpolated from plots of molecular weight against R_F for 4%, 5% and 10% gels using proteins of known molecular weights. The results are summarized in Table 5. Three main bands were usually ob-

Table 5. Molecular weights of hemocyanin subunits by polyacrylamide gel electrophoresis in SDS and dithiothreitol[a]

Gel percentage	*J. edwardsii*		*O. catharus*		*C. destructor*[b]	
	MW	%	MW	%	MW	%
4%	66,600		62,000		79,000 (M_2)	20
	77,500		68,200		84,000 (M_3)	10
			73,500		86,000 (M_1)	70
6%	64,000	12	69,000	9		
	98,000	13	101,000	27		
	110,000	75	112,000	64		
	160,000[c]		146,000[c]			
10%	89,000		67,250			
	96,000		83,250			
	105,000		102,000			
			119,000[c]			

[a]Temperature of incubation was 100°C for 4% and 6% gels and 65°C for 10% gels. Runs at 37°C gave multiple bands (usually 7) in the molecular weight range 28,000-177,000 for *Jasus* and 26,000-152,000 for *Ovalipes*.
[b]Data of Murray and Jeffrey (1974).
[c]Small amount only. Generally observed if no DTT present. Murray and Jeffrey (1974) also observed a dimer band at 190,000 if no DTT was present.

served, though a fourth band appeared at times, usually in the absence of DTT. This fourth band, which we consider to be a dimer, was quite transient, only just detectable by the eye, and difficult to pick up on a Joyce-Loebl Chromoscan.

With the exception of the band of lowest molecular weight, there seems to be agreement in the molecular weights of the monomers, both from electrophoresis of the native protein at pH 10.6, and from treatment with SDS. Since three bands are observed in both cases, it would appear that none of the native subunits are conformational isomers, as all three give rise to different bands on electrophoresis in SDS (Ressler, 1973).

Comparing the relative percentages of the monomer bands in the native protein at pH 10.6 (Table 4) with the percentages in SDS (Table 5), we found a shift in proportions had taken place, so that for both species, most hemocyanin was now found in the gel band of highest molecular weight in the SDS runs. A similar shift in proportions was also observed by Murray and Jeffrey (1974) with *C. destructor* hemocyanin, and they demonstrated that their largest band still corresponded to their monomer M_1. From their results and ours, there seems little doubt that SDS interacts differently with the different monomer units producing some anomalies in the observed molecular weights.

Reassociation Studies. The results are illustrated in Figures 6 and 7. For both *Jasus* and *Ovalipes* hemocyanins, when the pH has been lowered from 8.8 to 7.0, association to both the 16S and 25S particles is observed. Despite some breakdown of the protein to units smaller than 16S, reassociation to the higher forms on lowering the pH seems to be largely complete.

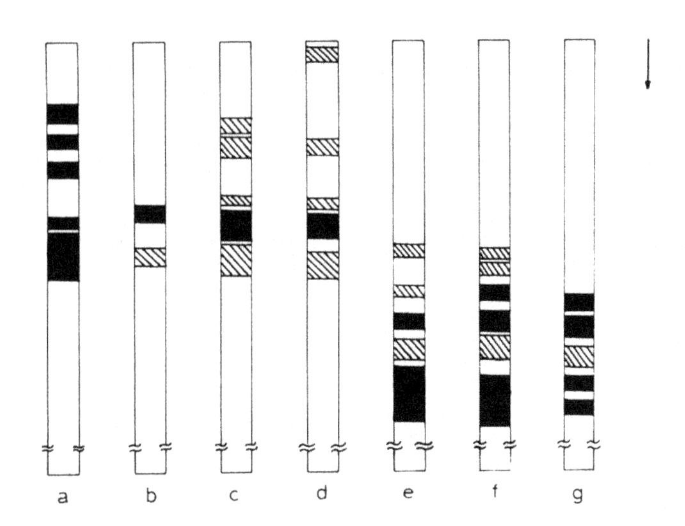

Fig. 6a-g. Reassociation of *J. edwardsii* hemocyanin. Native hemocyanin at pH 7.0 was dialyzed to pH 8.8 or 10.6, subsequently lowered back to pH 7.0 or 8.8, and then subjected to PAGE. (a) Native hemocyanin at pH 7.0 containing 0.035 M Ca^{2+}; (b) pH raised to 10.6, then lowered to 7.0; same bands for both 0.035 M Ca^{2+} present or no Ca^{2+} added; (c) pH raised to 8.8, then lowered to 7.0, no Ca^{2+} added; (d) Same as c, except 0.035 M Ca^{2+} added on lowering pH to 7.0; (e) Native hemocyanin at pH 8.8, no Ca^{2+}; (f) pH raised to 10.6, then lowered to 8.8 in the presence of 0.035 M Ca^{2+}; (g) same as f, but no Ca^{2+} present

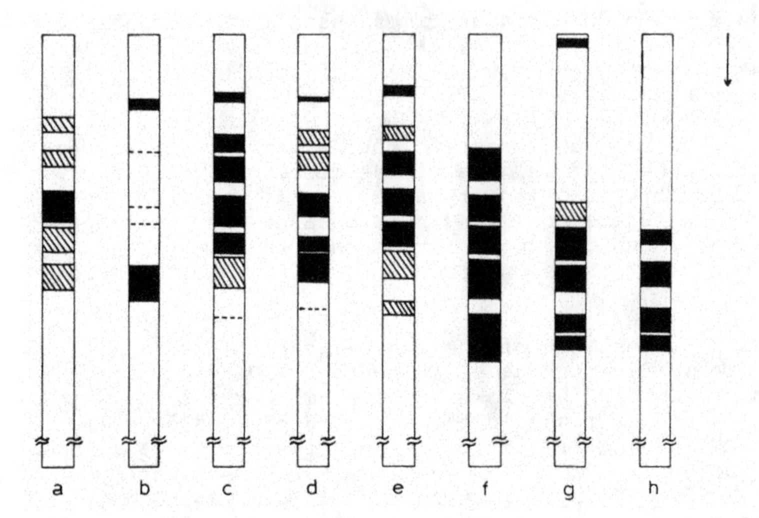

Fig. 7a-h. Reassociation of *O. catharus* hemocyanin. Similar treatment as for *Jasus* samples in Figure 6. (a) Native hemocyanin at pH 7.0 containing 0.01 M Ca^{2+}; (b) pH raised to 10.6, then lowered to 7.0, no Ca^{2+} added; (c) same as (b), but 0.01 M Ca^{2+} added on lowering pH to 7.0; (d) pH raised to 8.8, then lowered to 7.0, no Ca^{2+} added; (e) same as (d), but 0.01 M Ca^{2+} added on lowering pH to 7.0; (f) native hemocyanin at pH 8.8, no Ca^{2+}; (g) pH raised to 10.6, then lowered to 8.8, in the presence of 0.01 M Ca^{2+}; (h) same as (g), but no Ca^{2+} added

For samples that have been previously dialyzed to pH 10.16, there appears to be a reduced ability of the subunits to reassociate. Some reassociation to 16S is found on lowering the pH to 8.8 for *Ovalipes* with or without Ca^{2+}, but the presence of Ca^{2+} is required before this association can be observed with *Jasus*. On reduction of pH to 7.0, reassociation of *Jasus* hemocyanin to the 16S particle was observed both in the presence or absence of Ca^{2+}, but no 25S particle was obtained. For *Ovalipes* in the absence of Ca^{2+}, a narrow band of 25S protein was observed with the remainder not reassociated, but with the addition of divalent metal further reassociation took place, with a mixture of 25S, 16S and 5S forms being observed.

One observation was common to both species: a quantity of hemocyanin precipitated out of solution upon reduction of the pH from 10.6 to 8.8 or 7.0 in the presence of Ca^{2+}. The electrophoresis experiments reported here were performed on the supernatant after removal of the precipitate. The precipitate was soluble in sodium hydroxide and the solution thus obtained had a UV absorption spectrum typical of proteins, with an absorption peak at 280 nm. We have concluded that isoelectric precipitation of some or all of the protein bands has occurred on addition of the Ca^{2+}. This is consistent with the observation, under these conditions, of an extremely slow-moving band obtained in a number of experiments which we believe is attributable to a Ca^{2+}-induced aggregation of the protein.

The ability of Ca^{2+} to promote stabilization of these proteins was illustrated in separate experiments at pH 8.8. Samples of each species were dialyzed to this pH in the absence of Ca^{2+}, and then further dialysis was performed with Ca^{2+} added, whilst maintaining the same pH. The resulting electrophoreses with samples containing Ca^{2+} at pH 8.8 showed striking similarities to the native protein at pH 7.0. The 25S bands were all restored in *Ovalipes*, though with *Jasus* only the main 25S band was restored, along with the major 16S band.

Our experiments clearly illustrate that the extent of reassociation is species-dependent and Ca^{2+}-dependent; it is easier to obtain reassociation of *Ovalipes* than *Jasus*, and this reassociation is achieved more readily in the presence of Ca^{2+}.

Influence of EDTA on Absorption Spectra. In the course of studying hemocyanins which had been "stripped" with EDTA of bound divalent cations (Sullivan et al., 1974), we have observed changes in the UV absorption spectra of these proteins. Aliquots of stock protein were exhaustively dialyzed at either pH 7.0 or 10.6, in the presence or absence of Ca^{2+} or EDTA. Spectral scans were made from 190 to 300 nm on a Pye Unican SP1800 spectrophotometer after a suitable dilution of the dialyzed samples.

Jasus and *Ovalipes* have absorption maxima at 214 and 212 nm respectively at pH 7 and in the presence of Ca^{2+}. The bands are observed to undergo a red shift upon removal of Ca^{2+} during the dialysis, and in addition there is a change in the magnitude of the peak. These changes are enhanced by the presence of EDTA and by increasing pH. We are currently looking further into these observations.

Acknowledgments. The authors wish to thank the Applied Biochemistry Division, D.S.I.R., and the late Dr. J.W. Lyttleton, for the use of the analytical ultracentrifuge. We are also grateful to Dr. R.W. Boyle and G.E.M. Aslin, Geological Survey of Canada, for assistance with the Varian Techtron and Perkin Elmer atomic absorption analyses. Our acknowledgments are extended to Alan F. Bagnall for providing the crabs and to the Southern Cross Fisheries Company Limited, Wellington, for supplying the spiny lobsters. This work was supported by funds for equipment from the New Zealand University Grants Committee, and the internal Research Committee, Victoria University of Wellington, to whom grateful acknowledgment is made.

References

Busselen, P.: The electrophoretic heterogeneity of *Carcinus maenas* hemocyanin. Arch. Biochem. Biophys. 137, 415-420 (1970)

Carpenter, D.E., Van Holde, K.E.: Amino acid composition, amino-terminal analysis, and subunit structure of *Cancer magister* hemocyanin. Biochemistry 12, 2231-2238 (1973)

De Ley, M., Lontie, R.: The preservation of haemocyanin under carbon dioxide. FEBS LETT. 6, 125-127 (1970)

Di Giamberardino, L.: Dissociation of *Eriphia* hemocyanin. Arch. Biochem. Biophys. 118, 273-278 (1967).

Dunker, A.K., Rueckert, R.R.: Observations on molecular weight determinations on polyacrylamide gels. J. Biol. Chem. 244, 5074-5080 (1969)

Ellerton, H.D., Carpenter, E.E., van Holde, K.E.: Physical studies of hemocyanins. V. Characterization and subunit structure of the hemocyanin of *Cancer magister*. Biochemistry 9, 2225-2232 (1970)

Ellerton, H.D., Collins, L.B., Gale, J.S., Yung, A.Y.P.: The subunit structure of the hemocyanin from the crayfish, *Jasus edwardsii*. 1976 (in press)

Everall, P.H., Wright, G.H.: Low-pressure ultrafiltration of protein-containing fluids. Med. Lab. Tech. 15, 209-213 (1958)

Fairbanks, G., Steck, T.L., Wallach, D.F.H.: Electrophoretic analysis of the major polypeptides of the human erythrocyte membrane. Biochemistry 10, 2606-2616 (1971)

Fenner, C., Trant, R.R., Mason, D.T., Wikman-Coffelt, J.: Quantification of Coomassie Blue-stained proteins in polyacrylamide gels based on analysis of eluted dye. Analyt. Biochem. 63, 595-602 (1975)

Ferguson, K.A.: Starch gel electrophoresis application to the classification of pituitary proteins and polypeptides. Metabolism 13, 985-1002 (1964)
Hedrick, J.L., Smith, A.J.: Size and charge isomer separation and estimation of molecular weight of proteins by disc gel electrophoresis. Arch. Biochem. Biophys. 126, 155-164 (1968)
Holde, K.E. van, Bruggen, E.F.J. van: In: Subunits in Biological Systems. Timasheff, S.N., Fasman, G.D. (eds.). New York: Marcel Dekker, 1971, pp. 1-53.
Kapitany, R.A., Zebrowski, E.J.: A high resolution PAS stain for polyacrylamide gel electrophoresis. Analyt. Biochem. 56, 361-369 (1973)
Loehr, J.S., Mason, H.S.: Dimorphism of *Cancer magister* hemocyanin subunits. Biochem. Biophys. Res. Commun. 51, 741-745 (1973)
Long, C. (ed.): In: Biochemistrs' Handbook. Princeton: Van Nostrand, 1961, p. 28
Maguire, G.B., Fielder, D.R.: Disc electrophoresis of the haemolymph proteins of some portunid crabs (decapoda: Portunidae) - I. Effects of storage. Comp. Biochem. Physiol. 52A, 39-42 (1975)
Moore, C.H., Henderson, R.W., Nichol, L.W.: Estimation of the polymerization behaviour of *Jasus lalandii* haemocyanin and its relation to the allosteric binding of oxygen. Biochemistry 7, 4075-4085 (1968)
Murray, A.C., Jeffrey, P.D.: Hemocyanin from the Australian freshwater crayfish, *Cherax destructor*. Subunit heterogeneity. Biochemistry 13, 3667-3671 (1974)
Pickett, S.M., Riggs, A.F., Larimer, J.L.: Lobster hemocyanin: properties of the minimum functional subunit and of aggregates. Science 151, 1005-1007 (1966)
Reisner, A.H., Nemes, P., Bucholtz, C.: The use of Coomassie Brilliant Blue 250 perchloric acid solution for staining in electrophoresis and isoelectric focusing on polyacrylamide gels. Analyt. Biochem. 64, 509-516 (1975)
Ressler, N.: A systematic procedure for the determination of the heterogeneity and nature of multiple electrophoretic bands. Analyt. Biochem. 51, 589-610 (1973)
Roxby, R., Miller, K., Blair, D.P., van Holde, K.E.: Subunits and association equilibria of *Callianassa californiensis* hemocyanin. Biochemistry 13, 1662-1668 (1974)
Salvato, B., Sartore, S., Rizzotti, M., Ghiretti-Magaldi, A.: Molecular weight determination of polypeptide chains of molluscan and arthropod hemocyanins. FEBS Lett. 22, 5-7 (1972)
Shapiro, A.L., Vinuela, E., Maizel, J.V., Jr.: Molecular weight estimation of polypeptide chains by electrophoresis in SDS-polyacrylamide gels. Biochem. Biophys. Res. Commun 28(5), 815-820 (1967)
Sugita, H., Sekiguchi, K.: Heterogeneity of the minimum functional unit of hemocyanins from the spider (*Argiope bruennichii*) the scorpion (*Heterometrus* sp.), and the horseshoe crab (*Tachypleus tridentatus*). J. Biochem. (Tokyo) 78, 713-718 (1975)
Sullivan, B., Bonaventura, J., Bonaventura, C.: Functional differences in the multiple hemocyanins of the horseshoe crab, *Limulus polyphemus* L. Proc. Natl. Acad. Sci. 71, 2558-2562 (1974)
Weber, K., Osborn, M.: The reliability of molecular weight determinations by dodecyl sulfate-polyacrylamide gel electrophoresis. J. Biol. Chem. 244, 4406-4412 (1969)
Weber, K., Ringle, J.R., Osborn, M.: In: Methods in Enzymology. Hirs, C.H.W., Timasheff, S.N. (eds.). New York-London: Academic Press, 1972, Vol. 26, pp. 3-27

Crystals of *Limulus* Hemocyanin and Its Subunits

K. A. MAGNUS AND W. E. LOVE

Introduction

Hemocyanin from the arthropod *Limulus polyphemus* weighs about 3×10^6 daltons and can be dissociated into subunits (Sullivan et al., 1974) that each have one oxygen-binding site and two copper atoms. The subunits, each weighing about 7×10^4 daltons, retain their ability to bind oxygen reversibly and they can be chromatographically separated into at least seven fractions (Sullivan et al., 1974).

We have begun an X-ray crystallographic analysis of the structure of *Limulus* hemocyanin. Thus far, we have obtained crystals both of whole molecules (60S) and from the five fractions I-V of Sullivan et al. (1974). The crystal forms from which preliminary X-ray data have been obtained are believed to be unsuitable for high resolution structure analysis. Therefore conditions are being sought for the growth of more suitable crystals.

Results and Discussion

Representative examples of crystal forms obtained to date are shown in Figures 1-7. With the exception of those shown in Figure 4, all crystals were grown using polyethylene glycol[1] (PEG 6000 J. T. Baker Chemical Co., Phillipsburg, N.J. 08865) as the precipitating agent. All of the crystals were grown at room temperature and are thought to be oxygenated. Unsuitably small crystals of whole molecules grew from solutions that were 0.05 M Tris, 0.05 M glycine, 0.5 M NaCl, 0.005 M $NaCl_2$, pH 8.4, 5% w/v PEG 6000 and approximately 3% w/v hemocyanin. These conditions tend to prevent the dissociation of whole molecules into subunits (J. Bonaventura and C. Bonaventura, personal communication). Recently some progress has been made toward growing larger single crystals.

The crystals thus far obtained from fractions I, III and V, Figures 2, 5, and 7 respectively, have all been too small. Note in Figure 5 that there are two distinct crystal forms. The slightly smaller needles are birefrigent whereas the larger plate-like crystals are not.

Of the three forms from which preliminary X-ray data have been taken, form II A, Figure 3, has been most studied. These crystals from fractions II have the symmetry of the triclinic space group P1 and the lattice constants are: a = 157 Å, b = 183 Å, c = 149 Å, $\alpha = 90^{\circ}$, $\beta = 110^{\circ}$, $\gamma = 108^{\circ}$. From Matthews' (1968) tabulation of the fractional solvent content of other protein crystals it is estimated that this unit cell could hold between 16 and 32 molecules weighing 7×10^4 daltons. If we assume this subunit is a sphere, with a partial specific

[1]Dr. Joseph Bonaventura, Duke University Marine Laboratory, first discovered polyethylene glycol will crystallize whole molecules of *Limulus* hemocyanin.

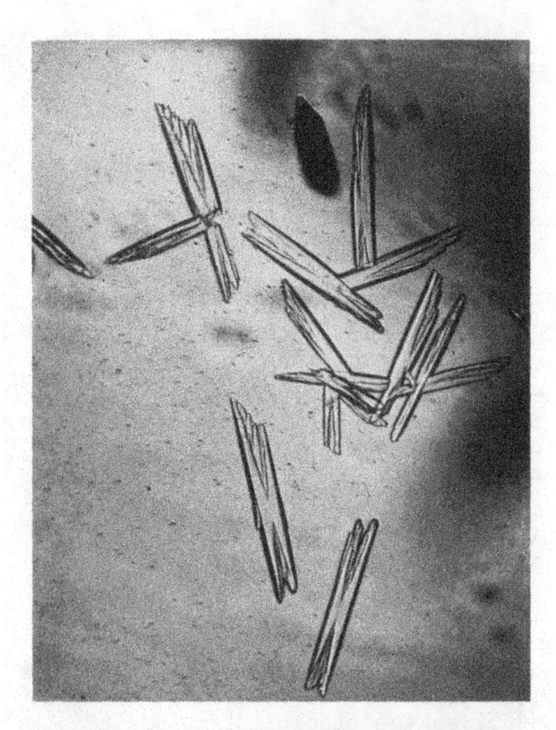

Figs. 1-7 are photographs of crystals of whole molecules and of crystals grown from fractions I through V of Sullivan et al. (1974). Fig. 1 shows crystals obtained from a solution of whole molecules

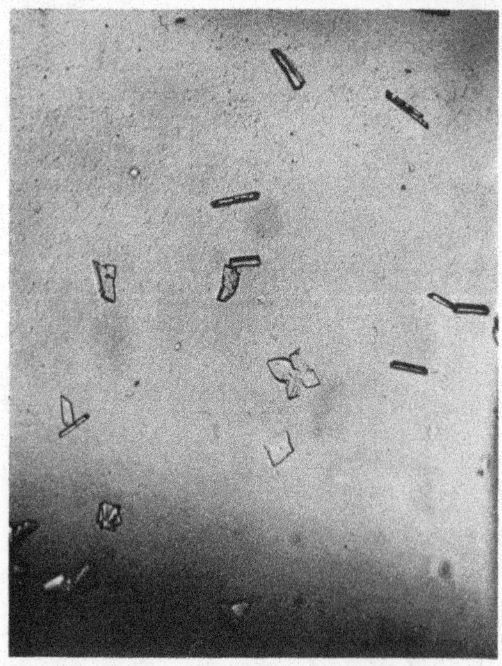

Fig. 2. Crystals from fraction I

volume of 0.75 and that the subunits are cubic closest packed, then this unit cell can hold 34 molecules. The interstices between the molecules will be 24.95% of the unit cell volume (Kasper and Lonsdale,

Figs. 3 and 4. Crystals from fractio
II

Fig. 4

1959) and such crystals would be relatively unhydrated. These unit cells thus probably contain somewhat less than 30 molecules. Current X-ray techniques are incapable of determining the structures of crystals with this many molecules and of this size in the asymmetric unit.

Figs. 5, 6, and 7. Crystals from fractions III, IV and V respectively

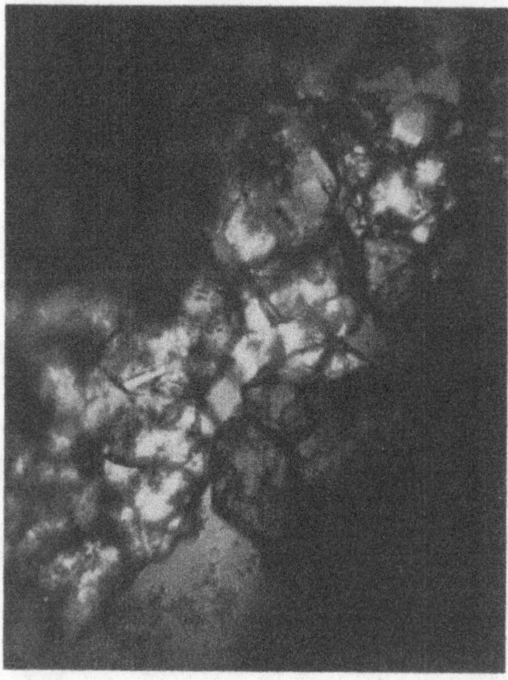

Fig. 6

The other crystal form from fraction II, grown from Drabkin's buffer (Drabkin, 1946) is less well characterized. Although form IIB, Figure 4, is highly symmetrical with beautifully formed facets, it diffracts

Fig. 7

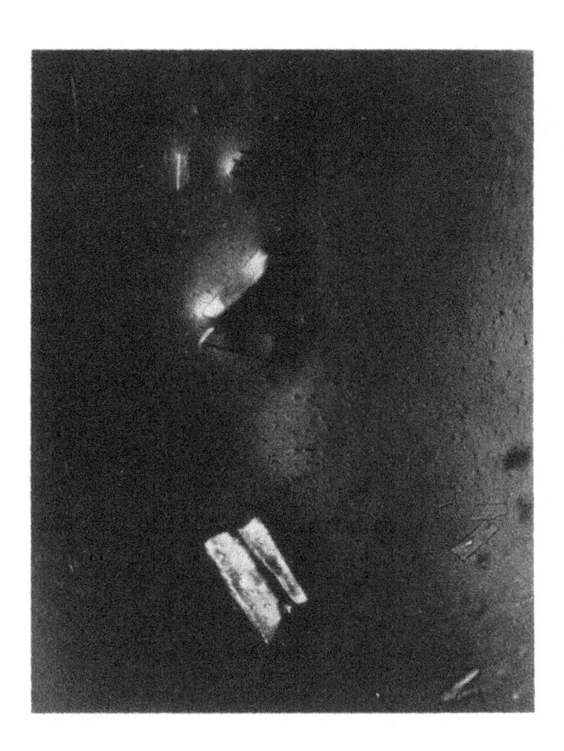

X-rays poorly. Form IIB is short-lived in the X-ray beam and so far, it has only been possible to determine that one cell dimension is about 300 Å.

Crystals grown from fraction IV are shown in Figure 6. The clusters of crystals are representative of the most frequent mode of growth in this fraction. However, single crystals have been obtained and shown to diffract.

Ionic strength, NaCl and $CaCl_2$, and pH are factors that influence the degree of association of subunits differently depending upon the fraction in which they are eluted (E.F.J. Van Bruggen, personal communication; Bonaventura et al., 1976). Subunits may have aggregated to 16S hexamers prior to crystallizing in fractions II and III, given the conditions employed. Crystals of subunits from fraction IV grew in the presence of 0.01 M Na_2EDTA, a condition that favors dissociation (Sullivan et al., 1974; Bonaventura et al., 1976; Schutter et al., this vol.). However, in none of these cases is the aggregation state within the crystal known.

The column fractions used in crystallization experiments were prepared according to Sullivan et al. (1974). As they have already noted, the five fractions eluted from the DEAE-Sephadex column show evidence of heterogeneity. We have recently used a shallower gradient than Sullivan et al. (1974), i.e., 0.25 to 0.50 M NaCl rather than 0 to 0.5 M NaCl. By this means fractions III and V each split into two resolvable peaks. We confirm Sullivan et al. (1974) that polyacrylamide disc gel electrophoresis on fractions I through V indicates that fraction II, as well as III and V, is composed of at least two polypeptides. As mentioned above the existence of III_a and III_b is suggested by the appearance of two crystal forms growing in the same tube, Figure 5.

The heterogeneity of the various column fractions raises questions about the composition of the whole 60S molecules in vivo. Do the column fractions reflect the stoichiometry of each whole molecule or are there different kinds of whole molecules in the hemolymph of an individual animal, or is there variation between the hemocyanins from different animals? We have measured the percent of 280 nm absorbing material in fractions I through to V on ten separate elution profiles of different samples of blood pooled from a small number of specimens, and the results are shown in Table 1. The small standard deviations suggest that hemocyanins from different animals are similar in their subunit conditions.

Table 1. Mean percentage of 280 nm absorbance in fraction I through V

Fraction	I	II	III	IV	V
Percent	9.6 ± 2.4	24.5 ± 3.9	31.3 ± 1.6	16.6 ± 2.4	18.1 ± 2.4

The numbers are averages of 10 different elution profiles ± the standard deviations of the individual observations.

This paper is a progress report. We have found conditions under which whole molecules of *Limulus* hemocyanin and some of its subunits can be crystallized. To date, no crystals have been found upon which further X-ray study is planned. Nevertheless, under conditions that favor aggregation we are attempting to grow X-ray quality crystals of whole 60S molecules, and under conditions that favor dissociation we are attempting to grow X-ray quality crystals of purified 5S subunits.

The 60S whole molecule consists of about fifty subunits, of which there are at least nine different kinds present in unknown ratio. Obviously, a prerequisite to the establishment of the structure of *Limulus* hemocyanin is an accounting of its subunit stoichiometry.

Acknowledgments. This work was supported by a grant, AM 02528, from the National Institutes of Health. We thank Dr. Joseph Bonaventura for generous gifts of *Limulus* hemocyanin and thank Drs. Van Bruggen, Sullivan, J. and C. Bonaventura and Schutter for their prepublication communications.

References

Bonaventura, J., Bonaventura, C., Sullivan, B.: Non-heme oxygen transport. In: Oxygen and Physiological Function. Jobsis, F. (ed.). Dallas: Professional Information Library, 1976 (in press)

Bonaventura, J., Bonaventura, C., Sullivan, B.: Hemoglobins and hemocyanins: comparative aspects of structure and function. J. Exptl. Zool. 194, 155-174 (1976)

Drabkin, D.L.: Spectrophotometric studies. XIV. Crystallographic and optical properties of the haemoglobin of man in comparison with those of other species. J. Biol. Chem. 164, 703-723 (1946)

Kasper, J.S., Lonsdale, K. (eds.): International Tables for X-Ray Crystallography. Birmingham, England: Kynoch Press, 1959, Vol. II, p. 343

Matthews, B.W.: Solvent content of protein crystals. J. Mol. Biol. 33, 491-497 (1968)

Sullivan, B., Bonaventura, C., Bonaventura, J.: Functional differences in the multiple hemocyanins of the horseshoe crab, *Limulus polyphemus* (L). Proc. Natl. Acad. Sci. 71, 2558-2562 (1974)

Subunits of Hemocyanin

Digestion of *Lymnaea stagnalis* Haemocyanin with Trypsin

E. J. WOOD

Abstract

Digestion of whole native molecules of *Lymnaea stagnalis* haemocyanin with trypsin resulted in the formation of "tubes" of digested haemocyanin molecules, some 20 to 30 times the length of the original molecule. Removal of the tubes by preparative ultracentrifugation left in the supernatant a fragment of tryptic digestion which bound oxygen. This fragment had a slightly different absorption spectrum to both native haemocyanin and to tubes. It appeared to be homogeneous in size with a molecular weight of approximately 124,000 and a polypeptide chain weight of approximately 45,000. The material was however heterogeneous in charge.

Introduction

Previous work with the haemocyanin from the fresh-water snail *L. stagnalis*, has established that this haemocyanin has the size and form typical of gastropod haemocyanins, and that it bound oxygen co-operatively (Hall et al., 1975). However it was observed that in the presence of high concentrations of salt, (e.g. 1-2 M NaCl) partial dissociation to one-half molecules appeared to take place, and it was difficult to decide whether the haemocyanin was of the α- or β-type, or whether the haemolymph in fact contained both of these. It is also of interest that recent studies on the carbohydrate and amino acid concent of *Lymnaea* haemocyanin (Hall and Wood, 1976) have shown that this haemocyanin contains less carbohydrate than many other gastropod haemocyanins, that it contains two unidentified carbohydrates and that it does not bind to concanavalin A. The haemocyanin contained elevated levels of alanine and lysine and a low histidine content compared with other gastropod haemocyanins whose amino acid compositions are on the whole rather similar to one another.

The haemocyanin has now been studied further and the present paper reports on the effect of tryptic digestion of the native protein in the presence of calcium ions in producing "tubes" similar to those observed when *Helix pomatia* β-haemocyanin is digested (Van Breemen et al., 1975).

Materials and Methods

Specimens of *L. stagnalis* were obtained from a canal near Leeds, U.K., and the haemocyanin was extracted and purified by preparative ultracentrifugation as described previously (Hall et al., 1975). Material was used freshly as far as possible in these studies. Analytical ultracentrifugation and electronmicroscopy were performed as described previously (Hall et al., 1975), and amino acid and carbohydrate analysis were performed as described by Hall and Wood (1976). Polyacrylamide gel electrophoresis in the presence of sodium dodecyl sulphate was carried out essentially as described by Weber and Osborn (1969).

Oxygen-binding curves were determined by the method of Rossi-Fanelli and Antonini (1958) using the appearance of the band at 345 nm as a measure of the extent of oxygenation of the haemocyanin.

Results

Digestion of Lymnaea Haemocyanin. Samples of purified haemocyanin dissolved in 0.1 M Tris-HCl, pH 7.8, containing 25 mM $CaCl_2$, were digested with 5% (w/v) trypsin (Sigma DCC-treated type XI) at 30°C for 2 h. At the end of this time the action of the trypsin was stopped by the addition of a suitable weight of either soya bean trypsin inhibitor (Sigma type 1-S) or of N-α-p-tosyl-L-lysin chloromethylketone. The solution by this time had a "silky" appearance due to the formation of extended tubular polymers of haemocyanin ("tubes") which were visualized in the electron microscope (Fig. 1). Preparative ultracentrifugation was used to separate the tubes from the supernatant which contained trypsin inhibitor, trypsin, and fragments resulting from the digestion of haemocyanin (see below).

The tubes and the supernatant solution were examined by detergent-gel electrophoresis (Fig. 2). The whole digest (i.e. before removal of the tubes by preparative ultracentrifugation) contained a characteristic set of three bands in the middle of the gel. A number of other bands were also present, both heavier and lighter, the former corrresponding to haemocyanin and the latter to trypsin, and trypsin inhibitor. It was observed that upon removal of the tubes by preparative ultracentrifigation, two of the three bands disappeared leaving the third and lightest in a relatively pure state. This band had a molecular weight close to that of ovalbumin (approximately 45,000) and no heavier bands were present. Some faster moving bands were present but were attributable to trypsin and trypsin inhibitor. They could be separated from the haemocyanin digestion product (hereafter referred to as "fragment") by gel filtration (see below).

Properties of Lymnaea Haemocyanin Tubes. The tubes formed by tryptic digestion of *Lymnaea* haemocyanin appeared to be very similar to those resulting from the digestion of *H. pomatia* β-haemocyanin (van Breemen et al., 1975). However as it is not known whether *Lymnaea* haemocyanin contains both an α- and a β-haemocyanin, and as we have not been able to separate the components observed when high concentrations of NaCl are added (Hall et al., 1975) it is uncertain whether or not the tubes result from the digestion of the total haemocyanin present. It may be significant that undigested *Lymnaea* haemocyanin molecules had some tendency to "stack" end-to-end, and Sminia and Boer (1973) have reported observing ordered rows of particles (believed to be haemocyanin molecules) in the pore cells of the animal.

The tubes themselves under the conditions of digestion, i.e. in the presence of 25 mM $CaCl_2$ had an enormous sedimentation coefficient, but it was clear that the material was heterogeneous, that is the tubes were different lengths.

Tube formation was reversible. If the sedimented tubes were taken up in 0.05 M sodium borate buffer, pH 9.2, containing 0.05 M NaCl, they went into solution easily and upon examination in the analytical ultracentrifuge, major components were found sedimenting at approximately 6.5S and 8.5S, and both were present in approximately equal amounts. Dialysis of this solution against 0.1 M Tris-HCl buffer pH 7.8 containing 25 mM $CaCl_2$ resulted in the re-formation of tubes which were indistinguishable from the original tubes in the electron microscope. The two light bands observed in detergent-gel electrophoresis had

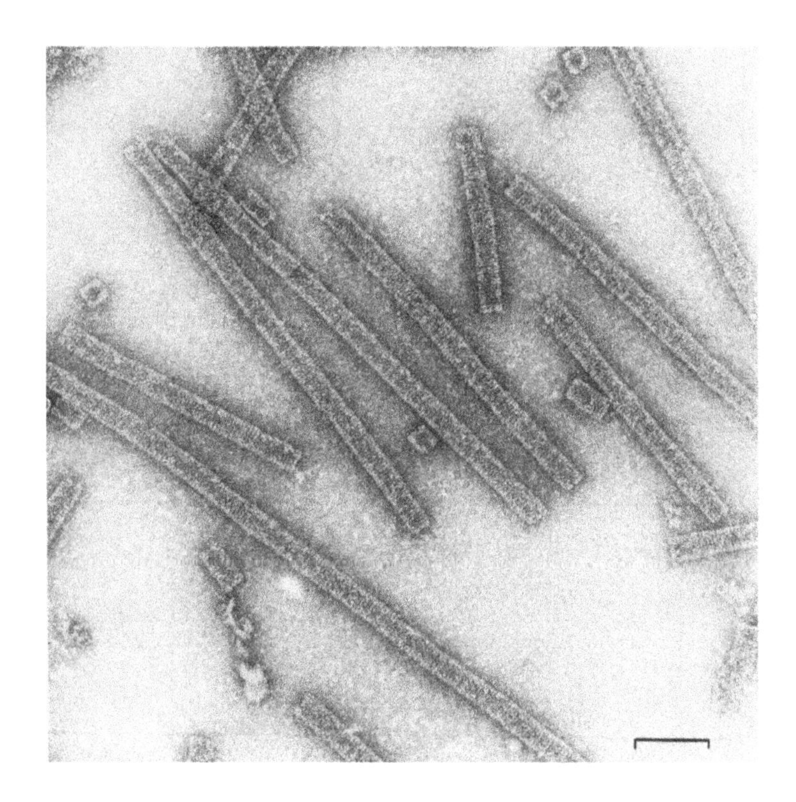

Fig. 1. Electron micrograph of negatively stained "tubes" resulting from tryptic digestion of *Lymnaea* haemocyanin. Length of bar approx. 0.1 μ

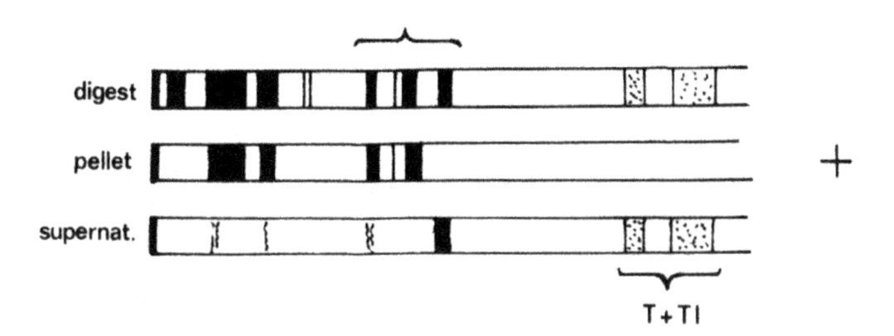

Fig. 2. Detergent-gel electrophoresis with digested *Lymnaea* haemocyanin, and tubes (= pellet) and fragment. *T* + *TI*: trypsin and trypsin inhibitor. See text for further details

approximate molecular weights of 55,000 and 65,000. It may be mentioned that dry weight measurements have shown that the absorbance at 280 nm of dissociated tubes is almost the same as that of undigested haemocyanin.

A partial fractionation of the dissociated tubes has been achieved on DEAE-cellulose (Fig. 3). Concentration of fractions of the major component yielded material which sedimented at 8.2S in the analytical ultracentrifuge as a homogeneous peak.

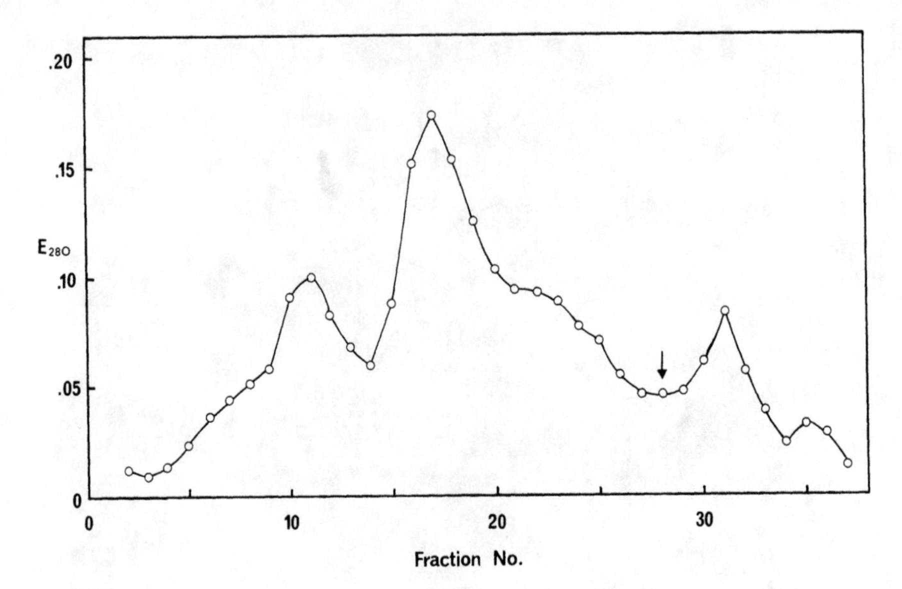

Fig. 3. Fractionation of dissociated tubes on DEAE-cellulose. Buffer 0.005 M Tris-HCl pH 7.5 with NaCl gradient tubes 0-28, 0-0.3 M. *Arrow*: addition of 1 M NaCl. Fraction vol. 5 ml

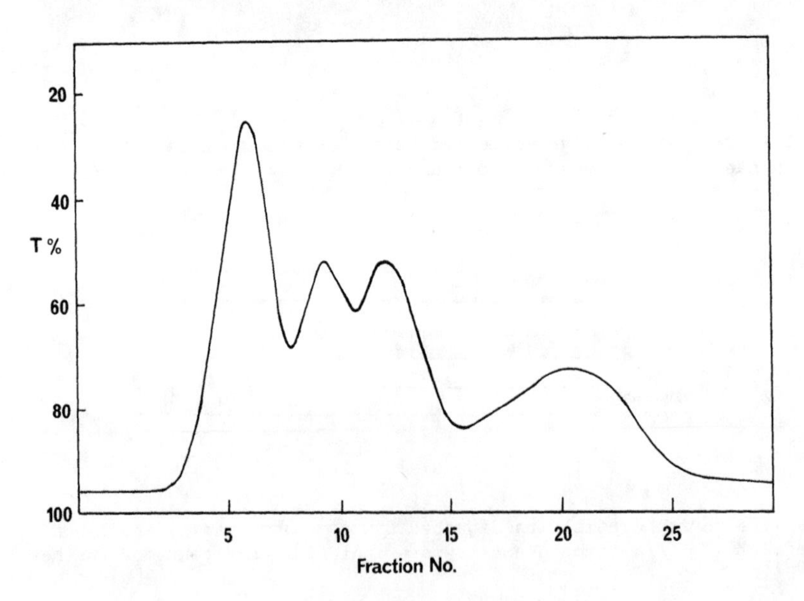

Fig. 4. Fractionation of supernatant after removal of tubes by ultracentrifugation on Sephadex G-100. Fraction vol. 3.7 ml, dimension of column 2 × 90 cm. Only the first peak contained material which absorbed at 346 nm as well as 280 nm

It has proved very difficult to obtain meaningful oxygen equilibrium curves with *Lymnaea* haemocyanin tubes because of reversible dissociation at low oxygen tensions. It was already known that the native haemocyanin bound oxygen highly co-operatively (Hall et al., 1975). Dissociated tubes at pH 7.4 bound oxygen nonco-operatively (see Fig. 6) with a moderate oxygen affinity (P_{50} = 7.5 mm Hg, n = 1). When it was tried to repeat this experiment with a dilute suspension of tubes (i.e. in the presence of 25 mM $CaCl_2$), the opalescence due to the tubes disappeared at low oxygen tensions but reappeared upon reoxygenation. As the light scattering contribution of the tubes to the absorption of the solution is considerable, no reliable results were obtained, and indeed work is in progress to study the kinetics of tube formation by menas of light-scattering measurements.

Properties of Lymnaea Haemocyanin Fragment. The fragment remaining after removing tubes by ultracentrifugation offered interesting possibilities depending upon its origin. This could be as a digestion product of a α-type haemocyanin (the tubes being produced by digestion of a β-type), but it was also possible that as the fragment only accounted for 10% or less of the total protein of native haemocyanin it might be a unit with some special function. It could for example be "collar" or "cap" (see Mellema and Klug, 1972), whose function might be to prevent long (and insoluble) haemocyanin polymers forming from the protomeric subunits.

The supernatant after removal of the tubes was concentrated and applied to a column of Sephadex C-100 (Fig. 4). The fragment, identifiable by its absorption at 340 nm in addition to that at 280 nm, emerged close to the void volume of the column and appropriate fractions were pooled, concentrated, and passed through the same column again to remove the last traces of trypsin and trypsin inhibitor. Calibration of the column revealed that the fragment had a molecular weight similar to or a little less than that of immunoglobulin G, that is 125-150,000. This was confirmed by the finding that the purified material sedimented as an apparently single peak with a sedimentation coefficient (at approx. 4 mg/ml) of 6.5S (uncorrected). Sedimentation equilibrium experiments indicated that the material was homogeneous and had a molecular weight of approx. 124,000 (Fig. 5 and Table 1). It could therefore perhaps be

Table 1. Molecular weight of tryptic fragment of *Lymnaea* haemocyanin by sedimentation equilibrium

Speed	Temperature	MW	Mz
17,947 r.p.m.	23.8°C	124,700	121,000
17,980	24.4	128,000	-
16,160	27.0	118,800	127,400
	Mean:	123,800	124,200

Fragment dissolved in 0.1 M Tris-HCl buffer, pH 7.8: protein concentration approx. 0.4 mg/ml, meniscus depletion experiments.

a trimer of the polypeptide chain (MW ∿45,000) found in detergent-gel electrophoresis, and would presumably contain three oxygen binding sites. The fragment was capable of binding oxygen (Fig. 6) but with a higher affinity than either haemocyanin or dissociated tubes.

It is interesting to note that the absorption bands due to the copper-oxygen complex were at approximately 340 nm and 545 nm rather than at 345 nm and 570 nm as is found with native haemocyanin or with tubes. An explanation for this shift is at present lacking.

An examination of the fragment on polyacrylamide gel electrophoresis revealed two major, rather diffuse bands poorly separated from each other, and a number of minor bands (Fig. 7). In an attempt to achieve

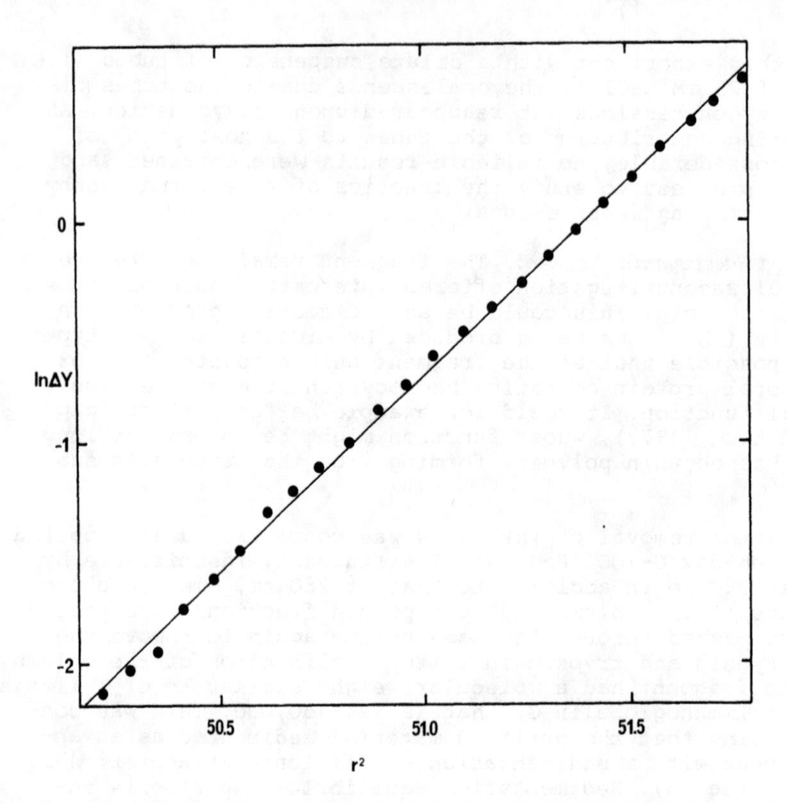

Fig. 5. Sedimentation equilibrium ultracentrifugation with purified *Lymnaea* haemocyanin fragment. Yphantis experiment, using Rayleigh interference optics, at 16,160 r.p.m. Line fitted by linear regression gives molecular weight 120,700

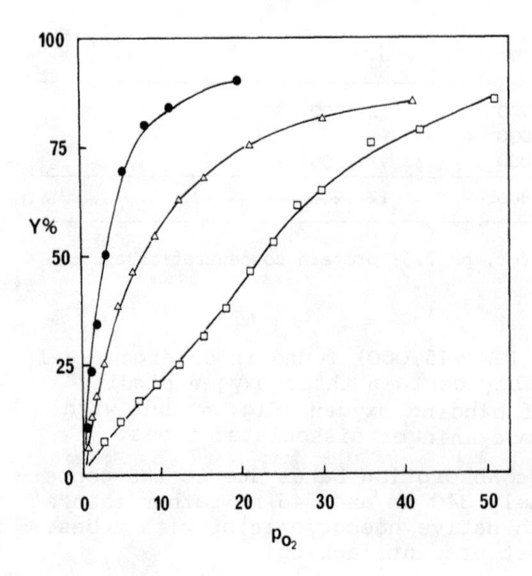

Fig. 6. Oxygen binding curves for native *Lymnaea* haemocyanin (□); dissociated tubes (Δ); fragment (●). Conditions: 25°C, 0.1 M Tris-HCl buffer, pH 7.8, containing 25 mM $CaCl_2$ (except in the case of tubes where the $CaCl_2$ was omitted)

Fig. 7. Polyacrylamide gel electrophoresis comparing dissociated tubes with the fragment derived by tryptic digestion of *Lymnaea* haemocyanin tubes

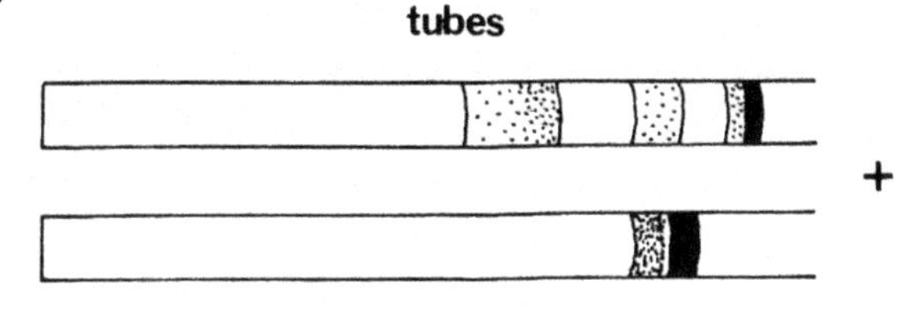

further separation of the components the fragment was subjected to ion-exchange chromatography on DEAE-cellulose. A separation was eventually achieved by both pH and salt gradients into two major fractions, but a number of minor components also appeared to be present. It is possible that these result from tryptic digestion to slightly different extents, but the possibility of microheterogeneity in the original haemocyanin cannot be ruled out.

Chemical Composition of Tubes and Fragments. Only preliminary data on amino acid and sugar compositions are available at present. We have already noted (Hall and Wood, 1976) that *Lymnaea* haemocyanin was similar in amino acid composition to other gastropod haemocyanins with the exception that the levels of alanine and lysine were elevated (ratios, alanine/glycine were 1.11 for *Helix* haemocyanin and 1.42 for *Lymnaea* haemocyanin, and for lysine/glycine 0.65 and 1.15), and those of histidine lowered (ratios 0.85 and 0.54). The composition of tubes was very similar to that of native *Lymnaea* haemocyanin except that the histidine content was raised (ratio histidine/glycine, 0.74). The fragment had similar lysine and histidine contents to native *Lymnaea* haemocyanin, but a lowered content of alanine.

As regards the sugar content of tubes and fragment the preliminary data show that most or all of the carbohydrates present in native *Lymnaea* haemocyanin are also present in tubes, but it is possible that the fragment lacks the two unidentified sugars (see Hall and Wood, 1976) and may also lack glucose.

Discussion

Lymnaea haemocyanin behaves in a very similar way to *H. pomatia* β-haemocyanin in its response to treatment with trypsin, but small differences are observed when the products of digestion are examined. The tubes formed appear to be identical with *Helix* haemocyanin tubes, which van Breemen et al. (1975) were able to study by optical image reconstruction and X-ray diffraction methods. However in detergent-gel electrophoresis they found that the tubes gave major bands at 290,000, 210,000 and 105,000 whereas *Lymnaea* tubes appeared to contain somewhat smaller polypeptide chains (e.g. 55,000 and 65,000) as well as larger chains. Furthermore the fragment obtained from the digestion of *H. pomatia* β-haemocyanin (= Peak 11 from gel filtration on Sepharose 2B, van Breemen et al., 1975) gave bands in SDS-gel electrophoresis at 96,000 and 73,000 compared with about 45,000 in the case of *Lymnaea* haemocyanin. The significance, if any, of these differences in polypeptide chain size after digestion remains to be seen.

It will be of great interest to study the fragment of *Lymnaea* haemocyanin further. It is intriguing to find that the spectral bands due to the copper-oxygen complex are shifted, and its oxygen-combining properties need to be studied further in comparison with those of the native haemocyanin and of dissociated tubes. It is interesting to speculate whether the fragment is an "end" of the native molecule, providing a signal to stop polymerizing, so that long, insoluble polymers are not formed, which would hamper blood flow. If this is the case, it remains to ascertain how it is incorporated into the native haemocyanin molecule.

Acknowledgments. I am grateful to Miss Lindsey Mosby for skilled technical assistance, to Mr. R. Hall for the carbohydrate analysis, to the Sequencing Unit, Biochemistry Department, University of Leeds, for the amino acid analysis, and to Mr. D. Kershaw for the electron micrography.

References

Breemen, J.F.L. van, Wichertjes, T., Muller, M.F.J., Driel, R. van, Bruggen, E.F.J. van: Tubular polymers derived from *Helix pomatia* β-haemocyanin. Europ. J. Biochem. 60, 129-135 (1975)

Hall, R.L., Pearson, J.S., Wood, E.J.: The haemocyanin of *Lymnaea stagnalis* L. (Gastropoda:pulmonata). Comp. Biochem. Physiol. 52B, 211-218 (1975)

Hall, R.L., Wood, E.J.: The carbohydrate content of gastropod haemocyanins. Trans. Biochem. Soc. 4, 307-309 (1976)

Mellema, J.E., Klug, A.: Quaternary structure of gastropod haemocyanin. Nature (London) 239, 145-150 (1972)

Rossi-Fanelli, A., Antonini, E.: Studies on the oxygen and carbon monoxide equilibria of human myoglobin. Arch. Biochem. Biophys. 77, 478-492 (1958)

Sminia, T., Boer, H.H.: Haemocyanin production in pore cells of the freshwater snail *Lymnaea stagnalis*. Z. Zellforsch. 145, 443-445 (1973)

Weber, K., Osborn, M.: The reliability of molecular weight determinations by dodecyl sulphate-polyacrylamide gel electrophoresis. J. Biol. Chem. 244, 4406-4412 (1969)

Structural Investigations on β-Haemocyanin of *Helix pomatia* by Limited Proteolysis

C. Gielens, G. Preaux, and R. Lontie

Introduction

The haemocyanin of the Roman snail (*Helix pomatia*) consists of an α- (75%) and a β-component (25%), both having a molecular weight of 9 × 10^6 (Wood et al., 1971) and a copper content of the order of 0.25% (2 Cu/ 50,000 on the average). The α-component dissociates into halves in the presence of 1 M NaCl or KCl (Heirwegh et al., 1961) or in slightly alkaline medium, pH 7.0-7.6 (Witters and Lontie, 1968), while the β-component does not. At higher pH values both haemocyanins successively dissociate into tenths and twentieths (Witters and Lontie, 1968; Siezen and van Driel, 1974). As the twentieths do not dissociate further under mild conditions, they most probably correspond to the polypeptide chains of the haemocyanin molecule.

Both the α- and the β-haemocyanin can be degraded by limited proteolysis into functional fragments with molecular weights of about 50,000 or a multiple (Gielens et al., 1973; Lontie et al., 1973; Brouwer, 1975). In the electron microscope twenthieths were observed as flexible chains of seven to eight globular units with a diameter of 55-60 Å (Siezen and van Bruggen, 1974). These domains were inferred to correspond to the 50,000 units containing one oxygen-binding site. The proteolytic cleavage seems to occur at easily accessible peptide bonds between the domains.

In order to determine the number and sequence of the domains in the polypeptide chain of β-haemocyanin a study was undertaken of the limited proteolysis of the tenths and twentieths by very selective enzymes. Trypsinolysis yielded five fragments: T1A, constituted of three domains, T1B and T1C, constituted each of two domains, and T2 and T3, consisting both of only one domain. From this it was concluded that twentieths are constituted of nine domains (Gielens et al., 1975). Moreover, the circular dichroic spectra of the fragments indicated the presence of two classes of copper groups according to their positive maximum at 455 or at 500 nm. More recently a single cleavage of the polypeptide chain could be obtained by the action of plasmin or of the protease of *Staphylococcus aureus*. Both enzymes are serine proteases. Plasmin (M_r 86,000), the enzyme with fibrinolytic activity, has the same specificity as trypsin but is much more selective. The staphylococcal protease (M_r 12,000), isolated by Drapeau et al. (1972), was shown to cleave peptide bonds at the carboxyl side of either aspartic or glutamic acid (Houmard and Drapeau, 1972). As both enzymes cleave the haemocyanin polypeptide chain between different domains, a comparative study including a further proteolytic degradation of the fragments enabled us to locate these in the polypeptide chain. From this study it also followed that the two domains of the tryptic fragment T1B must be identical and belong to two different polypeptide chains (associated twentieths). The twentieths thus seem to be constituted of only eight domains.

Material and Methods

Haemocyanin. Total haemocyanin was separated from the haemolymph of *H. pomatia* by salt precipitation and solubilized in 0.1 M sodium acetate buffer pH 5.7 (Heirwegh et al., 1961). The β-component was precipitated by dialysis at $4^{o}C$ against 10 mM sodium acetate buffer pH 5.3. After solubilization of the precipitate in phosphate buffer pH 6.5, I = 0.1 M, 0.02% NaN_3, a second isoelectric precipitation was performed at pH 5.3, I = 10 mM. The β-haemocyanin was redissolved in the phosphate buffer pH 6.5 and filtered through a Millipore filter type HAWP in a 90-mm Hi-Flux Cell (Millipore Corp., Bedford, MA, USA).

The concentration of the β-haemocyanin solution was determined at 278 nm in 0.1 M sodium tetraborate pH 9.2, using a specific absorption coefficient A (1%, 1 cm) = 14.16 (Heirwegh et al., 1961). The concentration of the proteolytic fragments was measured in sodium borate-HCl buffer pH 8.2, I = 0.1 M, 0.02% NaN_3, using the same absorption coefficient.

Proteolytic Enzymes and Inhibitors. Human plasmin was a lyphilized powder (lot no. 19221, 20 caseinolytic units per vial), graciously provided by AB KABI (Stockholm, Sweden). The protease of *S. aureus*, strain V8 (batch 275), was purchased from Miles (Slough, England). Crystalline bovine pancreatic trypsin (Batch 261-2) was a gift from Novo Industri A/S (Copenhagen, Denmark). Trypsin pancreatic inhibitor was a product of Worthington (Freehold, NJ, USA), phenylmethane sulphonyl fluoride was purchased from Merck (Darmstadt, Germany).

Proteolysis. The limited proteolysis of β-haemocyanin with plasmin and with the protease of *Staphylococcus aureus* was performed at room temperature in borate-HCl buffer pH 8.2, I = 0.1 M, 0.02% NaN_3 (on tenths) or in borate buffer pH 9.2, I = 0.1 M, 0.02% NaN_3 (on twentieths) at a protein concentration of 3-5%. The E/S ratio varied from 1/150 to 1/600 (w/w). The hydrolysis with plasmin was stopped by the addition of trypsin pancreatic inhibitor in a molar ratio of 2, compared to plasmin. The staphylococcal protease was inactivated by phenylmethane sulphonyl fluoride added in a molar ratio of 10.

The trypsinolysis of fragments P1 and Sp1P1 was carried out at room temperature in the borate buffer pH 8.2 at a substrate concentration of 13.5 mg/ml and of 4 mg/ml respectively, the E/S ratio amounting to 1/400 (w/w). The hydrolysis was stopped by denaturation with SDS (see below).

Chromatography on Ultrogel. The proteolytic hydrolysates of β-haemocyanin were fractionated on Ultrogel AcA 34 (batch 6905, Industrie Biologique Francaise, Gennevilliers, France), equilibrated with borate-HCl buffer pH 8.2, I = 0.1 M, 0.02% NaN_3. The dimensions of the column were 83 × 4.8 cm. Elution was carried out at 120 ml/h, fractions of 15 ml were collected. The absorbances at 278 and 346 nm were measured in a Beckman DU spectrophotometer (Munich, Germany).

The fractions corresponding to the elution peaks were characterized by P or Sp, referring to plasmin or to the staphylococcal protease, and by numbers indicating the elution order.

The liquid chromatograms of the plasmin hydrolysates were resolved into gaussian curves by a Curve Resolver Du Pont 310 (Du Pont de Nemours, Wilmington, DE., USA).

Sodium Dodecyl Sulphate - Polyacrylamide Gel Electrophoresis. The method of Weber et al. (1973) slightly modified was used. The samples were diluted to a protein concentration of about 1 mg/ml with borate-HCl buffer pH 8.2, I = 0.1 M, 0.02% NaN_3. Approximately 5 µl of these solutions were mixed with 50 µl of 10 mM phosphate buffer pH 7.2, containing 1% sodium dodecyl sulphate (SDS) and 1% 2-mercaptoethanol. After heating for 2 min at 100°C 10 µl tracking dye (0.05% Bromophenol Blue in 10 mM phosphate buffer pH 7.2) and 50 µl of a 40% solution of sucrose were added. The electrophoresis was performed on cylindrical 5% polyacrylamide (PAA) gels (90 × 4.6 mm) with a weight ratio acrylamide/N,N'-methylene-bisacrylamide of 37 in 0.1 M Tris-glycine, pH 9.0, containing 0.1% SDS (Vandekerckhove and van Montagu, 1974). The applied potential amounted to 110 V with a current of 2 mA/gel.

The gels were stained for 2 h with a 0.25% solution of Coomassie Brilliant Blue R-250 (Mann, New York, NY, USA) in acetic acid/methanol/water (1:5:5). Destaining was accomplished at 40°C with a mixture of methanol/acetic acid/water (2:3:5).

The following reference proteins were used for molecular weight determinations: bovine serum albumin and ovalbumin, prepared in the laboratory (Onkelinx et al., 1969), sperm whale myoglobin, obtained from Miles (Slough, England), β-galactosidase from Boehringer (Mannheim, Germany), and dog thyroglobulin, a generous gift from Prof. S. Lissitzky (Marseille, France).

The gels were scanned at 550 nm with a densitometer built in the laboratory and equipped with a recorder linear in transmission (model 8100, Philips, Eindhoven, The Netherlands). The mobilities were related to bromophenol blue.

Analytical Ultracentrifugation. The analytical ultracentrifugation of the fragments was carried out at 20.0°C in a Spinco Model E analytical ultracentrifuge, equipped with phase plate schlieren optics and an RTIC temperature control system (Beckman Instruments, Palo Alto, CA, USA). The experiments were performed at 50,740 r.p.m. in 12-mm single-sector (4°) cells at a protein concentration of 5 mg/ml.

The diagrams were measured with a Nikon profile projector model 6C (Nippon Kogaku K.K., Tokyo, Japan). For the conversion of the sedimentation coefficient of the fragments to the standard conditions of viscosity and density of water at 20.0°C ($s_{20,w}$) the partial specific volume 0.726 ml/g of β-haemocyanin (Wood et al., 1971) was used.

Results and Discussion

Proteolysis of β-Haemocyanin with Plasmin. The plasminolysis in borate buffer pH 8.2 of the tenths mainly yielded the fragments P1 and P3, which were separated by chromatography on Ultrogel. Their molecular weights were estimated by several methods (Table 1). The values based on the sedimentation coefficients were deduced by the equation of Stewart and Johns (1976) established for immunoglobins, also constituted of domains. This equation was preferred to that proposed by the same authors for globular proteins, as it yielded molecular weights of 930,000 and 410,000 respectively for loose tenths and twentieths (based on $s^{0}_{20,w}$ = 17.6S and 11.2S, values taken from Siezen and van Bruggen, 1974) against 565,000 and 275,000 obtained with the equation for globular proteins ($\log s^{0}_{20,w} = -2.339 + 0.623 \log M_r$). The elution volume in gel chromatography is also dependent on the shape of the protein, which has to be kept in mind in the evaluation of the molecular weights obtained by Sephadex and Ultrogel chromatography. The molecular weights from SDS-PAA electrophoresis may be overestimated (Bretscher, 1971), when there are bound carbohydrates like in haemocyanin (Dijk et al., 1970).

Table 1. Molecular weights of the fragments obtained from β-haemocyanin with plasmin or with the protease of *Staphylococcus aureus*

Fragment	10^{-3} × Molecular weight determinated by					Proposed value[e]
	Ultracentrifugation[a]		Ultrogel AcA 34[b]	Sephadex G-100[c]	SDS-PAA electrophoresis[d]	
	pH 8.2	pH 9.2	pH 8.2	pH 8.2		
P1	530	360	570		290	2 × 275
P3	175	180	130	140	170	165
Sp1	390	420	340		390	385
Sp4	145	120	90	110	65	2 × 55

[a]From $s_{20,w}$ values in borate-HCl or borate buffer, I = 0.1 M, using the equation $\log s^{o}_{20,w} = -2.026 + 0.548 \log M_r$ derived for immunoglobulins (Stewart and Johns, 1976).
[b]From K_{av} values (Laurent and Killander, 1964) versus log M_r calibrated with β-lactoglobulin, ovalbumin monomers, dimers, and trimers, serum albumin monomers, dimers, and trimers, alcohol dehydrogenase, aldolase, catalase, apoferritin, and tenths of *H. pomatia* haemocyanin.
[c]From V_e values versus log M_r (Andrews, 1964) on a column calibrated with myoglobin, ovalbumin, and serum albumin monomers and dimers.
[d]According to Shapiro et al. (1967) using the same reference proteins as described in Materials and Methods; for Sp1 a value of 390,000 is proposed as it migrated just in between haemocyanin twentieths (M_r 450,000) and thyroglobulin (M_r 330,000).
[e]On the basis of a molecular weight of 55,000 per domain.

From the values of Table 1 fragment P1 seems to be constituted of two identical non-covalently bound subfragments as, besides by SDS treatment in the presence of 2-mercaptoethanol, a dissociation also occurred simply by increasing the pH to 9.2. The subfragments reassociated when the pH was lowered from 9.2 to 8.2, as shown by the identity of the Ultrogel chromatographies at pH 8.2 of the hydrolysates obtained after plasminolysis either at pH 8.2 or 9.2. The fraction P1, separated at pH 8.2 after proteolysis at pH 9.2 (Fig. 1), showed a molecular weight of 570,000 and one of 375,000 after increasing the pH to 9.2, as deduced from $s_{20,w}$ = 13.46S and 10.70S respectively. The subfragments of P1 thus must originate from 2 twentieths, associated into tenths.

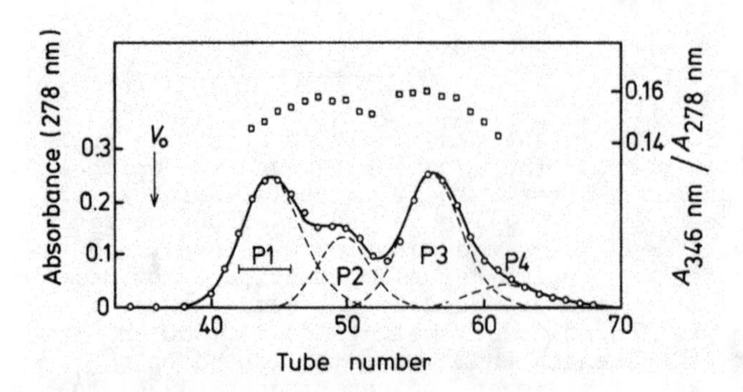

Fig. 1. Chromatography on Ultrogel AcA 34 at pH 8.2 of plasmin hydrolysate of the twentieths of β-haemocyanin. 120 mg haemocyanin in 4 ml borate buffer pH 9.2, I = 0.1 M, 0.02% NaN_3, E/S = 1/400 (w/w), 30 h at room temperature. (o) A (278 nm, 5 mm), (□) A (346 nm, 1 cm)/ A (278 nm, 1 cm), (----) analysis into Gaussian curves

The values for $A_{346\ nm}/A_{278\ nm}$ after plasminolysis at pH 9.2 (Fig. 1) were rather low, compared to the initial value of 0.2 found for the twentieths, indicating a slight decrease of the copper band. During plasminolysis of twentieths some copper groups were apparently destroyed, in contrast with the proteolysis at pH 8.2.

The molecular weight of P3 indicated this fragment to be composed of three domains. It showed, moreover, a CD spectrum very similar to that of the tryptic fragment T1A, also found to consist of three domains: two domains containing a copper group with a positive CD maximum at 455 nm and one with a copper group with a positive CD maximum at 500 nm. As P1 consisted of two associated subfragments, each originating from a twentieth, two P3 fragments must have been formed for each P1.

In addition to P1 and P3 two other fragments, P2 and P4, were also found in most plasmin hydrolysates. They were slowly formed by cleavage of P1, as shown by SDS-PAA electrophoresis performed after increasing times of hydrolysis. The constancy of the percentage of fragment P3 in the Ultrogel chromatography of the hydrolysates after plasminolysis under several experimental conditions (Table 2) also confirmed that this fragment was resistant to further proteolytic degradation. The

Table 2. Percentages of the fragments obtained by plasminolysis of tenths and twentieths of *H. pomatia* haemocyanin (Hc)

Proteolytic conditions				Percentage[a]				P3
pH	[Hc] (mg/ml)	E/S (w/w)	Hydrolysis time (h)	P1	P2	P3	P4	P1 + P2 + P4
8.2	56	1/400	26	41	17	36	6	0.56
8.2	46	1/150	5	36	20	37	7	0.59
8.2	53	1/150	24	26	28	39	7	0.64
9.2	30	1/400	30	30	17	37	7	0.59

[a]Obtained by resolution into Gaussian curves of the chromatograms on Ultrogel AcA 34 at pH 8.2.

percentages of the different fragments were estimated from the liquid chromatograms, resolved into gaussian curves (Fig. 1), assuming a constant extinction coefficient. In order to estimate the number of domains in each P1 subfragment the percentage of P3 was related to that of the sum of P1 and P2 + P4, since P2 and P4 originated from P1. This value corresponds to the ratio of the molecular weight of P3 to that of the P1 subfragment, if it is assumed that both are formed in equimolar amounts. The values from Table 2 are in good agreement with the ratio of 0.59, following from the molecular weights of the fragments estimated by SDS-PAA electrophoresis (Table 1). They strongly suggest that subfragment P1 must be constituted of five domains, provided they have similar dimensions, as P3 was considered to contain three domains. The twentieths must thus be constituted of eight domains.

Proteolysis with the Protease of Staphylococcus aureus. Limited proteolysis in borate buffer pH 8.2 of the tenths of β-haemocyanin with the protease of *S. aureus* yielded mainly fragments Sp1 and Sp4, although other minor fragments were formed. Fragments Sp1 and Sp4 were separated by Ultrogel chromatography at pH 8.2.

The molecular weights (Table 1) estimated for fragment Sp1 by the different methods agreed fairly well. No dissociation was observed by treatment with SDS, nor by increasing the pH to 9.2, indicating that this fragment was constituted of a single polypeptide.

Fragment Sp4 seemed to be constituted of two domains, which dissociated upon treatment with SDS, but not on increasing the pH to 9.2 (Table 1). Taking into account the molecular weight of Sp1, the two domains must originate from two polypeptide chains associated into tenths. Consequently each Sp4 domain would correspond to an end-domain of the polypeptide chain (side to be determined) and contain a region of fairly strong interaction where the association between two-twentieths occurs. As there was no dissociation of Sp4 at pH 9.2 the rest of the polypeptide chain must strongly contribute to the repulsion of the twentieths at pH 9.2.

Fragment Sp4 was characterized by the same molecular weight and the same behaviour in SDS electrophoresis as the tryptic fragment T1B. It also showed a very similar CD spectrum with positive maximum at 500 nm. From this it was inferred that both fragments were constituted of the same domains and that each domain of T1B should also be ascribed to a different polypeptide chain. This reduced the number of domains per polypeptide chain, estimated from the tryptic fragments from 9 to 8. This value agrees with that found by plasminolysis. Fragment Sp1 thus is considered to contain seven domains.

The action of the staphylococcal protease on twentieths at pH 9.2 appeared to be even more selective than on tenths at pH 8.2. Chromatography on Ultrogel AcA 34 at pH 8.2 yielded practically only Sp1 and Sp4 (Fig. 2), as confirmed by SDS-PAA electrophoresis of the fractions.

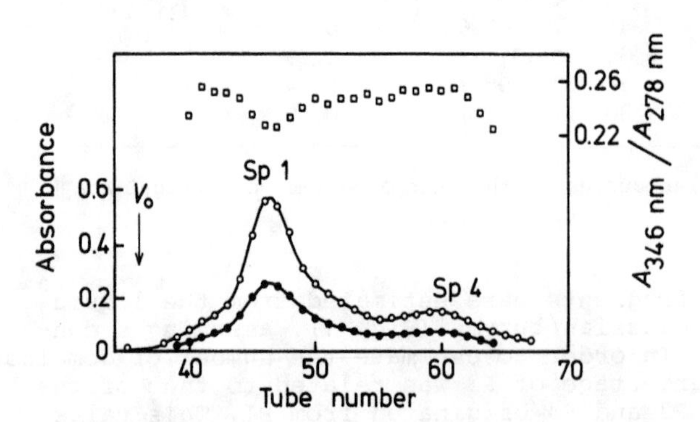

Fig. 2. Chromatography on Ultrogel AcA 34 at pH 8.2 of a hydrolysate of the twentieths of β-haemocyanin with staphylococcal protease.
120 mg haemocyanin in 4 ml borate buffer pH 9.2, I = 0.1 M, 0.02% NaN_3, E/S = 1/600 (w/w), 19 h at room temperature. (o) A (278 nm, 5 mm), (•) A (346 nm, 1 cm), (□) A (346 nm, 1 cm) / A (278 nm, 1 cm)

Reassociation of the two originally separated Sp4 domains must also have occurred as the same elution volume was observed as for Sp4 originating from tenths. There was no loss of activity of the copper groups (compare the $A_{346\ nm}/A_{278\ nm}$ values with those in Fig. 1).

Proteolysis of Fragments

1. With Plasmin and with the Protease of Staphyloccocus aureus. As mentioned P1 could slowly be split further by the action of plasmin into P2, most likely constituted of four domains, and P4, very similar to Sp4 and hence composed of two identical domains originating from two polypeptide chains. The staphylococcal protease cut fragment P1 mainly into the fragments P1Sp1, most probably containing four domains, and P1Sp2, identical with Sp4. Their molecular weights are given in Table 3 together with the fragments, to which they correspond according to their molecular weight and their CD spectra.

Sp1 could slowly be degraded further by the staphylococcal protease, mainly into fragments apparently containing three and four domains. Plasmin cleaved Sp1 exlcusively into fragments Sp1P1 and Sp1P2. Sp1P1 fully corresponded to P1Sp1, while Sp1P2 was similar to P3 and T1A consisting of three domains (Table 3).

Table 3. Molecular weight of the fragments obtained by further proteolysis of P1 with the staphylococcal protease and of Sp1 with plasmin and their correspondence with other fragments

Fragment	10^{-3} × Molecular weight estimated[a] by				Proposed value[b]	Corresponding fragments
	Ultracentrifugation pH 8.2	Ultrogel AcA 34 pH 8.2	SDS-PAA electrophoresis on purified fragments	SDS-PAA electrophoresis on hydrolysis mixtures		
P1Sp1	265	225	230	235	220	Sp1P1
P1Sp2	160	100	70	65	2 × 55	T1B, Sp4
Sp1P1	270		235	220	220	P1Sp1
Sp1P2	195		180	170	165	T1A, P3

[a] According to the methods described in Table 1.
[b] On the basis of a molecular weight of 55,000 per domain.

It thus appeared that the largest fragment resulting from the first action on the tenths of plasmin or of the staphylococcal protease remained susceptible to a further attack especially by the other enzyme. This cleavage occurred in the fragments between the same domains as in the intact polypeptide chain.

2. With Trypsin. Trypsinolysis of P1 and Sp1P1 yielded the same fragments with the exception of component c with an apparent molecular weight of 68,000 as determined by SDS-PAA electrophoresis, which was missing in the Sp1P1 hydrolysate (Fig. 3). The fragments could from their migration in the SDS-PAA electrophoresis be correlated to the fragments from trypsinolysis of the tenths (Gielens et al., 1975): component a corresponded to T1C, b to T2, c to T1B, and d to T3. The molecular weight deduced for component a (M_r 130,000) was higher than the earlier value for T1C (M_r 100,000). This mainly resulted from a different slope of the calibration curve for SDS-PAA electrophoresis, due to the inclusion of β-galactosidase (M_r 130,000) and thyroglobulin (M_r 330,000: Rolland and Lissitzky, 1976) as reference proteins. In addition the tryptic hydrolysate of the tenths also contained fragment T1A (M_r 165,000 against 140,000 earlier). Its absence in both the tryptic hydrolysate of P1 and of Sp1P1 confirmed that it must correspond to P3 and to Sp1P2.

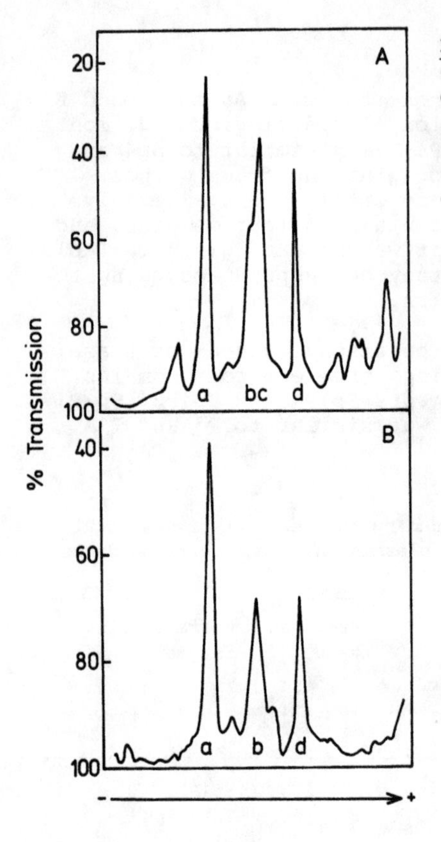

Fig. 3A and B. Densitometric scans of SDS-PAA gels after electrophoresis of tryptic hydrolysates of P1 and Sp1P1 and staining with Coomassie Brilliant Blue. (A) P1 after 15 min; (B) Sp1P1 after 1 h. Electrophoresis from left to right. The deduced molecular weights are *a*: 130,000; *b*: 78,000; *c*: 68,000; *d*: 42,000

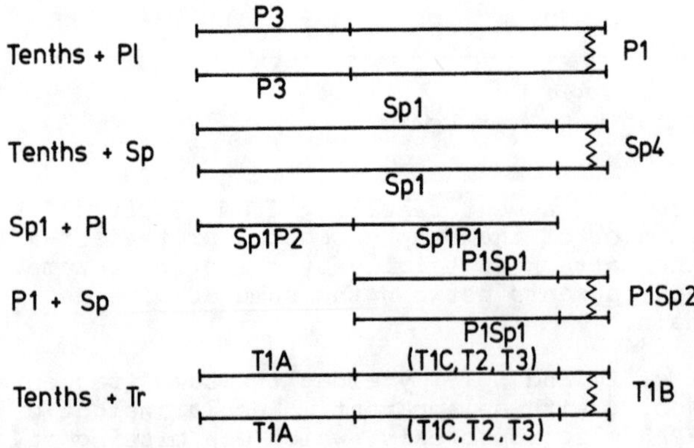

Fig. 4. Splitting scheme of the tenths of *H. pomatia* haemocyanin with plasmin (*Pl*), staphylococcal protease (*Sp*), and trypsin (*Tr*) and of the fragments *Sp1* and *P1* with plasmin and staphylococcal protease respectively

The absence of T1B in the hydrolysate of Sp1P1 was a further proof that it must correspond to Sp4.

Conclusions

The twentieths of β-haemocyanin of *H. pomatia*, even when associated into tenths, appear to contain two sites in between the domains very susceptible to proteolysis, one preferentially attacked by plasmin and one by the protease of *S. aureus*. During these cleavages the site responsible for the association of twentieths was preserved keeping the corresponding fragments together. This site is situated in a terminal domain of the polypeptide chain (side to be determined). The results obtained with both proteases used separately or in succession yielded the splitting scheme of Figure 4 and pointed to the presence of eight domains per polypeptide chain. Trypsin is not as selective and produced four cleavages in the polypeptide chain, yielding five fragments. From the trypsinolysis of P1 and Sp1P1 the fragments T1A and T1B could be located in the tenths (Fig. 4).

Acknowledgments. This work was supported by grants from the Fonds voor Collectief Fundamenteel Onderzoek (Contract Nr. 2.0016,76) and by the Fonds Derde Cyclus, Katholieke Universiteit te Leuven. We are grateful to Prof. P. Edman, Martinsried, Munich, for suggesting the study of the action of plasmin, to Prof. H. Koch, Leuven, for kindly allowing us to use his Du Pont 310 Curve Resolver, and to Prof. S. Lissitzky, Marseille, for the sample of dog thyroglobulin. We wish to thank AB KABI, Stockholm, for the sample of human plasmin and Novo Industri A/S, Copenhagen, for the crystalline trypsin. We express our gratitude to Miss J. Van de Weyer and Mr. L. Verschueren for their skilful assistance in performing the experiments and to Mr. G. van Aerschot for the ultracentrifugal runs. We are much indebted to Mr. J. Pauwels and Mr. G. Goemans for building the densitometer for polyacrylamide gels.

References

Andrews, P.: Estimation of the molecular weights of proteins by Sephadex gel-filtration. Biochem. J. 91, 222-233 (1964)

Bretscher, M.S.: Major human erythrocyte glycoprotein spans the cell membrane. Nature (New Biol.) 231, 229-232 (1971)

Brouwer, M.: Structural domains in *Helix pomatia* α-hemocyanin. Groningen: Veenstra-Visser Offset, 1975, pp. 32-45

Dijk, J., Brouwer, M., Coert, A., Gruber, M.: Structure and function of hemocyanins. VII. The smallest subunit of α- and β-hemocyanin of *Helix pomatia*: size, composition, N- and C-terminal amino acids. Biochem. Biophys. Acta 221, 467-479 (1970)

Drapeau, G.R., Boily, Y., Houmard, J.: Purification and properties of an extracellular protease of *Staphylococcus aureus*. J. Biol. Chem. 247, 6720-6726 (1972)

Gielens, C., Preaux, G., Lontie, R.: Isolation of the smallest functional subunit of *Helix pomatia* haemocyanin. Arch. Intern. Physiol. Biochem. 81, 182-183 (1973)

Gielens, C., Preaux, G., Lontie, R.: Limited trypsinolysis of β-haemocyanin of *Helix pomatio*. Characterization of the fragments and heterogeneity of the copper groups by circular dichroism. Europ. J. Biochem. 60, 271-280 (1975)

Heirwegh, K., Borginon, H., Lontie, R.: Separation and absorption spectra of α- and β-haemocyanin of *Helix pomatia*. Biochem. Biophys. Acta 48, 517-526 (1961)

Houmard, J., Drapeau, G.R.: Staphylococcal protease: a proteolytic enzyme specific for glutamoyl bonds. Proc. Natl. Acad. Sci. 69, 3506-3509 (1972)

Laurent, T.C., Killander, J.: A theory of gel filtration and its experimental verification. J. Chromatogr. 14, 317-330 (1964)

Lontie, R., De Ley, M., Robberecht, H., Witters, R.: Isolation of small functional subunits of *Helix pomatia* haemocyanin after subtilisin treatment. Nature (New Biol.) 242, 180-182 (1973)

Onkelinx, E., Meuldermans, W., Joniau, M., Lontie, R.: Glutaraldehyde as a coupling reagent in passive haemaglutination. Immunology 16, 35-43 (1969)

Rolland, M., Lissitzky, S.: Endogenous proteolytic activity and constituent polypeptide chains of shepp and pig 19S thyroglobulin. Biochem. Biophys. Acta 427, 696-707 (1976)

Shapiro, A.L., Vinuela, E., Maizel, J.V.: Molecular weight estimation of polypeptide chains by electrophoresis in SDS-polyacrylamide gels. Biochem. Biophys. Res. Comm. 28, 815-820 (1967)

Siezen, R.J., van Bruggen, E.F.J.: Structure and properties of hemocyanins. XII. Electron microscopy of dissociation products of *Helix pomatia* α-hemocyanin: quaternary structure. J. Mol. Biol. 90, 77-90 (1974)

Siezen, R.J., van Driel, R.: Structure and properties of hemocyanins. XIII. Dissociation of *Helix pomatia* α-hemocyanin at alkaline pH. J. Mol. Biol. 90, 91-102 (1974)
Stewart, G.A., Johns, P.: Empirical and theoretical relationships between the sedimentation coefficient and molecular weight of various proteins, with particular reference to the immunoglobulins. J. Immunol. Methods 10, 219-229 (1976)
Vandekerckhove, J., Van Montagu, M.: Sequence analysis of fluorescamine stained peptides and proteins purified on a nanomole scale. Europ. J. Biochem. 44, 279-288 (1974)
Weber, K., Pringle, J.R., Osborn, M.: Measurement of molecular weights by electrophoresis on SDS-acrylamide gel. Methods Enzymol. 26, 3-27 (1973)
Witters, R., Lontie, R.: Stability regions and amino acid composition of gastropod haemocyanins. In: Physiology and Biochemistry of Haemocyanins. Ghiretti, F. (ed.). London, New York: Academic Press, 1968, pp. 61-73
Wood, E.J., Bannister, W.H., Oliver, C.J., Lontie, R., Witters, R.: Diffusion coefficients, sedimentation coefficients and molecular weights of some gastropod haemocyanins. Comp. Biochem. Physiol. 40B, 19-24 (1971)

Composition and Structure of Glycopeptides Obtained by Cyanogen Bromide Cleavage of *Buccinum undatum* Haemocyanin

R. L. Hall and E. J. Wood

Introduction

Gastropod haemocyanins have been shown to be glycoproteins (Dijk et al., 1970; Waxman, 1975; Hall and Wood, 1976) but at present little is known of the structure or function of the carbohydrate moieties. Carbohydrate side chains have however been suggested as possibly being involved in linking together polypeptide chains in haemocyanin (Dijk et al., 1970) or perhaps having a functional role at the copper-binding site (van Holde and van Bruggen, 1971). Siezen and van Driel (1973) have in addition suggested that the carbohydrate may be the cause of the microheterogeneity that is found with *Helix pomatia* and probably other haemocyanins. An understanding of the role of the carbohydrate in gastropod haemocyanins is therefore important to studies of the structural and functional properties of these proteins.

The ability of gastropod haemocyanins to bind to the plant lectin concanavalin A through their carbohydrate groups has been reported (Hall and Wood, 1976), and Waxman (1975) used concanavalin A linked to Sepharose (Con A-Sepharose) to isolate small glycopeptides obtained after tryptic digestion of *Busycon canaliculatum* haemocyanin. In the present work, studies were made of the glycopeptides obtained from *Buccinum undatum* haemocyanin by cyanogen bromide cleavage and isolated on Con A-Sepharose.

Cyanogen bromide has been used previously to obtain large peptides from *H. pomatia* (Dijk, 1971) and *B. canaliculatum* (Waxman, 1975) haemocyanins, but separation of the resulting peptides was extremely difficult because of the formation of aggregates of the material, which were insoluble except in 8M-urea, 5% (w/v) sodium dodecyl sulphate, or 1M-formic acid. In this study, the problem of insoluble peptide aggregates was overcome not by altering the properties of the solvent, which would make separation by affinity chromatography impossible, but by altering the solubility properties of the peptides themselves by treatment with an alkylating agent. This did not prevent the binding of glycopeptides to Con A-Sepharose and therefore enabled them to be isolated specifically.

Methods

Preparations of Haemocyanin. B. undatum haemocyanin was isolated as described by Wood (1973) and stored under CO in 0.1 M-sodium acetate/acetic buffer, pH 5.7 containing 0.01% (w/v) NaN_3.

Cyanogen Bromide Cleavage. Haemocyanin was reduced and alkylated by stirring solutions containing 10 mg/ml haemocyanin in 0.05 M-sodium borate/HCl buffer, pH 8.7 containing 8 M-urea and 1% (v/v) β-mercaptoethanol at room temperature for 1 h, followed by the addition of a 10-fold molar excess of ethyleneimine with stirring for a further 3 h. After dialysis against 0.1 M-NH_4HCO_3 adjusted to pH 9.0 with 0.1 M-NH_4OH, the S-aminoethyl haemocyanin (SAE-haemocyanin) was freeze-dried.

Cyanogen bromide cleavage was carried out by dissolving the SAE-haemocyanin in 70% (v/v) formic acid to a concentration of 10 mg/ml and adding a 500-fold molar excess of CNBr to methionine residues and stirring in a fume cupboard at room temperature for 48 h. The resultant peptide-containing solution was diluted with 10 volumes of water and freeze-dried.

Solubilisation with Citraconic Anhydride. CNBr peptides were added to 0.05 M-sodium borate/HCl pH 9.0 buffer and stirred vigorously at $0^{o}C$ for 2 h to produce a homogeneous suspension. A 30-fold molar excess of citraconic anhydride (2-methylmaleic anhydride) to lysine residues was added dropwise in 1,4 dioxan (1:4 v/v) with constant stirring, and the pH maintained at 9.0 by the addition of 1 M-NaOH. After the final addition the solutions were stirred for a further 1 h. Removal of the citraconyl groups from the peptides was carried out when required by lowering the pH to 2.0 for 6 h (Gibbons and Perham, 1970).

Isolation of Concanavalin Binding Peptides. A Con A-Sepharose column (5.0 cm diam. × 6.0 cm) was equilibrated with binding buffer consisting of 10 mM-Tris HCl, pH 7.0, containing 0.1 M-NaCl, 0.7 mM-$CaCl_2$, 0.7 mM-$MgCl_2$, 0.7 mM-$MnCl_2$ and 0.01% (w/v) NaN_3. Solutions containing citraconylated cyanogen bromide peptides adjusted to pH 7.0 by the addition of 0.1 M-HCl were applied to the Con A-Sepharose column and the unbound peptides eluted with 200 ml binding buffer. The bound peptides were subsequently eluted from the column with 200 ml binding buffer containing 0.4 M-α-methyl D-glucoside and freeze-dried.

Gel Filtration of Peptides. Citraconylated glycopeptides were de-salted on a 2.8 cm × 11.0 cm column of Biogel P2 which had been equilibrated with 0.05 M-NH_4HCO_3. Fractionations of citraconyl glycopeptides were carried out on Biogel P30 (3.2 cm diam. × 19.0 cm) and Sephadex G-50 (2.2 cm diam. × 113 cm) columns equilibrated with 0.05 M-NH_4HCO_3 containing 0.01% (w/v) NaN_3.

Detergent-gel Electrophoresis. Detergent-gel electrophoresis was carried out on polyacrylamide gels containing 0.1% (w/v) sodium dodecyl sulphate essentially as described by Weber and Osborn (1969).

Isoelectric Focusing in Polyacrylamide Gels. Polyacrylamide gel isoelectric focusing was performed as described by Wrigley (1968) using 5% polyacrylamide gels containing 8 M-urea and 1% (w/v) Ampholine.

Carbohydrate Determinations. Methyl glycosides were prepared from glycopeptides by methanolysis under N_2 with 1 M-HCl in anhydrous methanol at $80^{o}C$ for 6 h. Neutral sugars were estimated as the trimethylsilyl derivatives of the methyl glycosides by g.l.c. on a Pye 104 gas chromatograph (Hall and Wood, 1976). Amino sugars were released by hydrolysis of the glycopeptides with 4 M-HCl in evacuated sealed tubes for 4 h at $105^{o}C$ and analysed on a Biocal BC 200 amino acid analyser.

Amino Acid Analysis. Peptide samples were hydrolysed with 6 M-HCl in the presence of 0.1% (w/v) phenol in evacuated sealed tubes for 24 h at $110^{o}C$. Amino acid analysis was performed with a Rank-Hilger Chromaspec amino acid analyser.

N-terminal Amino Acid Determination. Dansyl-amino acid derivatives were prepared by the method of Gros and Labouesse (1969) and separated by ascending chromatography on thin-layer polyamide plates.

Results

Hydrolysis of *B. undatum* haemocyanin with CNBr produced peptides of which only about 20% were soluble in water at 20°C. However the addition of either 1% (w/v) sodium dodecyl sulphate, or of 8 M-urea, rendered the peptides completely soluble. Examination of the peptides by detergent-gel electrophoresis revealed two broad bands in the molecular weight ranges 5000-10,000 and 10,000-14,000 (Fig. 1). Samples of haemocyanin,

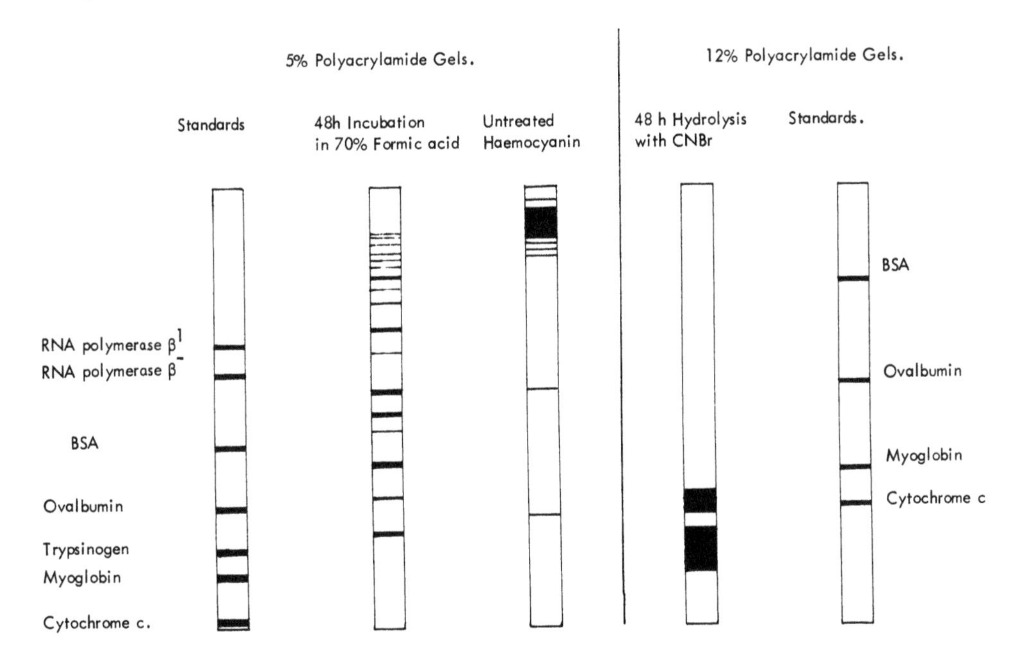

Fig. 1. Detergent-gel electrophoresis with *Buccinum* haemocyanin, formic acid-treated haemocyanin, and CNBr peptides from *Buccinum* haemocyanin

which had been reacted only with 70% (v/v) formic acid for 48 h, were also analysed by SDS gel electrophoresis, and numerous discrete bands ranging in molecular weight from 28,000 to 400,000 were produced. However no dansyl NH_2-terminal amino acids could be detected for haemocyanin that had been subjected to formic acid treatment only, which suggested that the peptide fragments were not produced as the result of hydrolysis of peptide bonds.

Attempts were made to isolate CNBr glycopeptides from the water-soluble fraction by separation on Con A-Sepharose, but no material was bound. This indicated that the carbohydrate residues responsible for binding native haemocyanin to concanavalin A were not present in the water soluble fraction. The total peptide mixture was treated with the alkylating agent citraconic anhydride (which reacts specifically with free primary amino groups: Dixon and Perham, 1968) and this treatment resulted in the solubilisation of almost all of the peptide material. A solution containing the citraconyl peptides, adjusted to pH 7.0, was applied to the Con A-Sepharose column. About 85% of the material was not bound by the column and passed straight through, and then the glycopeptide-containing fraction was eluted with buffer containing 0.4 M-α-methyl D-glucoside (Fig. 2). Glycopeptides in the eluate from this column were separated from small molecules by passage through a Biogel

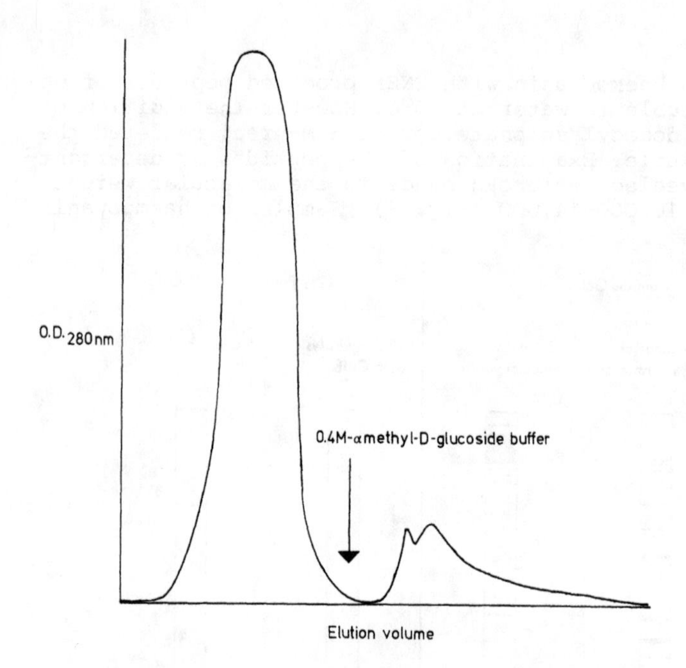

Fig. 2. Affinity chromatography of citraconylated CNBr glycopeptides of *Buccinum* haemocyanin on Con A-Sepharose. The unbound peptides were eluted first with 10 mM-Tris/HCl buffer pH 7.0 containing 0.1 M-NaCl, 0.7 mg-$CaCl_2$, 0.7 mM-$MgCl_2$, 0.7 mM-$MnCl_2$ and 0.01% (w/v) NaN_3. The glycopeptide fraction was then eluted with the same buffer containing 0.4 M-α-methyl D-glucoside

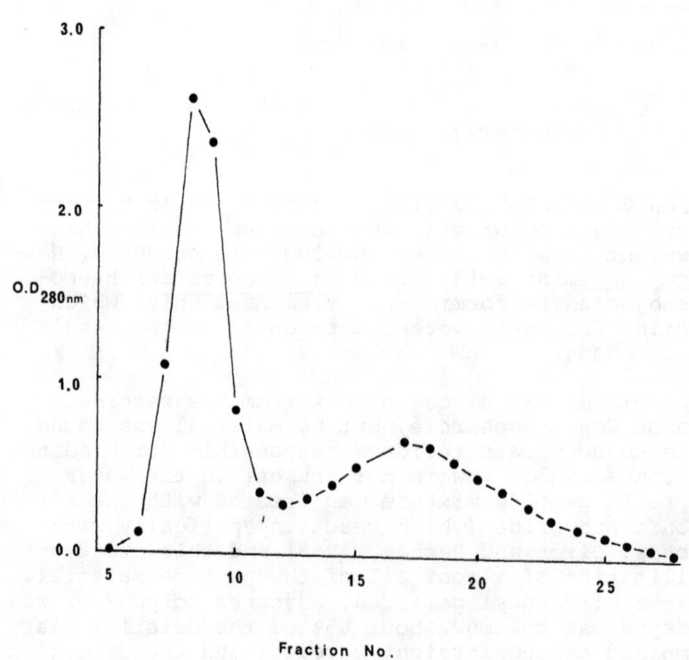

Fig. 3. Fractionation of citraconylated CNBr glycopeptides on Bio-gel P30 (column dimension 3.2 cm × 19.0 cm). The column was eluted with 0.05 M-NH_4HCO_3 at a flow rate of 5.7 ml/h h and 3.8 ml fractions were collected

P2 column equilibrated with 0.05 M-NH_4HCO_3. The material eluting in the void volume was pooled and concentrated by freeze-drying.

The glycopeptides were redissolved in 3 ml of 0.05 M-NH_4HCO_3 and were fractionated on a Biogel P30 column (Fig. 3). Amino acid and carbohydrate analyses were carried out on samples taken from different parts of each peak (Table 1). The results of these analyses showed that the

Table 1. Amino acid and carbohydrate compositions of glycopeptides of Biogel P30 column fraction 7 (peak containing large glycopeptides). The results are expressed as molar ratios using a value of 2.0 for alanine

Amino Acid	
Aspartic acid	2.9
Threonine	1.3
Serine	1.4
Glutamic acid	2.9
Proline	2.3
Glycine	2.1
Alanine	2.0
Valine	1.8
Methionine	trace
Isoleucine	1.3
Leucine	2.4
Tyrosine	1.2
Phenylalanine	1.8
Histidine	1.6
SAE-cysteine	0.3
Lysine	1.5
Arginine	1.6
Tryptophan	n.d.
Carbohydrate	
Glucosamine	0.36
Galactosamine	trace
Fucose	0.03
Mannose	0.29

amino acid composition of the large glycopeptide peak was broadly similar to that of the small glycopeptide peak with the exception that it contained slightly higher amounts of glutamic acid, proline, tyrosine and phenylalanine. Comparison of the amino acid contents of fractions taken at different positions in the small peptide peak also indicated a more or less uniform composition over the range of the peak (Table 2). However when carbohydrate compositions were compared, marked differences were found. The large glycopeptide peak had a glucosamine content slightly higher than the mannose content, but in the small glycopeptide peak the mannose content was higher than the glucosamine content. Furthermore, unlike the amino acid composition, the content of carbohydrate more particularly mannose varied in different fractions from the small peak. After removal of citraconyl groups by exposure to pH 2.0 for 6 h, the NH_2-terminal amino acids present in the glycopeptides of of both the large and small peaks were determined in order to try to ascertain the number of polypeptide chains present in each peak. However, no dansyl α-amino acid derivatives were found for peptides in the large glycopeptide peak, only dansyl ε-amino lysine and dansyl o-tyrosine being detected. This may indicate that the large glycopeptide(s) may contain the NH_2-terminal region of *B. undatum* haemocyanin since no free NH_2-terminal amino acid has been detected in the native protein (Hall, unpublished). NH_2-terminal analysis of the small peak revealed the presence of eleven dansyl α-amino acid derivatives, show-

Table 2. Amino acid and carbohydrate compositions of glycopeptides of Biogel P30 column fractions 14-23 (small glycopeptide peak)

	Column fraction 14	15	17	18	19	20	22	23
Amino acid								
Aspartic acid	2.5	2.6	2.7	2.9	2.9	2.5	2.3	2.5
Threonine	1.0	1.0	1.1	1.1	1.1	1.2	0.9	1.1
Serine	1.3	1.4	1.3	1.4	1.3	1.6	1.5	1.6
Glutamic acid	2.5	2.4	2.5	2.5	2.2	2.2	2.0	2.3
Proline	1.1	1.1		1.2		1.3		
Glycine	2.2	2.0	2.1	1.8	2.0	2.2	2.7	2.7
Alanine	2.0	2.0	2.0	2.0	2.0	2.0	2.0	2.0
Valine	1.6	1.8	1.6	1.5	1.6	1.5	1.4	1.8
Methionine		0.3						
Isoleucine	1.1	1.4	1.1	1.2	1.2	1.0	1.1	1.0
Leucine	1.9	2.1	1.9	2.2	2.3	1.9	1.9	2.0
Tyrosine	0.8	1.0	0.8	0.8	0.8	0.6	0.8	0.7
Phenylalanine	1.1	1.4	1.0	1.0	1.0	1.1	1.0	0.9
Histidine	1.2	1.5	1.2	1.2	1.3	1.4	1.1	1.0
SAE-cysteine	0.3					0.3		
Lysine	1.4	1.5	1.2	1.0	1.0	1.3	1.1	1.3
Arginine	1.4	1.9	1.4	1.3	1.3	1.8	1.7	2.0
Tryptophan								
Carbohydrates								
Glucosamine	0.91	1.16	1.31	1.24	0.99	0.83	0.46	Trace
Galactosamine	Trace	Trace	Trace	Trace	Trace	Trace	Trace	Trace
Fucose	0.20	0.23	0.12	0.12	0.31	0.28	0.07	Trace
Mannose	1.65	2.20	1.77	3.84	4.22	4.01	0.81	0.55

The results are expressed as molar ratios using a value of 2.0 for alanine.

ing that considerable heterogeneity existed in the peptide portions of the glycopeptides in this peak.

Glycopeptides from which the citraconyl groups had been removed were subjected to isoelectric focusing in polyacrylamide gels in the presence of 8 M-urea. Four bands staining with Coomassie Blue G were found in the large glycopeptide fraction, and six bands in the small glycopeptide fraction. Preparative column isoelectric focusing in the presence of 8 M-urea was attempted but was unsuccessful for fractionating the glycopeptides because of isoelectric precipitation.

Further attempts to fractionate the small citraconylated glycopeptides on the basis of size were made using a Sephadex G-50 column. The eluate from the column was divided into five fractions (Fig. 4) and each pooled fraction was freeze-dried. Each of the fractions A, B, C and D was passed separately through the column twice, and each time material eluted between the limits of the original fraction pooled and freeze-dried. The NH_2-terminal amino acids present in each of the resultant fractions was determined, but all the eleven dansyl α-amino acids found for the small glycopeptide peak from the Biogel P30 column were present in each of the final fractions re-eluted from the Sephadex G-50 column. This may well indicate that the glycopeptides are heterogeneous in size because they contain different amounts of carbohydrate attached to the polypeptide chains.

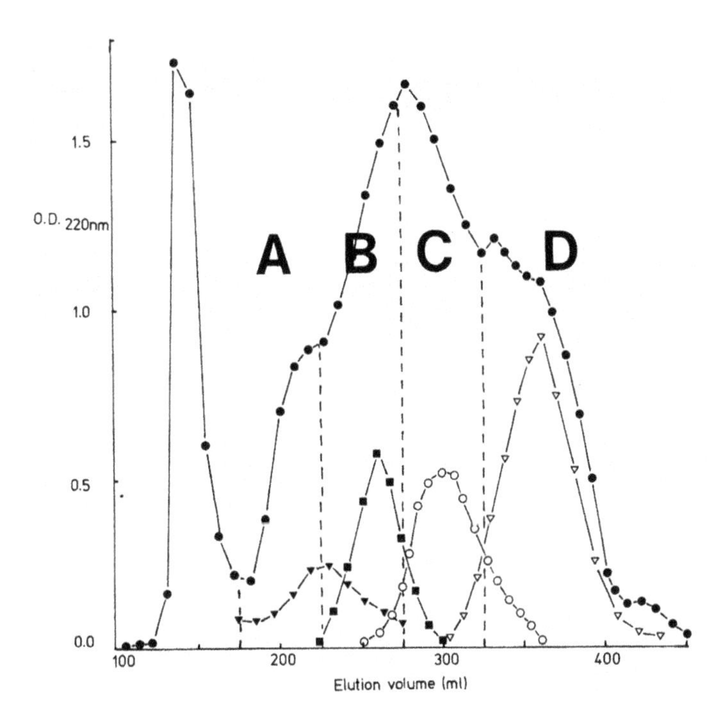

Fig. 4. Fractionation of small citraconyl CNBr glycopeptides on Sephadex G-50 (column dimensions 2.2 cm × 113 cm). The column was eluted with 0.05 M-NH_4HCO_3 at a flow rate of 12 ml/h. (●) Total small citraconyl CNBr glycopeptides from Biogel P30 column; (▼) fraction A 175-225 ml eluent, three times chromatographed; (■) fraction B, 225-275 ml eluent, three times chromatographed; (O) fraction C, 275-325 ml eluent, three times chromatographed; (∇) fraction D, 325-450 ml eluent, three times chromatographed

Discussion

Studies of the products of cleavage at specific residues in denatured gastropod haemocyanins have previously been hampered by the formation of aggregates of material which were insoluble except in strongly denaturing solvents (Gruber, 1968; Dijk, 1971; Waxman, 1975). In the present work peptide products were solubilised by treatment with citraconic anhydride. This method may prove to be of general use in solubilising peptides from other proteins, particularly when it is intended that glycopeptides are to be isolated by affinity chromatography.

Attempts to isolate tryptic glycopeptides from *H. pomatia* (Dijk, 1971) and *B. canaliculatum* (Waxman, 1975) haemocyanins have met with only a limited success. The procedures used gave a mixture of peptides which could not be resolved even after several steps of purification. This suggested that heterogeneity exists among the peptide portions of the glycopeptides obtained from these proteins. No fractionation of insoluble peptides was made in either of these studies. Recently it has been reported (Brower et al., 1976) that limited trypsinolysis of undenatured *H. pomatia* α-haemocyanin gave rise to a number of discrete polypeptide products. When analysed for carbohydrate content these products showed large variation among the fragments, again indicating that heterogeneity exists in the peptide portions of the glycopeptides from this protein. The presence of eleven N-terminal end groups in the small CNBr glycopeptide fraction from *B. undatum* haemocyanin makes it likely that numerous sites of carbohydrate attachment are also present in this haemocyanin. The number and size of the peptides found would also appear to exclude any explanation that they are derived from a single 50,000 molecular weight subunit.

Brouwer et al. (1976) put forward the hypothesis that the primary structure of *H. pomatia* α-haemocyanin consisted of single polypeptide chains of molecular weight about 360,000 containing six to eight structurally similar domains having molecular weights of 50,000. This hypothesis is not supported by the results of incubation of gastropod haemocyanins with formic acid as reported here and by Dijk et al. (1970) in which no NH_2-terminal amino acids were found for the resultant polypeptide fragments. However this might be consistent with carbohydrate, amide or sidechain peptide bonds linking together structural domains formed from separate polypeptide chains. Such an explanation would not have the requirement for six to eight closely linked genes, which would be a necessary corollary to the Brouwer hypothesis.

Acknowledgments. R.L.H. is grateful to the Science Research Council for a studentship.

References

Brouwer, M., Walters, M., van Bruggen, E.F.J.: Proteolytic fragmentation of *Helix pomatia* α-haemocyanin: Structural domains in the polypeptide chain. Biochemistry 15, 2618-2623 (1976)

Dijk, J., Brouwer, M., Coert, A., Gruber, M.: Structure and functions of haemocyanins. VII. The smallest subunit of α- and β-haemocyanin of *Helix pomatia*: Size, composition, N- and C-terminal amino acids. Biochim. Biophys. Acta 221, 467-479 (1970)

Dijk, J.: Alpha and beta-haemocyanins of *Helix pomatia*. Thesis, Univ. Groningen, Netherlands (1971)

Dixon, H.B.F., Perham, R.N.: Reversible blocking of amino groups with citraconic anhydride. Biochem. J. 109, 312-314 (1968)

Gibbons, I., Perham, R.N.: The reaction of aldolase with 2-methylmaleic anhydride. Biochem. J. 116, 843-849 (1970)

Gros, C., Labouesse, B.: Study of the dansylation reaction of amino acids, peptides and proteins. Europ. J. Biochem. 7, 463-470 (1969)

Gruber, M.: Structure and function of *Helix pomatia* haemocyanin. In: Physiology and Biochemistry of Haemocyanins. Ghiretti, F. (ed.). London, New York: Academic Press, 1968, pp. 49-59

Hall, R.L., Wood, E.J.: The carbohydrate content of gastropod haemocyanins. Biochem. Soc. Trans. 4, 307-309 (1976)

Siezen, R.J., van Driel, R.: Structure and properties of haemocyanins. VIII. Microheterogeneity of α-haemocyanin of *Helix pomatia*. Biochim. Biophys. Acta 295, 131-139 (1973)

van Holde, K.E., van Bruggen, E.F.J.: The haemocyanins. In: Subunits in Biological Systems: Part A. Timasheff, S.N., Fasman, G.D. (eds.). New York: Marcel Dekker, 1971, Vol. V, pp. 1-53

Waxman, L.: The structure of arthropod and mollusc haemocyanins. J. Biol. Chem. 250, 3796-3806 (1975)

Weber, K., Osborn, M.: The reliability of molecular weight determinations by dodecyl sulphate - polyacrylamide gel electrophoresis. J. Biol. Chem. 244, 4406-4412 (1969)

Wood, E.J.: Gastropod haemocyanins: Dissociation of haemocyanins from *Buccinum undatum, Neptunea antiqua* and *Colus gracilis* in the region pH 7.5-9.2. Biochim. Biophys. Acta 328, 101-106 (1973)

Wrigley, C.W.: Gel electrophoresis. A technique for analyzing multiple protein samples by isoelectric focusing. Science Tools 15, 17-23 (1968)

Breakdown of *Murex trunculus* Haemocyanin into Subunits

J. V. BANNISTER, J. MALLIA, A. ANASTASI AND W. H. BANNISTER

Introduction

The minimum functional unit in molluscan haemocyanins has a molecular weight of about 50,000 (Ghiretti-Magaldi et al., 1966). Various attempts have been made to prepare a 50,000 molecular weight subunit from these haemocyanins by chemical or enzymatic means. However, Brouwer and Kuiper (1973) and Waxman (1975) have reported that molluscan haemocyanins do not dissociate into polypeptide chains of less than 265,000-290,000 molecular weight under strongly denaturing and reducing conditions. The results of reduction and carboxymethylation and proteolytic digestion of haemocyanin from the whelk *Murex trunculus* (L.) are described in this paper.

Materials and Methods

Haemocyanin was prepared as previously described (Bannister and Wood, 1971). Apohaemocyanin was prepared according to Fernandez-Moran et al. (1966). Protein concentrations were estimated spectrophotemetrically at 280 nm using an $E_{1\ cm}^{1\%}$ at pH 9.2 of 13.9 for oxyhaemocyanin and 12.8 for apohaemocyanin (Bannister et al., 1973).

Reduction and carboxymethylation was carried out on apophaemocyanin according to Crestfield et al. (1963) except that reduction was allowed to proceed for 48 or 120 h at 45°C.

Subtilisin digestion of haemocyanin was carried out as described by Bannister et al. (1975). Trypsinolysis in the presence of Ca^{2+} was performed under the conditions described by Van Breemen et al. (1975). Trypsin digestion in the absence of Ca^{2+} was carried out for 3 h at 37°C using 0.5 mg trypsin per 100 mg haemocyanin at pH 8.2.

Digestion of haemocyanin with papain was carried out by the method of Porter (1959) for γ-globulin, whilst pronase digestion of haemocyanin was performed as described by Ambler (1963) for *Pseudomonas* cytochrome c-551. In both cases the incubation was carried out at 37°C for 16 h.

The products obtained as a result of reduction and carboxymethylation or proteolysis were separated by gel filtration. Analytical polyacrylamide gel electrophoresis on 5% gel rods was carried out according to Davis (1964). Sodium dodecylsulphate (SDS) gel electrophoresis was performed by the method of Fairbanks et al. (1971). Samples were incubated with 4 M urea and 1% SDS, with or without 1% (v/v) 2-mercaptoethanol, for 1 h at 45°C before electrophoresis. All gels were stained with Coomassie Brilliant Blue R.

Sedimentation analyses were carried out at 20°C in a Beckman Model E analytical ultracentrifuge. Carboxymethylcysteine was determined by amino acid analysis of 24-h acid hydrolysates. N-terminal amino acids were determined by the method of Percy and Buchwald (1972).

Results

Reduction and carboxymethylation products of apohaemocyanin were separated on a Bio-Gel A-15m column. As many as six peaks were obtained after 48-h reduction (Fig. 1), whilst only two peaks were obtained after 120-h reduction (Fig. 2). Amino acid analysis showed four carboxy-

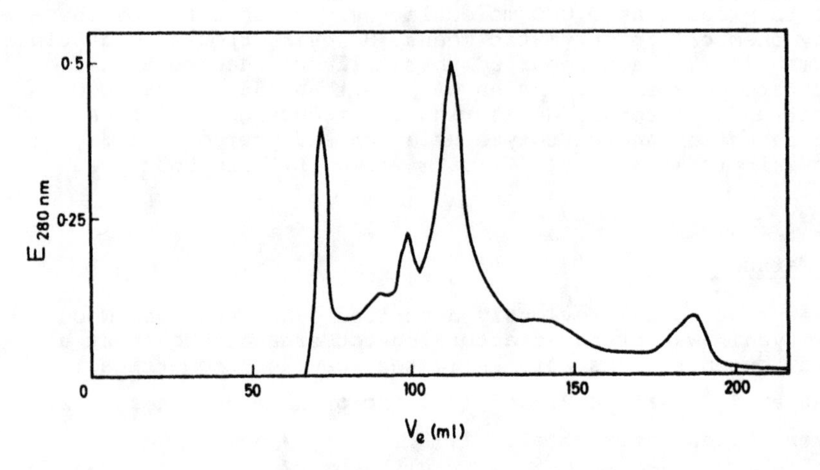

Fig. 1. Gel filtration of 48-h reduced and carboxymethylated apohaemocyanin on a column of Bio-gel A-15m (90 × 1.5 cm) equilibrated with 0.1 M Tris-HCl buffer, pH 8.0, containing 8 M urea. Fractions of 1.8 ml were collected at a flow rate of about 18 ml/h

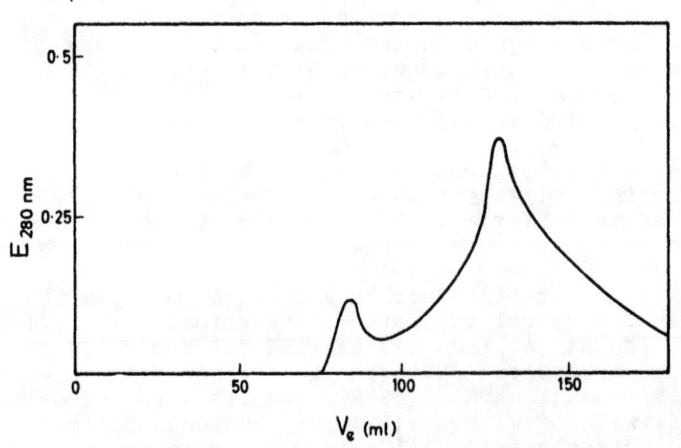

Fig. 2. Gel filtration of 120-h reduced and carboxymethylated apohaemocyanin on Bio-gel A-15m. Column and conditions as in Figure 1

methylcysteine residues per 50,000 molecular weight in both preparations. SDS-gel electrophoresis of the 48-h reduced protein revealed bands of molecular weight ranging from 15,000 to 200,000. In the 120-h reduced protein five bands of molecular weight varying from 12,800 to 60,000 were obtained. No marked differences in molecular weights were obtained when the samples for SDS electrophoresis were incubated in the presence or absence of 2-mercaptoethanol before electrophoresis. The molecular weights obtained for the different samples are given in Table 1.

Table 1. Molecular weights of bands seen in sodium dodecylsulphate-gel electrophoresis of reduced and carboxymethylated apohaemocyanins

With 2-mercaptoethanol[a] Band	MW	Without 2-mercaptoethanol[a] Band	MW
(a) *48-h reduction*			
I	13,500	I	15,000
II	19,000	II	21,000
III	25,000	III	27,000
IV	36,000	IV	39,000
V	41,600	V	43,000
VI	47,500	VI	49,000
VII	58,800	VII	60,000
VIII	76,000	VIII	76,000
IX	91,000	IX	87,000
X	141,000	X	135,000
XI	170,000	XI	166,000
XII	209,000	XII	190,000
XIII	224,000	XIII	200,000
(b) *120-h reduction*			
I	15,500	I	12,800
II	27,000	II	26,000
III	41,000	III	43,600
IV	50,000	IV	53,700
V	57,000	V	60,000

Removal of urea from the treated haemocyanin was achieved by sequential dialysis against buffers containing decreasing concentrations of urea until finally the protein was dialysed against 0.05 M borate buffer, pH 8.2. The urea-free carboxymethylated protein was chromatographed on a Bio-gel A-15m column. Two peaks were obtained (Fig. 3). The faster

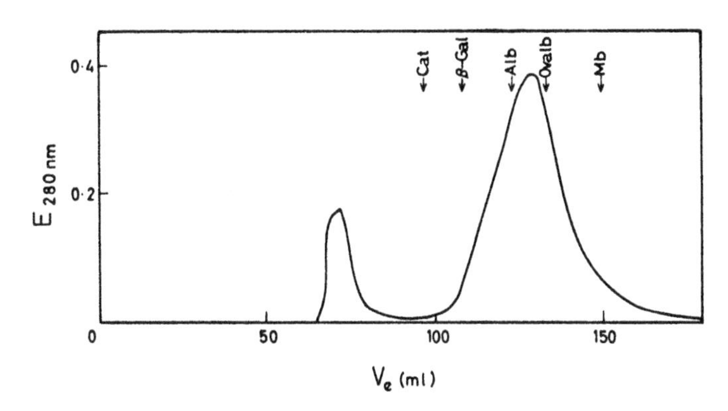

Fig. 3. Gel filtration of 120-h reduced and carboxymethylated apohaemocyanin following removal of urea. Bio-gel A-15m column (90 × 1.5 cm) equilibrated with 0.05 M borate buffer, pH 8.2. Fractions of 1.8 ml were collected at a flow rate of about 12 ml/h

eluting peak had a molecular weight of 690,000, whilst the slower peak had an approximate molecular weight of 50,000. However, concentration of the latter material and its application on a Bio-gel P-200 column resulted in the protein eluting in the excluded volume, indicating a molecular weight higher than 200,000 and reaggregation.

N-terminal amino acid analysis of both 48-h and 120-h reduced and carboxymethylated protein showed the presence of free N-terminal amino acids only in the latter. Five N-terminal amino acids were identified: alanine, isoleucine, leucine, phenylalanine and valine.

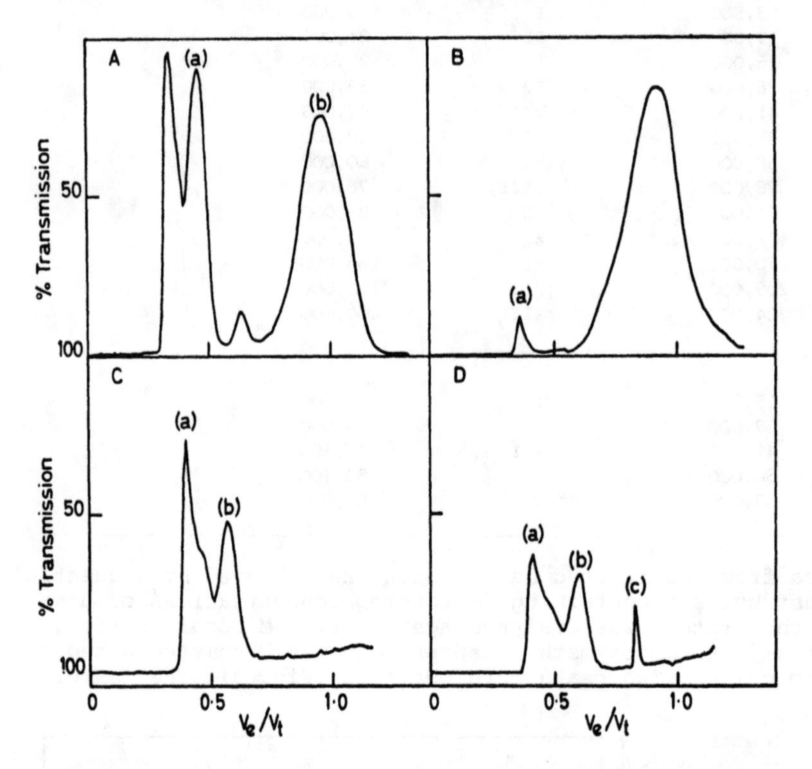

Fig. 4A-D. Gel filtration of fragments after subtilisin digestion of haemocyanin. (A) Sephadex G-75 column (81 × 5 cm) equilibrated with 0.1 M borate buffer, pH 8.2, loaded with digestion mixture containing approximately 3 g haemocyanin. (B) Sephadex G-50 column (87 × 5 cm) equilibrated with 0.1 M borate buffer, pH 8.2, loaded with column (A), peak (*b*) fractions from approximately 12 g haemocyanin. (C) Sephadex G-75 column (93 × 1.5 cm) equilibrated with 0.1 M ammonium bicarbonate, pH 8.0, loaded with peak (*a*) from column (B). (D) Sephadex G-75 column (93 × 1.5 cm) equilibrated with 0.1 M ammonium bicarbonate, pH 8.0, containing 1% (v/v) 2-mercaptoethanol, loaded with peak (*a*) from column (C)

The fractionation of a subtilisin digest on Sephadex G-75 is shown in Figure 4A. Peak (a) corresponds to the 50,000 molecular weight fragments characterized in a previous investigation (Bannister et al., 1975). This material showed seven N-terminal groups - alanine, aspartic acid, glutamic acid, glycine, leucine, phenylalanine and valine - while no N-terminal group was detected in native haemocyanin or the apoprotein. Peak (b) corresponds to small molecular weight material with an elution volume around the total bed volume of the column. The elution volume is approximately the same as that of cytochrome c, indicating an apparent molecular weight of about 12,000. The material in peak (b) reacted positively for copper. Between peaks (a) and (b) a small peak is evident. The material eluted here did not give a positive reaction for copper. Its elution corresponds with that expected for subtilisin (molecular weight 27,500) in the digestion mixture.

The low molecular weight, copper-positive material of peak (b) was applied to a Sephadex G-50 column. Its fractionation is shown in Figure 4B. The leading fraction, peak (a) was investigated. This material was reapplied to a column of Sephadex G-75. The result is shown in Figure 4C. A peak eluting around the void volume of the Sephadex G-75 column followed by a peak with an elution volume corresponding to that for 50,000 molecular weight material was observed [peak (b)]. In fact a molecular weight of 50,200 ± 700 was calculated for this peak after calibration of the column. It is clear that the original small molecular weight material from the subtilisin digestion mixture [peak (b), Fig. 4A] had aggregated after its initial separation on Sephadex G-75.

The material from peak (a), Figure 4C partially disaggregated when it was treated with 2-mercaptoethanol. After treatment this material was applied to a column of Sephadex G-75. Material eluting around the void volume still appeared [peak (a), Fig. 4D]. However, material of approximately 50,000 molecular weight [peak (b), Fig. 4D] and smaller material of about 15,500 molecular weight [peak (c), Fig. 4D] were also observed.

Polyacrylamide gel electrophoresis of the material from peak (a), Figure 4B showed three major and several minor components. The 50,000 molecular weight material from peak (b), Figure 4C (reaggregated material) showed two major and some indistinct trailing bands. Sedimentation analysis of this reaggregated material revealed a single broad band with a sedimentation coefficient of 3.8S and the material was found to have a molecular weight of 54,000 ± 4,000 in SDS-gel electrophoresis.

Haemocyanin incubated with trypsin in the absence of Ca^{2+} was fractionated on a column of Bio-gel P-200. Two peaks (I and II) were observed (Fig. 5). Peak I eluted in the void volume, whilst peak II eluted in the included volume of the column. Polyacrylamide gel electrophoresis of peaks I and II showed various bands. Peak I gave two major bands

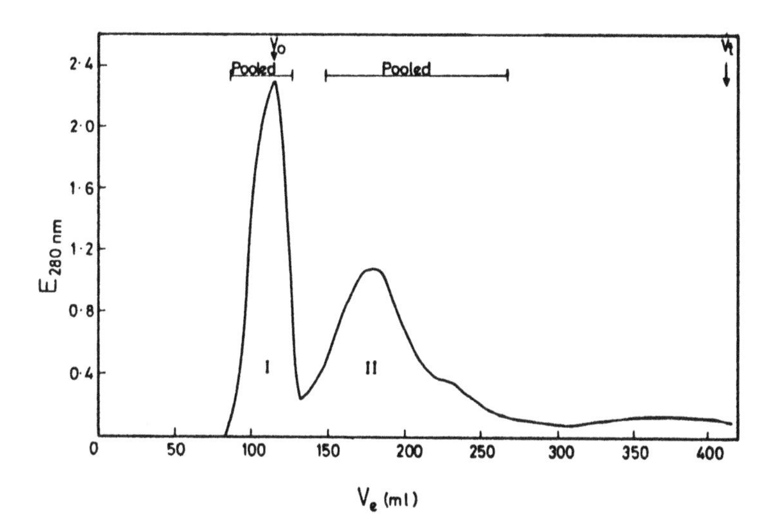

Fig. 5. Gel filtration of fragments after tryptic digestion of haemocyanin in absence of Ca^{2+}. Bio-gel P-200 column (82 × 2.5 cm) equilibrated with 0.1 M borate buffer, pH 8.2. Fractions of 8.3 ml were collected at a flow rate of 14 ml/h

together with two minor ones, whilst peak II gave a series of bands of varying intensity. SDS-gel electrophoresis without use of 2-mercaptoethanol indicated a single component with a molecular weight of 346,000 for peak I. After incubation with 2-mercaptoethanol, peak I gave six bands with molecular weights of 390,000, 355,000, 316,000, 200,000, 141,000 and 87,000. Peak II gave six bands of molecular weight 145,000, 135,000, 81,000, 56,000, 47,000 and 31,000 without exposure to 2-mercaptoethanol, whilst nine bands of molecular weight 218,000, 190,000, 158,000, 135,000, 104,000, 83,000, 60,000, 49,000 and 39,000 were obtained after incubation with 2-mercaptoethanol. N-terminal analysis gave the same end groups for peaks I and II. The following end groups were identified: alanine, arginine, glycine, leucine, phenylalanine, serine and valine. In the ultracentrifuge, peak I gave a sedimentation coefficient of 120S, whilst peak II gave a value of 22S.

The fractionation on Sephadex G-100 of haemocyanin digested with trypsin in the presence of 10 mM Ca^{2+} is shown in Figure 6. The bulk of the

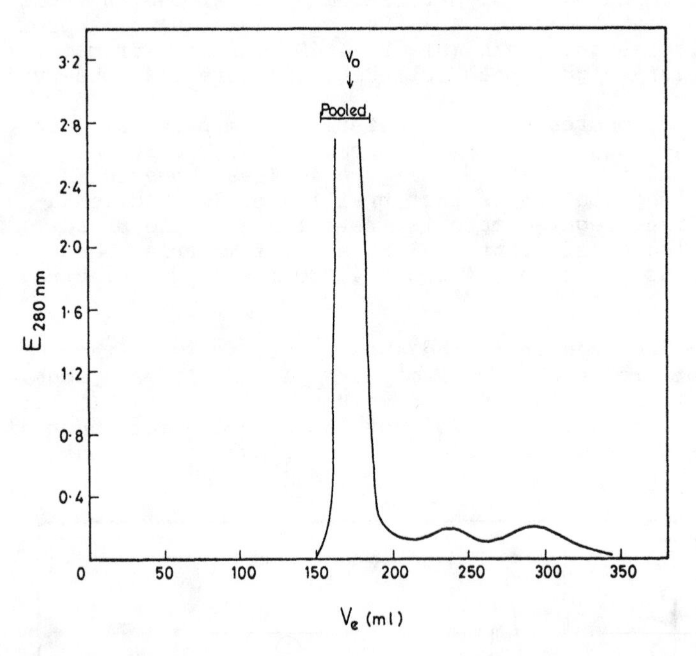

Fig. 6. Gel filtration of haemocyanin after tryptic digestion in the presence of Ca^{2+}. Sephadex G-100 column (95 × 2.5 cm) equilibrated with 0.1 M Tris-HCl buffer, pH 8.2, containing 10 mM Ca^{2+}. Fraction volume and flow rate as in Figure 5

protein eluted in the void volume, whilst a very small fraction (less than 5% of the total material) eluted within the fractionation range of the column. Ultracentrifugal analysis of the tryptic digest gave two sedimenting peaks with sedimentation coefficients of 83S and 6S, respectively. The main component of the digest therefore had a sedimentation coefficient very close to that of the whole molecule.

Fractionation of papain digested haemocyanin on Sephadex G-75 gave four peaks, three of which were in the included volume (Fig. 7). Peak I eluted in the void volume and probably represented partially digested haemocyanin. Peak II had a molecular weight of 50,000 and was copper-positive with an oxyhaemocyanin absorption spectrum. Peak III and an-

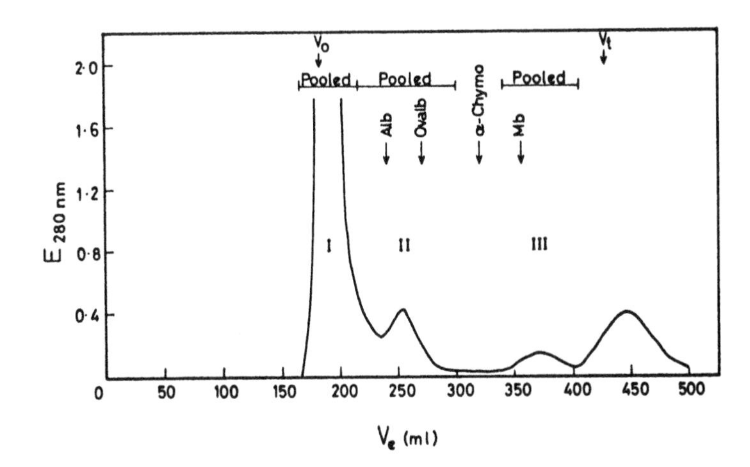

Fig. 7. Gel filtration of haemocyanin after digestion with papain. Sephadex G-75 column (87 × 2.5 cm) equilibrated with 0.1 M phosphate buffer, pH 7.0. Fractions of 5 ml were collected at a flow rate of 45 ml/h

other trailing peak just outside the included volume were copper negative. Further separation of peak II from peak III was achieved on Sephadex G-100.

SDS-gel electrophoresis confirmed the presence of a 50,000 molecular weight fragment following papain digestion of haemocyanin. A molecular weight of 59,000 was obtained without use of 2-mercaptoethanol. Incubation with 2-mercaptoethanol produced no further breakdown and a molecular weight of 50,000 was found. Whilst SDS-gels indicated homogeneity in size for this fragment, polyacrylamide gel electrophoresis gave a series of bands indicating heterogeneity in charge. Only one N-terminal amino acid was found - glutamic acid. N-terminal analysis of the partially digested protein (Peak I, Fig. 7) indicated the presence of six amino acids, namely, alanine, isoleucine, leucine, phenylalanine, proline and valine.

Digestion of haemocyanin with pronase and fractionation of the digest on Sephadex G-75 resulted in the appearance of a peak with a very broad shoulder on its trailing edge containing material of molecular weight 42,000 extending to 25,000. Further resolution of the components in the shoulder was difficult to achieve.

Discussion

Prolonged reduction and carboxymethylation of *Murex trunculus* haemocyanin appears to produce a 50,000 subunit. Polypeptide chains with a high molecular weight as obtained by Brouwer and Kuiper (1973) and Waxman (1975) could have been the result of incomplete dissociation. No N-terminal amino acids were detected by Brouwer and Kuiper (1973) after reduction carboxymethylation of *Helix pomatia* haemocyanin, whilst Waxman (1975) found one N-terminal amino acid (aspartic acid) for *Busycon canaliculatum* haemocyanin. Reduction and carboxymethylation of *M. trunculus* haemocyanin has revealed the presence of five N-terminal amino acids. This investigation has also shown that gradual removal of urea does not lead to the insolubilisation of the modified protein as would happen if the urea was abruptly removed. However, when the concentration of the urea-free modified protein is increased above 12 mg/ml, association takes place to higher molecular weight products.

Lontie et al. (1973), Pearson and Wood (1974) and Bannister et al. (1975) obtained fragments of about 50,000 molecular weight after subtilisin digestion of gastropod haemocyanin. Bannister et al. (1975) postulated that the fragments consisted of four polypeptide chains of about 12,500 molecular weight. The presence of such low molecular weight material following subtilisin digestion has been demonstrated in the present investigation. This material has also been shown to aggregate and give rise to 50,000 molecular weight material. Previous investigations have shown that after treatment of *Helix pomatia* haemocyanin with formic acid (Dijk et al., 1970), or alkali followed by succinylation (Cox et al., 1972), fragments of 25,000 and 22,300 molecular weight, respectively, are obtained. Rizzotti (1974) demonstrated the presence of both 23,500 and 12,000 molecular weight material after reduction of *Carcinus maenas* apohaemocyanin. In this investigation, 12,500 molecular weight material has been observed after reduction and carboxymethylation. The results obtained with reduction and carboxymethylation and subtilisin digestion therefore indicate that the minimum functional subunit of *M. trunculus* haemocyanin might be an aggregate of four polypeptide chains of about 12,500 molecular weight each.

The two components obtained on Bio-gel P-200 following trypsinolysis of *M. trunculus* haemocyanin at pH 8.2 in the absence of Ca^{2+} were shown to be heterogeneous by polyacrylamide gel electrophoresis. SDS-gel electrophoresis of the two components indicated that they contained fragments of varying molecular weights. Identical N-terminal amino acids were found for both components. This suggests that the higher molecular weight component is an aggregate of the second, lower molecular weight component. The sedimentation coefficient of 120S for the higher molecular weight component obtained after tryptic digestion, is somewhat greater than that of the whole molecule (100S) thus confirming the tendency of the tryptic digestion products to aggregate. Previous experiments by Gielens et al. (1973, 1975) have shown that small molecular weight material can be obtained by the action of trypsin on *H. pomatia* β-haemocyanin. Recently a limited trypsinolysis of *Helix pomatia* α-haemocyanin described by Brouwer (1975) yielded fragments of 200,000, 150,000, 100,000 and 50,000 molecular weight. In the present investigation small fragments were obtained only under the conditions of SDS-gel electrophoresis. Following trypsinolysis of *M. trunculus* haemocyanin in the absence of Ca^{2+}, the smallest functional component (with oxyhaemocyanin absorption maxima) that was obtained had a sedimentation coefficient of 22S. This suggests that the fragment (Fig. 5, peak II) corresponded to a tenth molecule, with a molecular weight of approximately 900,000.

When Ca^{2+} was present very little trypsinolysis of *M. trunculus* took place. Van Breemen et al. (1975) digested *H. pomatia* β-haemocyanin with trypsin in the presence of Ca^{2+}. Limited trypsinolysis of this haemocyanin was also observed. However, the digested material was found to form polymers with a sedimentation coefficient of about 700S. The digested material obtained from *Murex trunculus* does not form these polymers. The large material obtained had a sedimentation coefficient of 83S which approximates that of the whole molecule (100S). Trypsinolysis of *M. trunculus* haemocyanin in the presence of Ca^{2+} appears to leave the bulk of the molecule intact.

In contrast to trypsinolysis, digestion of *M. trunculus* haemocyanin with papain resulted in the production of 50,000 molecular weight subunits. Although only a single N-terminal amino acid was identified, the material was heterogeneous in polyacrylamide gel electrophoresis.

These results are the first demonstration that papain produces functional 50,000 molecular weight fragments in a similar way to subtilisin digestion. The main distinction between the two digests is that whilst papain digestion produced a single N-terminal amino acid, seven N-terminal groups were found in the 50,000 molecular weight subunits obtained by subtilisin digestion. Also whilst partially digested haemocyanin (as evidenced by the presence of six N-terminal amino acids) of high molecular weight is obtained with papain, the subtilisin digestion mixture contains a significant amount of undigested haemocyanin with no free N-terminal groups.

The present investigation has shown that appropriate adjustment of the conditions for reduction and carboxymethylation can lead to the production of 50,000 molecular weight subunits from molluscan haemocyanin. However, the resistance of various haemocyanins to dissociation, which may be related to their carbohydrate content and protein-sugar linkages, remains to be assessed. The identity of the subunits with the 50,000 molecular weight fragments obtainable by proteolysis also remains to be established, and the polypeptide chain structure of the subunits remains to be clearly elucidated.

Acknowledgment. W.H.B. thanks the Wellcome Trust for financial support.

References

Ambler, R.P.: Purification and amino acid composition of *Pseudomonas* cytochrome C-551. Biochem. J. 89, 349-378 (1963)

Bannister, J.V., Galdes, A., Bannister, W.H.: Isolation and characterization of two-copper subunits from *Murex trunculus* haemocyanin. Comp. Biochem. Physiol. 51B, 1-4 (1975)

Bannister, W.H., Camilleri, P., Chantler, E.N.: Relationship of tryptophan to the copper-oxygen complex in oxyhaemocyanin. Comp. Biochem. Physiol. 45B, 325-333 (1973)

Bannister, W.H., Wood, E.J.: Ultraviolet fluorescence of *Murex trunculus* haemocyanin in relation to the binding of copper and oxygen. Comp. Biochem. Physiol. 40B, 7-18 (1971)

Brouwer, M.: Structural domains in *Helix pomatia* α-hemocyanin. Thesis, Croningen (1975)

Brouwer, M., Kuiper, H.A.: Molecular weight analysis of *Helix pomatia* α-hemocyanin in guanidine hydrochloride, urea and sodium dodecylsulphate. Europ. J. Biochem. 35, 428-435 (1973)

Cox, J., Witters, R., Lontie, R.: The quaternary structure of *Helix pomatia* haemocyanins as determined by alkali treatment and succinylation. Intern. J. Biochem. 3, 283-293 (1972)

Crestfield, A.M., Moore, S., Stein, W.H.: Preparation and enzymic hydrolysis of reduced and S-carboxymethylated proteins. J. Biol. Chem. 238, 622-627 (1963)

Davis, B.J.: Disk electrophoresis. II. Method and application to human serum proteins. Ann. N.Y. Acad. Sci. 121, 404-427 (1964)

Dijk, J., Brouwer, M., Coert, A., Gruber, M.: Structure and function of hemocyanins. VII. The smallest subunit of α- and β-hemocyanin of *Helix pomatia*: Size, composition, N- and C-terminal amino acids. Biochem. Biophys. Acta 221, 467-479 (1970)

Fairbanks, G., Steck, T.L., Wallack, D.F.H.: Electrophoretic analysis of the major polypeptides of the human erythrocyte membrane. Biochemistry 10, 2606-2616 (1971)

Fernandez-Moran, H., Van Bruggen, E.F.J., Ohtsuki, M.: Macromolecular organisation of hemocyanins and apohemocyanins as revealed by electron microscopy. J. Mol. Biol. 16, 190-207 (1966)

Ghiretti-Magaldi, A., Nuzzolo, C., Ghiretti, F.: Chemical studies on haemocyanins - I. Amino acid composition. Biochemistry 5, 1943-1951 (1966)

Gielens, C., Preaux, G., Lontie, R.: Active subunits of crystalline β-haemocyanin of *Helix pomatia* obtained by trypsin treatment. Arch. Intern. Physiol. Biochim. 81, 5 (1973)

Gielens, C., Preaux, G., Lontie, R.: Limited trypsinolysis of β-haemocyanin of *Helix pomatia*. Europ. J. Biochem. 60, 271-280 (1975)

Lontie, R., De Ley, M., Robberecht, H., Witters, R.: Isolation of small functional subunits of *Helix pomatia* haemocyanin after subtilisin treatment. Nature (New Biol.) 242, 180-182 (1973)

Pearson, J.S., Wood, E.J.: Attempts to obtain small functional subunits of the haemocyanin from *Buccinum undatum* and *Neptunea antiqua*. Trans. Biochem. Soc. 2, 333-336 (1974)

Percy, M.E., Buchwald, B.M.: A manual method of sequential Edman degradation followed by dansylation for the determination of protein sequences. Analyt. Biochem. 45, 60-67 (1972)

Porter, R.R.: Hydrolysis of rabbit γ-globulin and antibodies with crystalline papain. Biochem. J. 73, 119-126 (1959)

Rizzotti, M.: On the quaternary structure of *Carcinus maenas* (Arthropoda) hemocyanin. Experientia 30, 1201 (1974)

Van Breemen, J.F.L., Wichterjes, T., Muller, M.F.J., van Driel, R., van Bruggen, E.F.J.: Tubular polymers derived from *Helix pomatia* β-hemocyanin. Europ. J. Biochem. 60, 129-135 (1975)

Waxman, L.: The structure of arthropod and mollusc hemocyanins. J. Biol. Chem. 250, 3796-3806 (1975)

The Minimal Subunit of Arthropod Hemocyanin

B. SALVATO AND F. RICCHELLI

Introduction

At high pH and in the absence of divalent cations arthropod hemocyanin dissociates into subunits with sedimentation coefficients of about 5S (van Holde and van Bruggen, 1971). This subunit is generally considered to be the minimal dissociation product of arthropod hemocyanin. For these subunits, a MW of 75,000-80,000 daltons has been proposed, essentially on the basis of ultracentrifugation measurements (Ellerton et al., 1970) and sodium dodecyl sulfate (SDS) polyacrylamide-gel electrophoresis (Loehr and Mason, 1973; Waxman, 1975). In our opinion only, it is mere coincidence that these determinations give values in agreement with and corresponding to that calculated from the copper/protein ratio. Thus, as pointed out by van Holde and van Bruggen (1971) the 5S subunit appears heterogeneous in the ultracentrifuge with even smaller components.

Moreover, the evaluation of the molecular weight by SDS polyacrylamide gel technique, requires some caution when information about the amount of SDS linked to the protein is lacking (Weber and Osborn, 1975). In this communication we report data of a different kind, which tend to demonstrate that arthropod hemocyanin is built up of subunits of about 50,000 daltons. The experiments were carried out mainly with *Limulus polyphemus* hemocyanin.

Methods

Materials. Hemocyanin was prepared by double $(NH_4)_2SO_4$ precipitation from 1:1 water diluted hemolymph collected from living animals. The procedure was previously described (Salvato and Tallandini, Chap. 28, this vol.).

All reagents were of analytical grade and were used without further purification. The magnesium salt of 1-anilino-8-naphthalenesulphonate (ANS) was kindly provided by Professor A. Azzi.

Protein and Copper Concentration. Protein concentration was determined spectrophotemetrically. For *L. polyphemus* Hcy the values of $E_{278}^{1\%} = 1.07$ in Tris-HCl μ 0.1, pH 8.0 buffer and $E_{288}^{1\%} = 1.24$ in KOH, 0.3 M were used. Copper was determined with the Perkin-Elmer 300 Atomic absorption spectrophotometer equipped with the multiple lamp.

Equilibrium of Hcy with SDS. One ml of protein solution at 10 mg/ml in 8 M urea, 1% mercaptoethanol, Tris-HCl 0.01 M, pH 7.2 buffer was exposed at 100°C per 3-5 min; then 4 ml of 0.75% SDS in Tris-acetate 0.01 M, pH 7.2 was added, and the mixture was incubated for 12 h at 37°C. About 1 ml of this solution was chromatographed on a column (0.8 × 10 cm) of Sephadex G 25 Superfine equilibrated with Tris-acetate 0.01 M, pH 7.2, SDS 0.1% buffer. The protein and SDS concentrations were determined in the recovered solution. The amount of linked SDS was calculated from the total SDS concentration by subtracting the concentration of SDS in the equilibration buffer alone. Measures were performed on *Carcinus maenas* and *L. polyphemus* hemocyanin, lysozyme, chytrotrypsinogen A, ovoalbumin and bovine serum albumin.

SDS Determination. SDS was determined colorimetrically. To 3 ml of 0.02% solution of basic fucsine in Tris-acetate 0.1 M, pH 7.0 buffer, 10-50 µl of sample (containing 5-50 µg of SDS) is added; the solution is then extracted with 5 ml of 10% v/v octanol in carbon tetrachloride and the absorbance of organic phase is read at 545 nm against the blank. The absorbance of the blank is about 0.030-0.040 OD. The amount of extracted dye is proportional to the added SDS and no interference of the protein was detected. The calibration curve is reported in Figure 1.

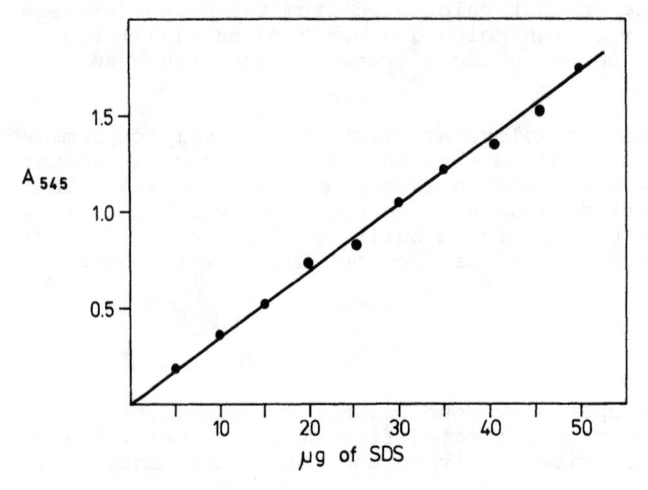

Fig. 1. Calibration curve of the method for SDS determination

Polyacrylamide Gel Electrophoresis. This was essentially performed as described by Weber and Osborn (1975). Gel concentration of 5% and 7.5% were used. Electrophoresis buffer was 0.025 M $NaH_2PO_4 \cdot H_2O$ - 0.07 M $Na_2HPO_4 \cdot 2H_2O$, pH 7.2 containing 0.1% SDS. Gels were stained in a solution of 0.25% Comassie Brilliant Blue R250 in 22.7:4.6:50 v/v methanol-glacial acetic acid-water. Samples were prepared as described for the SDS-binding experiments. Lysozyme, chymotrypsinogen A, ovoalbumin, bovine serum albumin were used as calibration standard. The data are presented as mobility of protein bands versus the log of their molecular weight.

Sedimentation Velocity. Sedimentation coefficients of *L. polyphemus* Hcy in Tris-acetate 0.1 M, methyl-ammonium-HCl 0.1 M,urea 2.5 M, pH 7.5 buffer were determined in a Beckman model E analytical ultracentrifuge using schlieren and absorption obtics at 52000 r.p.m. and at 20°C.

Gel Filtration. Gel filtration was performed on a column (1.8 × 110 cm) of Bio-gel A 1.5 m (200-400 mesh) equilibrated with Tris-HCl 0.1 M, $CH_3NH_2 \cdot HCl$ 0.1 M, urea 3 M, pH 7.5 buffer, and on a column of the same dimensions of Sephadex G 150 Superfine equilibrated with Tris-HCl 0.05 M EDTA 0.01 M pH 8.9 buffer.

As standards blue-dextran 2000, horse-γ-globulin, dansylated-BSA, ovoalbumin, chymotrypsinogen A and cytochromec were used. Sample and standards were put in the column together, dissolved in 1 ml of elution buffer and at a concentration of about 6-7 mg/ml. Fractions of 50 drops were collected and read at 280 and 350 nm. Dansylated-bovine serum albumin was read fluorimetrically (λex = 340 nm, λem = 500 nm).

Elution data were reported as V_e/V_o versus log of MW. The elution data of the Bio-gel column was elaborated according to Grover and Kapoor (1973).

Reconstitution with $K_3Cu(CN)_4$. Reconstitution experiments were effected on both native and apo-Hcy in the presence of different concentrations of urea. To 2 ml of 5 mg/ml Hcy solution in Tris-HCl 0.1 M, $CH_3NH_2 \cdot$ HCl 0.1 M, pH 7.5 buffer containing 1-5 molar of urea, 10-50 µl of a solution of $K_3Cu(CN)_4$ (8 µg/ml of Cu^+) was added. The mixture was dialyzed against the same buffer containing 20-50 µg of Cu^+ for 24 h at room temperature, and then against the buffer containing EDTA 0.01 M for 48 h. The spectrum of the recovered solution between 240 and 390 nm and its copper content was determined.

1-8 ANS Binding. The binding of 1-8 ANS to Hcy was measured with the MPF-4 Perkin-Elmer spectrofluorimeter equipped with the thermostated cell assembly. The measurements were effected at 17 ± 0.1°C, with the excitation wavelength of 390 nm (excitation slit 7 nm, emission slit 8 nm). The spectra were recorded in the ratio mode for minimizing any errors arising from fluctuations of the light source intensity. Experiments were carried out on native and apo-Hcy associated (80-90% as 16S component) in Tris-HCl µ 0.1, pH 7.5 buffer; and dissociated 95% 4-5S component in 0.1 M tetraborate, 0.01 M EDTA, pH 8.9 buffer (Sullivan et al., 1974) and in Tris-HCl 0.1 M, $CH_3NH_2 \cdot$HCl 0.1 M, urea 2.5 M, pH 7.5 buffer.

We used also Hcy reassociated (80-90% as 24 and 34S components) at pH 8.0. The protein was dialyzed against borate buffer, EDTA 0.01 M, pH 8.9 for 96 h in the cold; and then against borate 0.1 M, NaCl 1 M, pH 8.0 buffer. Prior to use the aggregation state of Hcy was controlled in the ultracentrifuge for all the experimental conditions.

The number of binding sites per mole of protein was determined as described by Azzi (1974).

Approximate values of binding constants were obtained with the Scatchard method (Scatchard, 1949). Quantum yields of protein bounds ANS were calculated from the equation

$$Q = \frac{0.4\ F_{max}}{F_{ethanol}}$$

were 0.4 is the fluorescence quantum yield of ANS in ethanol and $F_{ethanol}$ and F_{max} are the fluorescence emitted by ANS solutions at the same concentration in ethanol and in the presence of an excess of protein.

Results

SDS Binding and Polyacrylamide Gel Electrophoresis. The Hcy binds much less SDS than the standard proteins. In Table 1 the SDS bound per mg of protein is reported for both hemocyanin and standards. The values found for standards satisfactorily agree with that reported for reduced proteins (Pitt-Rivers and Ambesi-Impiombato, 1968). In Figure 2 the plot of mobilities of standard proteins versus the logarithm of their molecular weight is shown.

Table 1. SDS bound per mg protein for hemocyanin and standards

Protein	SDS:protein ratio mg/mg
Limulus polyphemus Hcy	0.70
Carcinus maenas Hcy	0.76
Lysozyme	1.37
Chymotrypsinogen A	1.39
Ovoalbumin	1.43
Bovine serum albumin	1.38

The Hcy samples showed the presence of one or two close protein bands with a MW of 70,000-77,000 daltons. This result is in good agreement with that reported elsewhere (Loehr and Mason, 1973; Waxman, 1975).

Sedimentation Velocity. The *L. polyphemus* Hcy in 2.5 M urea moves in the ultracentrifuge as a single peak which rapidly spreads. The slope of the plot of log of the radial distance versus time shows a monotonic drift. Indicative S values were obtained from the mean slope calculated with the least squares method. In Figure 3 the values of S are reported versus protein concentration. With increasing protein concentration the S values show an upward trend.

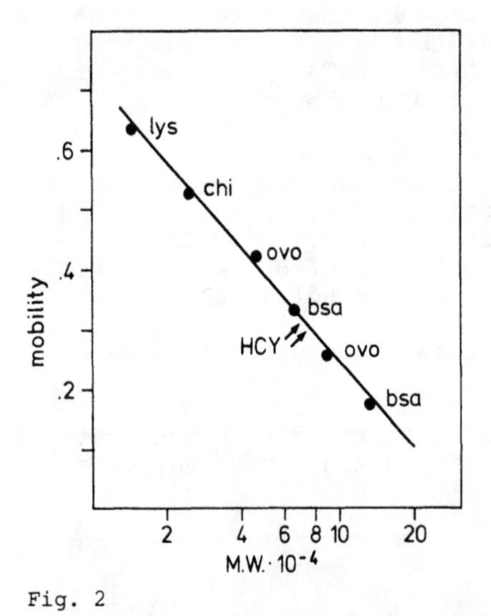

Fig. 2

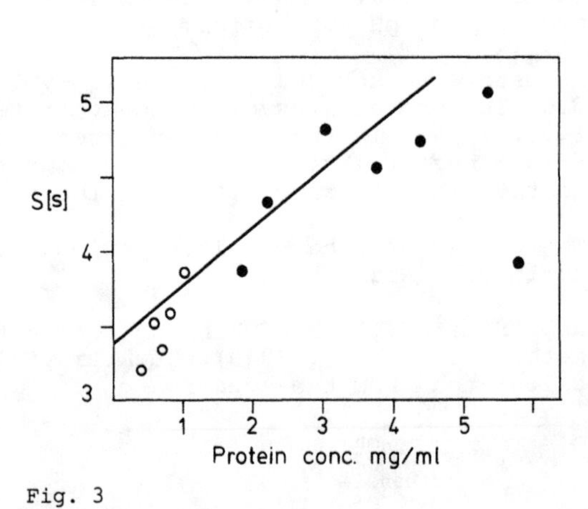

Fig. 3

Fig. 2. Polyacrylamide gel electrophoresis. Standard protein mobilities are reported versus their MW in a semilogarithmic plot. *Arrows*: Hcys

Fig. 3. Sedimentation velocity. The S values for Hcy in 2.5 M urea are reported vs. protein concentration. (●) Schlieren optics, (o) absorption optics

Gel Filtration. In Figures 4 and 5 the elution profile of Hcy and some standards for the Bio-gel and Sephadex column, respectively, are reported. In both cases hemocyanin elutes as a single peak between the dansylated-bovine serum albumin and ovoalbumin. In Figure 6 the plot of V_e/V_o versus the logarithm of MW is shown. In both cases the MW of Hcy is 56,000 daltons.

Reconstitution with $K_3Cu(CN)_4$. In Figure 7 the Cu/P ratio of reconstituted Hcy are reported against the urea concentration. Up to 1.5 M urea the

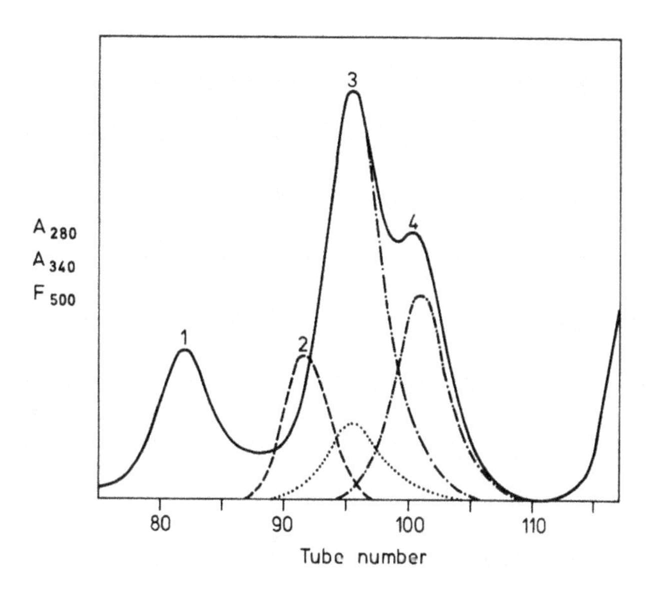

Fig. 4. Elution profile from Bio-gel A 1.5 M column. In the ordinate scale absorption at 280 nm ——— ; 340 nm ....; fluorescence emission intensity at 500 nm (λ_{ex} 340) --- in arbitrary units are reported. *Numbers* correspond to horse γ-globulin (*1*); BSA (*2*); Hcy (*3*); ovoalbumin (*4*): Resolution between the curves 3 and 4 was obtained by calculating absorption at 280 nm of Hcy from the absorption at 340 nm using the ratio $\frac{A_{340}}{A_{280}} = 0.20$

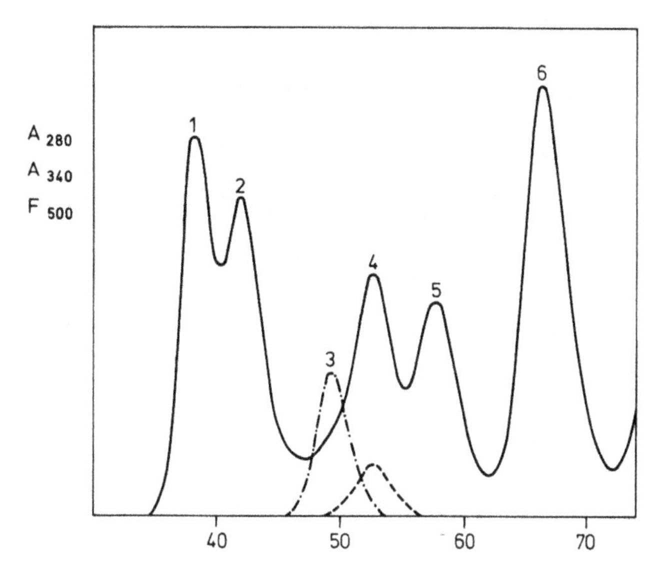

Fig. 5. Elution profile from Sephadex G 150 column. In the ordinate scale absorption at 280 nm ———; 350 nm ---; fluorescence emission intensity at 500 nm (λ_{ex} 340) -•-•-• in arbitrary units are reported. The numbers correspond to Blue Dextran 2000 (*1*); horse γ-globulin (*2*); BSA (*3*); Hcy (*4*); ovoalbumin (*5*); chymotrypsinogen A (*6*)

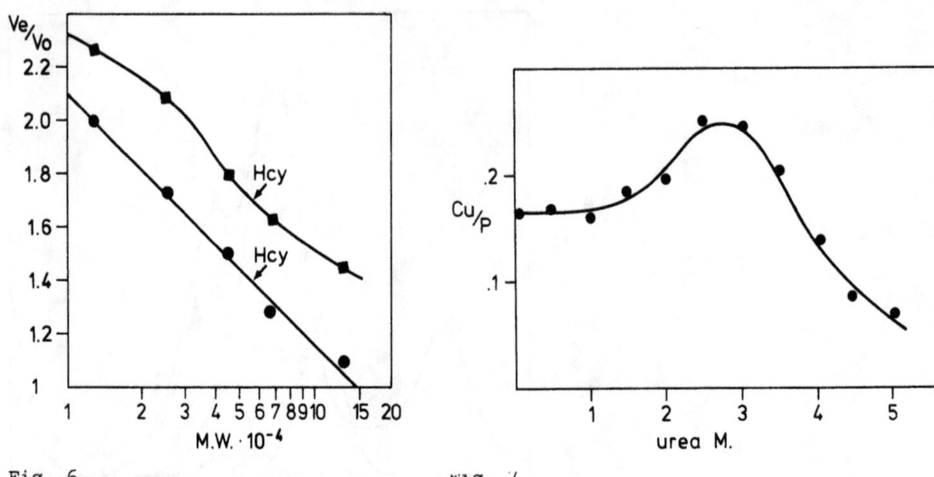

Fig. 6

Fig. 7

Fig. 6. Semilogarithmic plot of V_e/V_o vs. the MW • Sephadex G 150 column, ■ Bio-gel A 1.5 M column

Fig. 7. Reconstitution of Hcy with $K_3Cu(CN)_4$ and in the presence of various urea concentrations. The copper band per 100 mg of protein versus urea concentration is shown

copper/protein ratio remains equal to that of the native proteins. When the urea concentration increases up to 3 M, the copper/protein ratio reaches the value of 0.25%. For higher urea concentration a gradual decrease of copper/protein ratio occurs. No substantial differences in the results were evident using the native or apo-hemocyanin.

ANS Binding. Native associated Hcy does not bind the ANS. The fluorescence emission spectrum shows a maximum at 520 corresponding to that of ANS in buffer alone. The quantum yield is also very low (~0.003). We obtained the same result with *Carcinus maenas* and *Octopus vulgaris* Hcy. In all other conditions a strong increase of ANS fluorescence occurs together with an auxochromic shift of the emission maximum toward 470-460 nm.

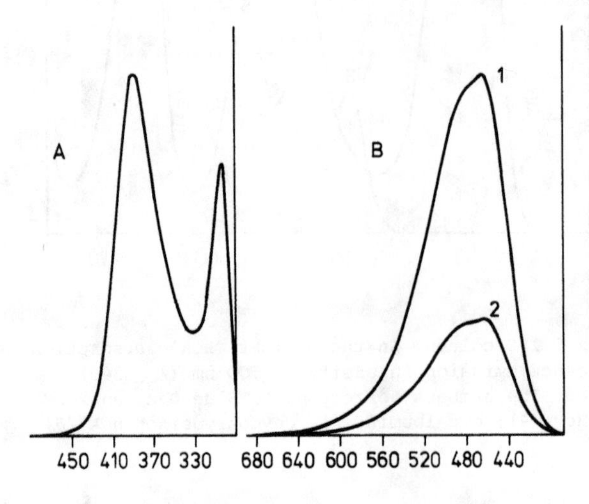

Fig. 8. (A) Excitation spectrum of ANS in presence of native reassociated Hcy in NaCl 1 M pH 8.0 (λ_{em} = 460 nm). Hcy = 6.4 mg/ml and ANS = 20 μM. (B) Emission spectra *1*: λ_{ex} = 390 nm; *2*: λ_{ex} = 340 nm

In Figures 8 and 9 excitation and emission spectra of ANS in presence of dissociated native and apo-Hcy are shown. The excitation spectrum of native-Hcy clearly shows the inner-filter effect of the 340 nm copper-oxygen band.

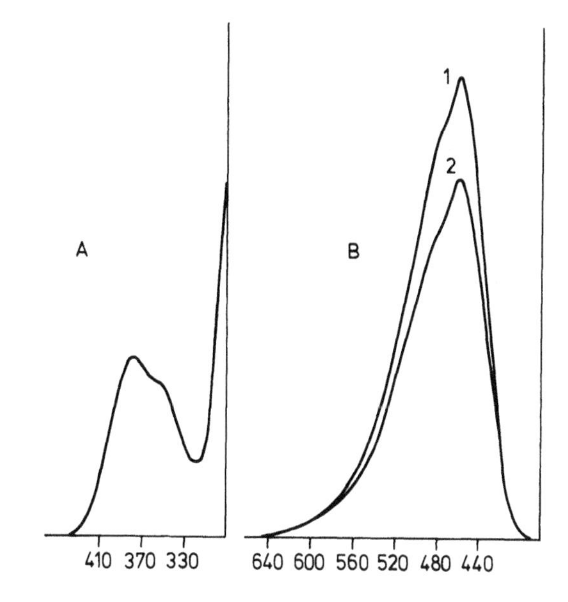

Fig. 9. (A) Excitation spectrum of dissociated apo-Hcy in the presence of EDTA 0.01 M pH 8.9. Hcy = 0.98 mg/ml, ANS = 20 µM. λ_{em} = 460 nm. (B) Emission spectra *1*: λ_{ex} = 390 nm; *2*: λ_{ex} = 340 nm

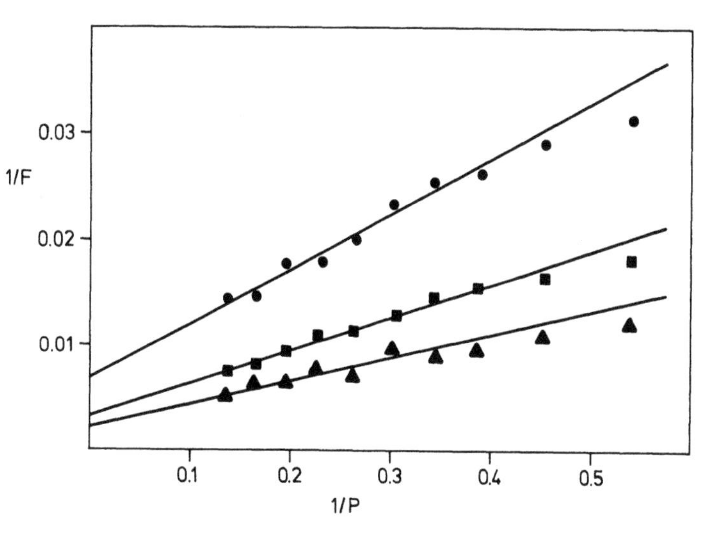

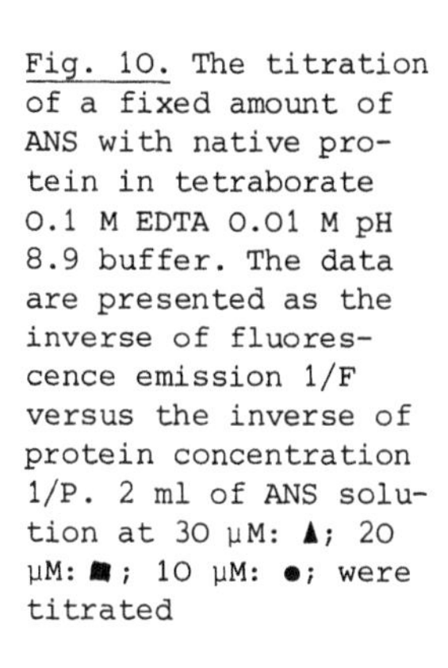

Fig. 10. The titration of a fixed amount of ANS with native protein in tetraborate 0.1 M EDTA 0.01 M pH 8.9 buffer. The data are presented as the inverse of fluorescence emission 1/F versus the inverse of protein concentration 1/P. 2 ml of ANS solution at 30 µM: ▲; 20 µM: ■; 10 µM: ●; were titrated

In Figures 10 and 11 typical ANS-protein titrations are reported. In Table 2, the approximate ANS binding constants values, the number of binding sites per 450,000 daltons the MW of 16S component and the quantum yields of ANS emission in the conditions we used are summarized. The native dissociated and reassociated hemocyanin bind about 3 ANS molecules per 450,000 daltons, the associated apo-Hcy binds about 6 and the dissociated apo-hemocyanin about 9 ANS molecules.

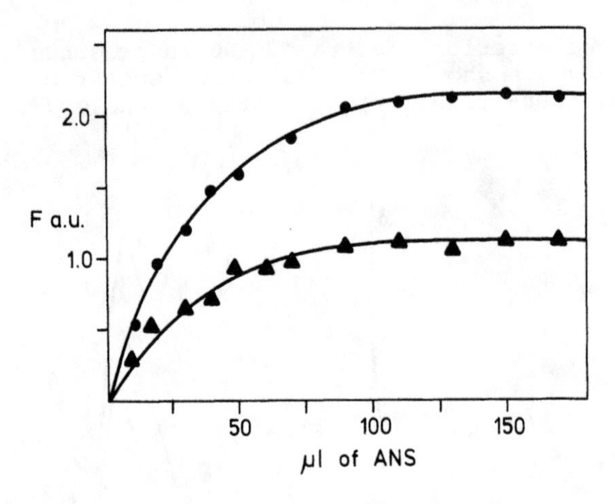

Fig. 11. The titration of 2 ml of native protein in 0.01 M tetraborate pH 8.9 buffer containing 0.1 M EDTA at 0.98 mg/ml: ▲, and 2.01 mg/ml: ● with a solution of ANS 5×10^{-3} M. The fluorescence (in arbitrary units) are reported versus the µl of added ANS solution

Table 2. Sedimentation coefficients (S), ANS binding constants (K), number of ANS binding sites per 450,000 daltons (N), quantum yields (Q), and wavelengths of maximum emission

	S	K	N	Q	λ_{em}
Hcy native in urea 2.5 M	4.5	$\sim 5.4 \times 10^{-5}$ [a]	3.2	0.052	460
Hcy native in EDTA 0.01 M	4.3	$\sim 4.7 \times 10^{-5}$ [a]	3.3	0.062	460
Hcy reassociated in 1 M NaCl	24 and 34	$\sim 5.0 \times 10^{-5}$ [a]	2.16	0.048	460
Hcy-apo in Tris-HCl buffer pH 7.5	16	$\sim 5.25 \times 10^{-5}$ [b]	6.25	0.103	460
Hcy-apo in urea 2.5 M	4.8	$\sim 5.5 \times 10^{-5}$	8.63	0.14	460
Hcy-apo in EDTA 0.01 M	4.6	5.7×10^{-5}	8.75	0.2	460

[a] Calculated using 1/3 of protein concentration.
[b] Calculated using 2/3 of protein concentration.

Discussion

The reported results clearly require the revision of some fundamental concepts on arthropod hemocyanin. The MW of about 75,000 daltons for the minimal arthropod subunit is thus inconsistent with all presented data: the MW of apo-hemocyanin evaluated by SDS gel electrophoresis in not reliable since the amount of SDS bound per mg of hemocyanin is about one half of that bound per mg of standard proteins. The ultracentrifugation data (van Holde and van Bruggen, 1971) indicate that the 5S component is actually heterogeneous, probably as a consequence of concentration-dependent association-dissociation phenomena (Teller, 1973).

The molecular weight arising from gel filtration experiments is lower than that of bovine serum albumin and close to 50,000 daltons. The maximum amount of reconstitution and of ANS binding is in agreement with the latter MW. We propose that the 16S arthropod subunit is actually built-up by nine subunits of about 50,000 daltons, three of these are copper-free and are arranged into the core of the aggregate, while the six external subunits contain the copper and can bind oxygen.

On this basis our reconstitution and ANS binding results can be readily interpreted. When hemocyanin is in the associated state, the three internal copper-free subunits are not accessible to the copper or to the ANS. Only the six external subunits of the apo-protein can bind it. After dissociation the interaction occurs with all nine subunits in apo-protein and only with the copper-free subunits in native protein. When the native proteins are dissociated and reassociated, the three copper-free subunits are distributed between the three internal and the six external positions at least in the ratio 1:2 and then accessible sites are available although in a minor number with respect to the dissociated protein.

A relevant biological implication of this model is that the copper is bound to the Hcy after the 16S component is built-up.

References

Azzi, A.: The use of fluorescent probes for the study of membranes. In: Methods in Enzymology. Fleischer, S., Packer, L. (eds.). London-New York: Academic Press, 1974, Vol. XXXII, Part B

Ellerton, H.D., Carpenter, D., van Holde, K.E.: Physical studies of hemocyanins. V. Characterization and subunit structure of the hemocyanin of *Cancer magister*. Biochemistry 9, 2225-2232 (1970)

Grover, A.K., Kapoor, M.: An improved treatment of data for estimation of molecular weights by Sephadex gel filtration. Anal. Biochem. 51, 163-172 (1973)

Holde, K.E. van, Bruggen, E.F.J. van: The hemocyanins. In: Subunits in Biological Systems. Timasheff, S.N., Fasman, G.D. (eds.). New York: M. Dekker, 1971, pp. 1-53

Loehr, J.S., Mason, H.: Dimorphism of *Cancer magister* hemocyanin subunits. Biochem. Biophys. Res. Commun. 51, 741-745 (1973)

Pitt-Rivers, R., Ambesi-Impiombato, F.S.: The binding of sodium dodecyl sulphate to various proteins. Biochem. J. 109, 825-830 (1968)

Scatchard, G.: The attraction of proteins for small molecules and ions. Ann. N.Y. Acad. Sci. 51, 660-672 (1949)

Sullivan, B., Bonaventura, J., Bonaventura, C.: Functional differences in the multiple hemocyanins of the horse shoe crab *Limulus polyphemus*. Proc. Nat. Acad. Sci. USA 71, 2558-2561 (1974)

Teller, D.C.: Characterization of proteins by sedimentation equilibrium in the analytical ultracentrifuge. In: Methods in Enzymology. Hirs, C.H.W., Timasheff, S.N. (eds.). London-New York: Academic Press, 1973, Vol. XXVII, Part D

Waxman, L.: The structure of arthropod and mollusc hemocyanins. J. Biol. Chem. 250, 3796-3806 (1975)

Weber, K., Osborn, U.: Proteins and sodium dodecyl sulphate: Molecular weight determination on polyacrylamide gels and related procedures. In: The Proteins. Neurath, H., Hill, R.L. (eds.). London-New York: Academic Press, 1975, Vol. I

Morphology of *Helix pomatia* Hemocyanin and Its Subunits

J. F. L. VAN BREEMEN, G. J. SCHUURHUIS AND E. F. J. VAN BRUGGEN

Introduction

At the 3rd Hemocyanin Meeting in Louvain, Mellema showed a model for a gastropod hemocyanin based on three-dimensional reconstruction from electron micrographs (Mellema and Klug, 1972). The cylindrical molecule has 52 symmetry and consists of a wall, two collars and two caps. The wall contains 60 morphological units with dyad symmetry bounded by two sets of helical grooves. The morphological units are of six crystallographic distinct types, but they are similar in size, shape and orientation. Each collar contains five blobs of material. The cap has no substructure.

At the 4th Hemocyanin Meeting in Padua the model of Siezen was shown for the dissociation of such a cylindrical hemocyanin into 1/2-, 1/10, and 1/20-molecules (Siezen and van Bruggen, 1974). The dissociation of 1/2- into compact 1/10-molecules occurs along one set of the helical grooves resulting in a piece of the cylinder wall with one blob of the collar attached to it. This compact 1/10-molecule can change its conformation to a loose 1/10-molecule, an irregular cluster of more than ten globules. Finally, the 1/20-molecules are observed as clusters of seven or eight globules, sometimes arranged as a necklace.

Today, at the 5th Hemocyanin Meeting at Malta we want to propose a refinement of the Mellema and Klug model based on the first preliminary results of a new three dimensional reconstruction map and on more detailed information obtained about the structure and single 1/10- and 1/20-molecules and their intermediates.

Is it Possible to Visualize by Electron Microscopy Eventual Structural Changes During Oxygenation? Van Driel and van Bruggen (1974) showed the possibility of cross-linking the whole cylindrical molecule in either the high or the low oxygen affinity state. Electron microscopy of single cross-linked molecules indicated small structural changes during oxygenation, but it was not possible to obtain a statistically meaningful result. Attempts for alignment of the molecules were undertaken and this resulted in the formation of long tubular polymers of *Helix pomatia* β-hemocyanin after treatment with trypsin (van Breemen et al., 1975). Although trypsin removed the collar part of the molecule, the polymers were still fully active in co-operative oxygen binding. A surface lattice was drawn based on electron microscopy and optical diffraction. As in the Mellema and Klug model it consists of 10 (5.9-) helices showing an angle of 54° with the tube axis. From the intensity of the spots we concluded that one set of helices is more protruding than the other. This is also clearly demonstrated on Figure 1 by freeze-etching (Kwak, personal communication). Recently we obtained the first three-dimensional reconstruction map of part of a tubular polymer. Cross-sections perpendicular to the cylinder axis at z = 0, 27.5 and 55 Å are shown on Figure 2. We see that the penetration of the heavy metal stain into one set of the helical grooves fully divides the tube into five separate helices. The other set of helical grooves is only shallow with holes

Fig. 1. Freeze-etched tubular polymers of *H. pomatia* β-hemocyanin. Only one set of helical grooves (= the more protuding set) is observed

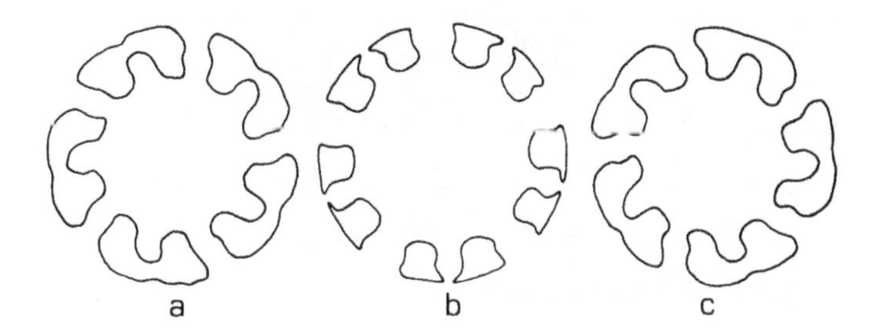

Fig. 2. Low-resolution results of the three-dimensional reconstruction of a *H. pomatia* β-hemocyanin tube. Shown are three cross-sections at intervals of 27.5 Å. Note the change in morphology and the rotation

fully penetrating the wall at 55 Å intervals. A more detailed analysis of this 3D-map is undertaken at this moment.

Structure of Compact 1/10-Molecule. Compact 1/10-molecules of *H. pomatia* α-hemocyanin are observed on electron micrographs as "boats" and as "parallelograms" (Siezen and van Bruggen, 1974). We continued this study and we analyzed a very large number of these structures. Figure 3 is a collection of different boat-, parallelogram- and intermediate structures. Our interpretation of the average boat-structure is given in Figure 5. The collar piece contains four domains of 55 Å diameter. It is located in between the top and the middle layer of morphological units of the 1/2-molecule where it forms a bridge, probably between two 1/20-molecules inside one 1/10-molecule. This collar piece seems rather flexible, as we concluded from the very diverse profiles of the parallelogram structures.

Arrangement of Two 1/20- Within One 1/10-Molecule. Compact 1/10-molecules were crosslinked with 0.1-0.4 mg/ml D.S.I. (= dimethylsuberimidate) at pH 8.4 and ionic strength 1.0. Then the molecules were dialyzed against a high ionic strength buffer of pH 10.5, a condition where normally only 1/20-molecules occur. EM-analysis of this solution showed all kinds of structures intermediate between compact 1/10- and completely separated 1/20-molecules. Some of these structures are collected in Figure 5. The interpretation is difficult, but it is clear to us that the dissociation of 1/10 into 1/20-molecules is a *length-wise* splitting (see also Fig. 6).

Folding of the Beaded Chain of the 1/20-Molecule. Figure 6 shows 1/20-molecules that were formed in different solutions. The three top rows show 1/20-

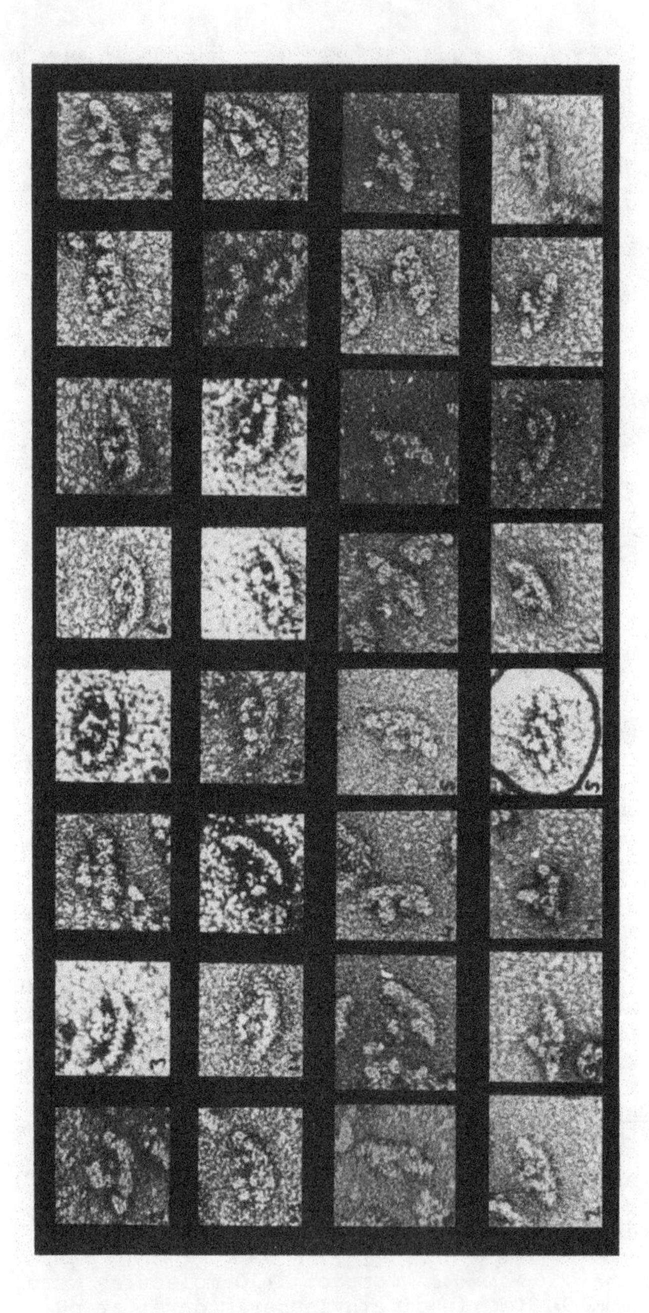

Fig. 3. Electron micrographs of *H. pomatia* α-hemocyanin compact-1/10-molecules: *First and second vertical row*: boat-structures, the collar piece often shows four domains. *Third vertical row*: parallelogram structures, the position of the collar piece varies. *Fourth vertical row*: intermediate structures, sometimes a "view in the boat"

molecules at pH 10.5 and ionic strength 0.1. The rows 4-6 were taken from electron micrographs of cross-linked compact 1/10-molecules after increasing pH. Rows 7 and 8 were collected from electron micrographs of cross-linked loose 1/10-molecules after increase of the pH. Gen-

Fig. 4. Schematic drawing of the boat-structure and its average dimensions; *p*: 142 Å; *q*: 177 Å

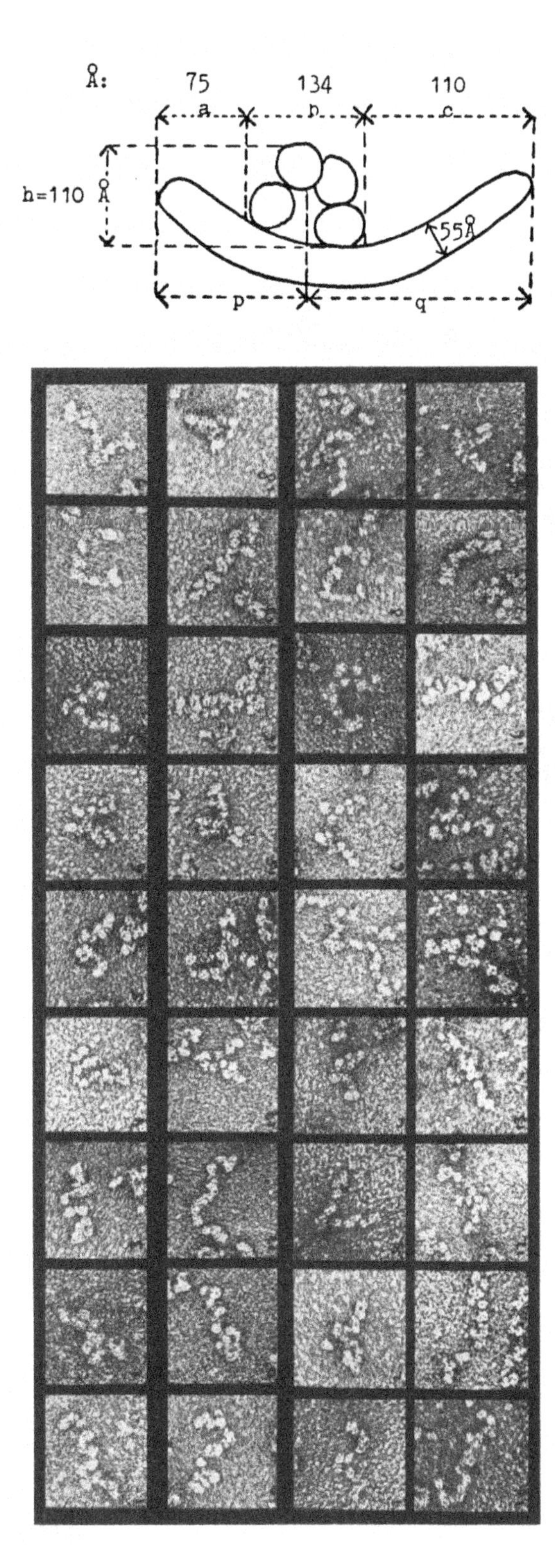

Fig. 5. Electron micrographs of *H. pomatia* α-hemocyanin structures intermediate between compact-1/10 and loose 1/10-molecules

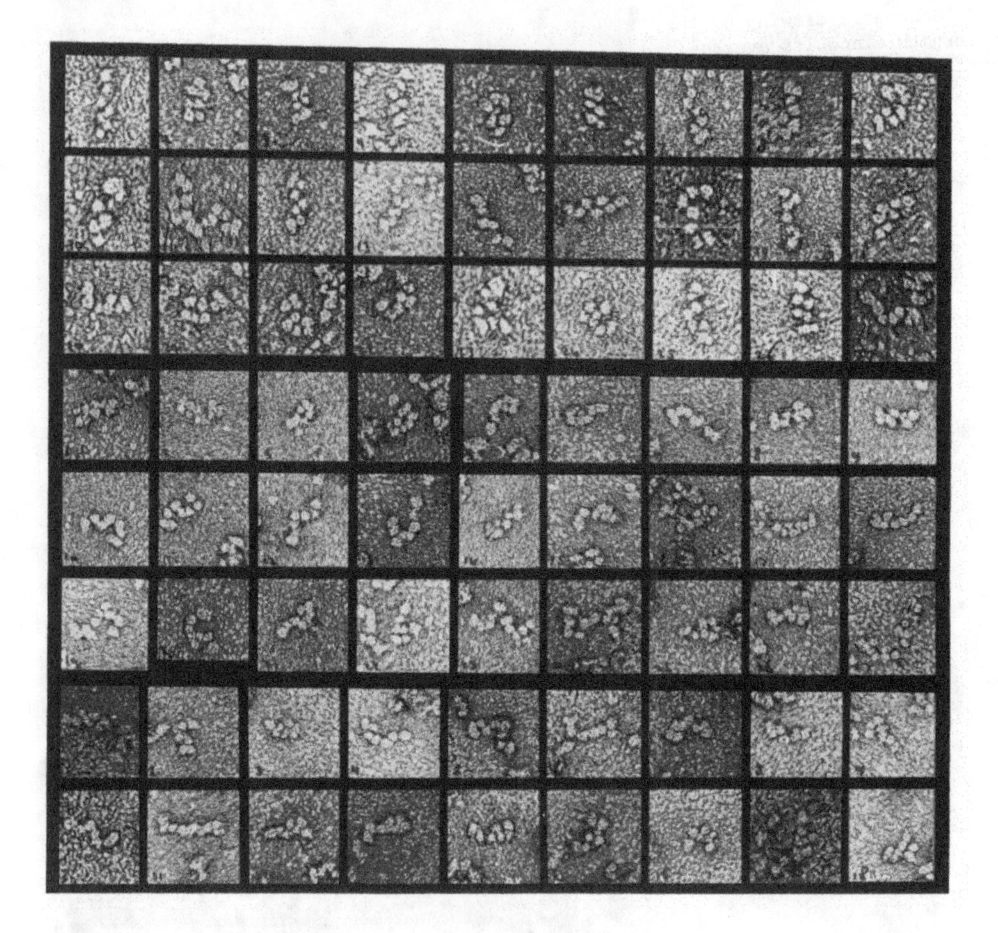

Fig. 6. Electron micrographs of *H. pomatia* α-hemocyanin 1/20-molecules

erally, we observe 1/20-molecules with seven or eight globules in equal amounts. Starting from compact 1/10-molecules also nine and ten globules are rarely seen. A definitive conclusion about the folding of the beaded chain is still impossible. However, analysis of all the observed structures in relation to all the theoretically possible structures makes the arrangement schematically shown on Figure 7 very probable.

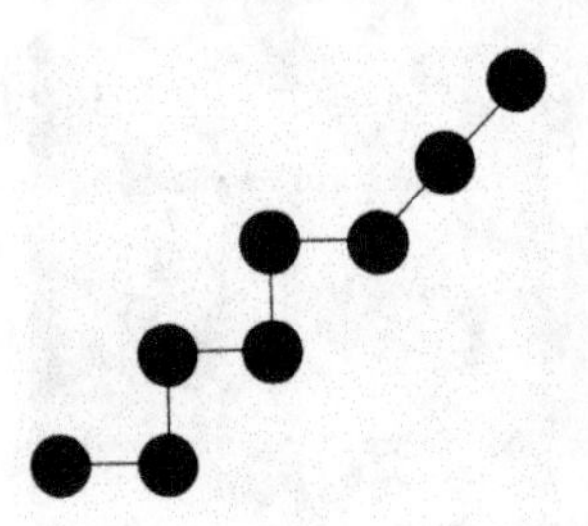

Fig. 7. Schematic drawing of domain arrangement within 1/20-molecules

Tentative Model for the Arrangement of Two 1/20-Molecules Within One Compact 1/10-Molecule. We assume for each 1/20-molecule a polypeptide chain of eight domains; six domains belong to the wall of the cylinder and two domains form part of the collar.

The arrangement of two 1/20-molecules within a compact 1/10-molecule is schematically drawn in Figure 8. The 12 wall domains correspond with six morphological units of the Mellema and Klug model while the four collar domains correspond with one blob of a collar.

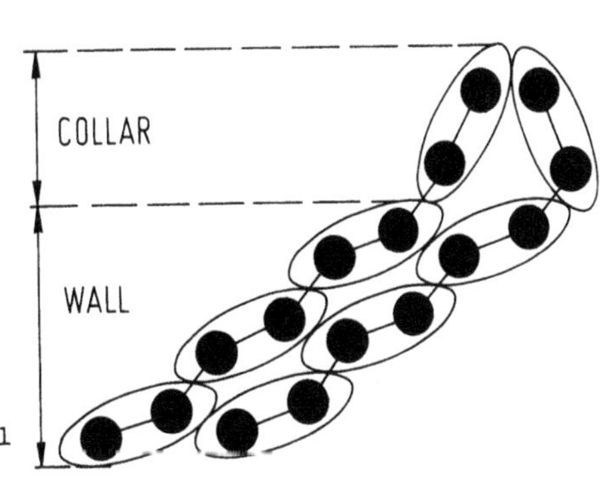

Fig. 8. Schematic drawing of arrangement of two 1/20-molecules within one compact-1/10-molecule. Two domains form one morphological unit of the Mellema and Klug model

This partly tentative model is in agreement with all our new data:

1. It explains the alternating depths of the helical grooves and the holes occurring in the most shallow groove.

2. It corresponds with the observation on the boats of a collar piece as a bridge of four globules.

3. It explains the lengthwise dissociation of compact 1/10-molecules into 1/20-molecules.

4. It is in correspondence with our best knowledge about the way of folding of the beaded chain.

Acknowledgments. This work was partly supported by the Netherlands Foundation for Chemical Research (S.O.N.) with financial aid from the Netherlands Organization for the Advancement of Pure Research (Z.W.O.).

We would like to thank Klaas Gilissen for printing and mounting the photographs and Ans van Rijsbergen for typing the manuscript.

References

Breemen, J.F.L. van, Wichertjes, T., Muller, M.F.J., Driel, R. van, Bruggen, E.F.J. van: Tubular polymers derived from *Helix pomatia* β-hemocyanin. Europ. J. Biochem. 60, 129-135 (1975)

Driel, R. van, Bruggen, E.F.J. van: Functional properties of chemically modified hemocyanin. Fixation of hemocyanin in the low and the high oxygen affinity state by reaction with a bifunctional imido ester. Biochemistry 14, 730-735 (1975)

Mellema, J.E., Klug, A.: Quaternary structure of gastropod hemocyanin. Nature (London) 239, 146-150 (1972)

Siezen, R.J., Bruggen, E.F.J. van: Electron microscopy of dissociation products of *Helix pomatia* α-hemocyanin: Quaternary structure. J. Mol. Biol. 90, 77-89 (1974)

Physical Properties of Hemocyanin

The Investigation of the Proton Magnetic Resonance Spectra of Some Copper Proteins

A.E.G. CASS, H.A.O. HILL, AND B. E. SMITH

Introduction

Of all the physical methods currently applied to the study of the structure and function of proteins in solution only nuclear magnetic resonance spectroscopy has the power to probe each and every part of the protein. Normally this results in a surfeit of information, but recent technical advances (Campbell et al., 1973; Campbell and Dobson, 1975) have made it possible to exploit the inherent potential of the method. With proteins of moderate weight, $\leq$ 50,000 daltons, sensitivity is no longer a problem, adequate signal-to-noise being efficiently achieved with samples having a concentration $\geq 5 \times 10^{-4}$M. Sufficiently resolved spectra can now be obtained by the employment of techniques, based on some discriminatory principle. Assignment may also be aided by such techniques but it is still laboriously achieved in most instances. For the full exploitation of the method it is desirable to assign individual resonances in the spectra to specific amino acid residues in the protein. This is currently the major problem. Although no general method has yet emerged such detailed assignments can be made in a number of instances. In this paper we describe the problems and the promise of the method by its application to azurins, plastocyanins and superoxide dismutase.

Results and Discussion

The ^{1}H NMR spectrum (Hill et al., 1976) of azurin from *P. aeruginosa* is shown in Figure 1. The resonances appearing at high field are those of methyl groups which are so positioned in the protein as to lie over and above the rings of neighbouring aromatic residues. At lower field are the resonances of relatively unperturbed methyl, methylene and methine protons. It is common preactice to utilize ^2H_2O, rather than ^1H_2O as solvent which leads to the replacement of the exchangeable protons by deuterons, with a consequent simplification of the spectrum at low field. It is in this region of the spectrum that the protons of the aromatic residues occur.

The calculation of convolution difference spectra (Campbell et al., 1973), as illustrated in Figure 2, is a very useful method of enhancing the apparent spectral resolution which can be further improved by making measurements at the highest possible temperature (Fig. 3). A more dramatic improvement results from making use of the different inherent relaxation times of different protons. By employing the pulse sequence, 90-τ-180-τ- collect, the resonances can (Hill et al., unpublished) be readily differentiated (Fig. 4). This not only improves the resolution but aids the 'type'-assignment of the spectra. The power of these methods can be appreciated more readily from the results of their application to larger proteins. In Figure 5, the conventional, convolution difference and spin echo spectra (Cass et al., unpublished) of bovine erythrocyte superoxide dismutase are compared. The spin-echo spectrum is wonderfully well-resolved and some assignments follow easily. The assignment to specific residues is trivial in this case

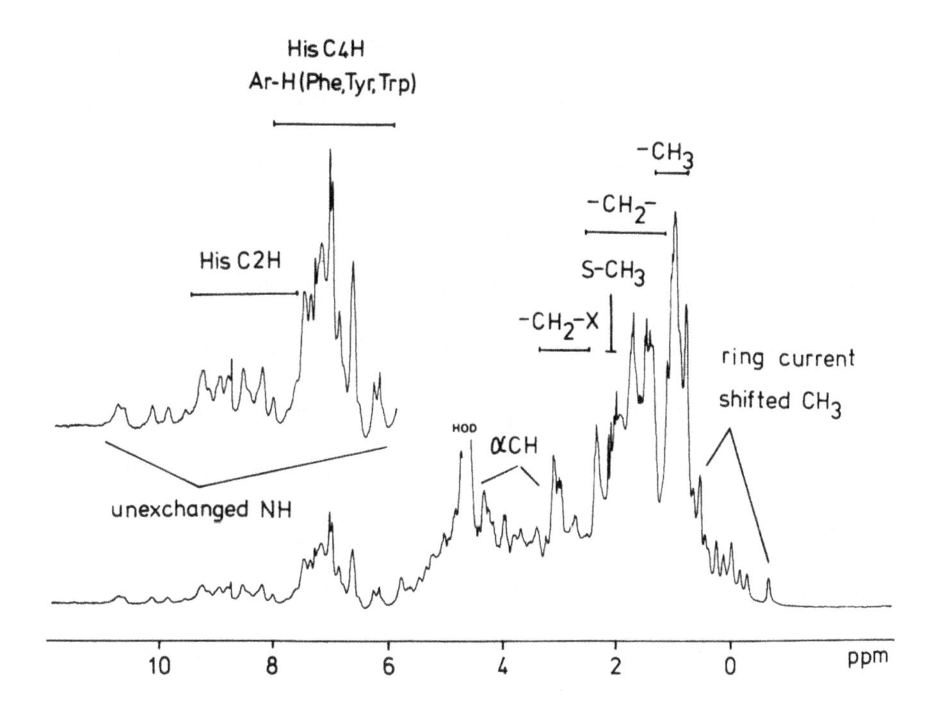

Fig. 1. The 270 MHz ^1H NMR spectrum of copper(I) azurin from *P. aeruginosa* at 45°C

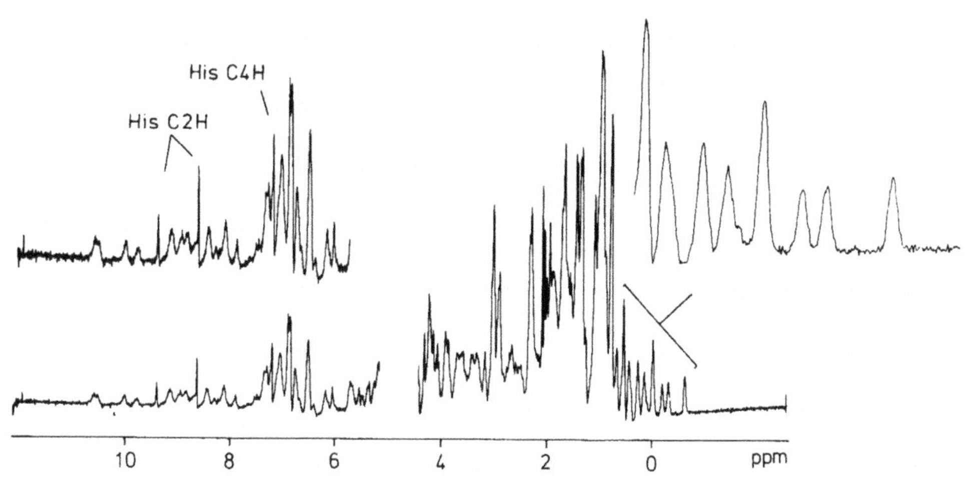

Fig. 2. The convolution difference spectrum of the copper(I) form of *P. aeruginosa* azurin at 25°C

since the protein contains only one tyrosine and one methionine. Such specific assignments can also be easily made in favourable cases. For example, the azurin from *Alcaligenes denitrificans* contains one more histidine than that from *P. aeruginosa*. This is readily apparent in a comparison of the difference spectra (Hill et al., unpublished) derived from measurements at two pH-values for each protein (Fig. 6).

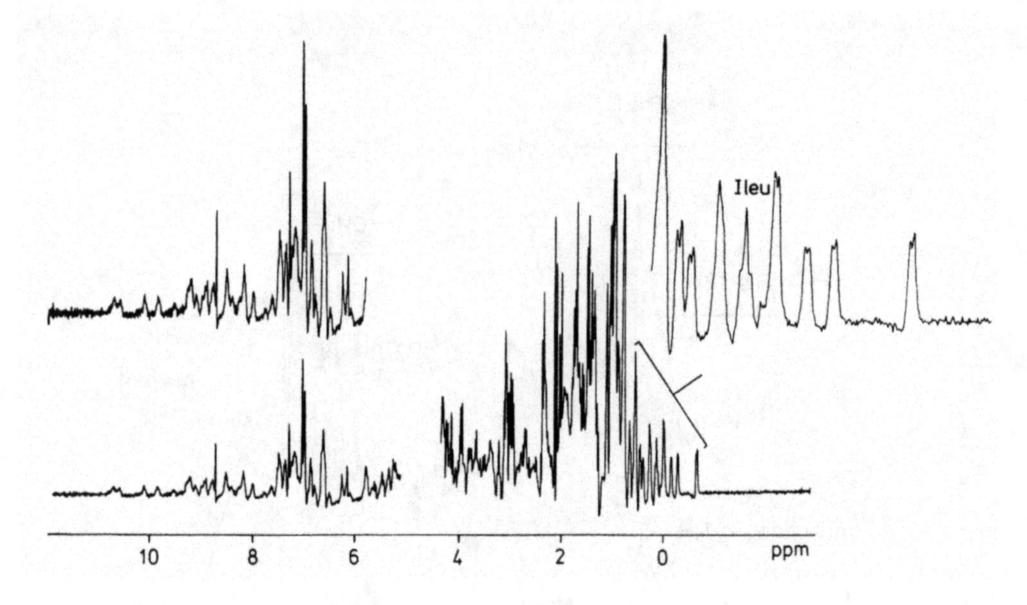

Fig. 3. The convolution difference spectrum of the copper(I) form of azurin from *P. aeruginosa* at 45°C

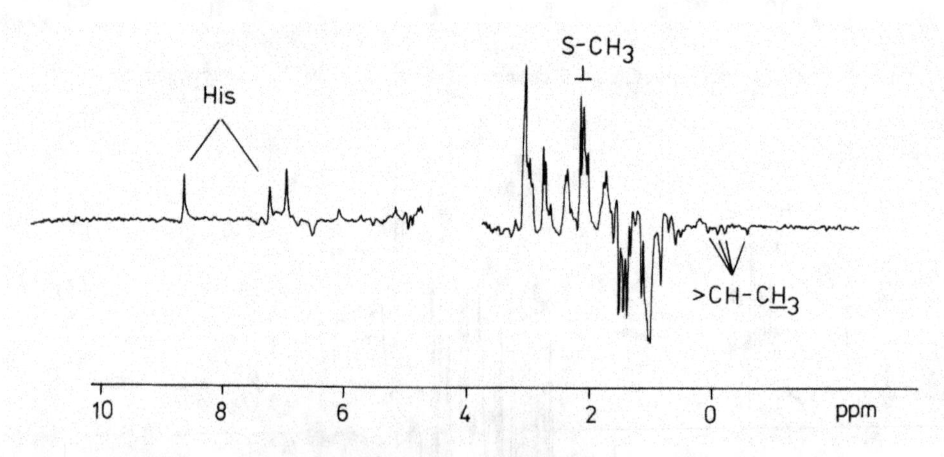

Fig. 4. The spin echo 270 MHz ^1H NMR spectrum of the copper(I) form of azurin from *P. aeruginosa* at 45°C, obtained using a pulse sequence 90-τ-180-τ collect where τ = 60 ms

The chemical shift of a given proton is sensitive not only to its immediate electronic environment but also to the influence of neighbouring groups. Thus, the ^1H NMR spectrum (Hill et al., unpublished) of a protein such as *P. aeruginosa* azurin is not that of a simple summation of the ^1H NMR of the component amino acids (Fig. 7). The resonances of those methyl groups situated near aromatic residues and shifted by the consequent ring currents provide a particularly sensitive indication of changes in tertiary structure. The spectra of the blue copper proteins as illustrated in Figure 8 by the plastocyanins from *Anabaena variabilis, Chlorella fusca,* spinach (Hill et al., unpublished) and french bean (Beattie et al., 1975) are relatively rich in such resonances. A comparison between the spectra of reduced copper (I) and oxidized

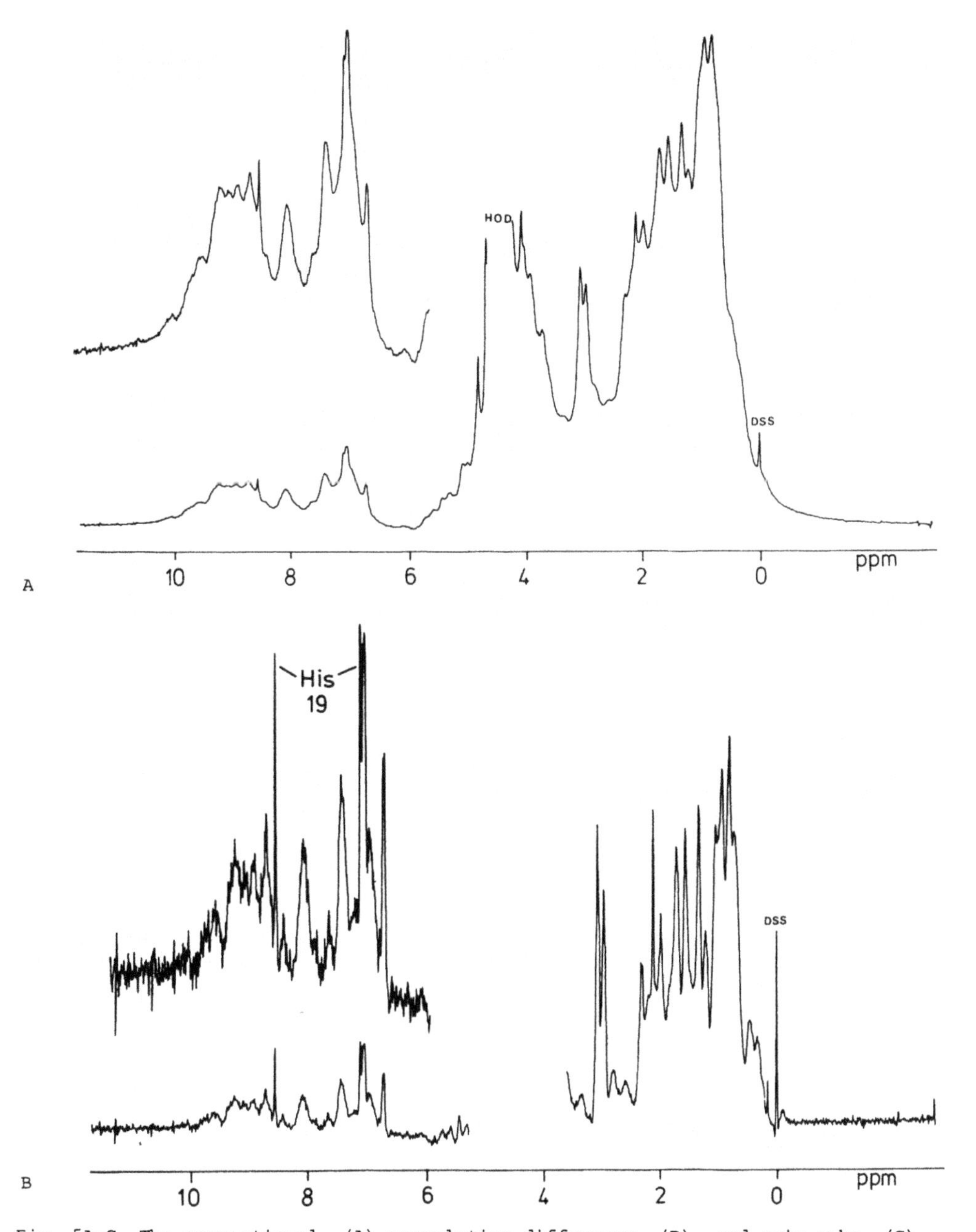

Fig. 5A-C. The conventional, (A), convolution difference, (B), and spin echo, (C), 270 MHz ^{1}H NMR spectra of the copper (II)-zinc (II) form of bovine erythrocyte superoxide dismutase

copper (II) forms of these proteins shows that there are no gross conformational changes as a consequence of the change in oxidation state. Those spectral changes which are observed arise (see below) from line-broadening caused by the paramagnetic copper (II) ion. Similarly, in *P. aeruginosa* azurin, there are no gross corformational changes

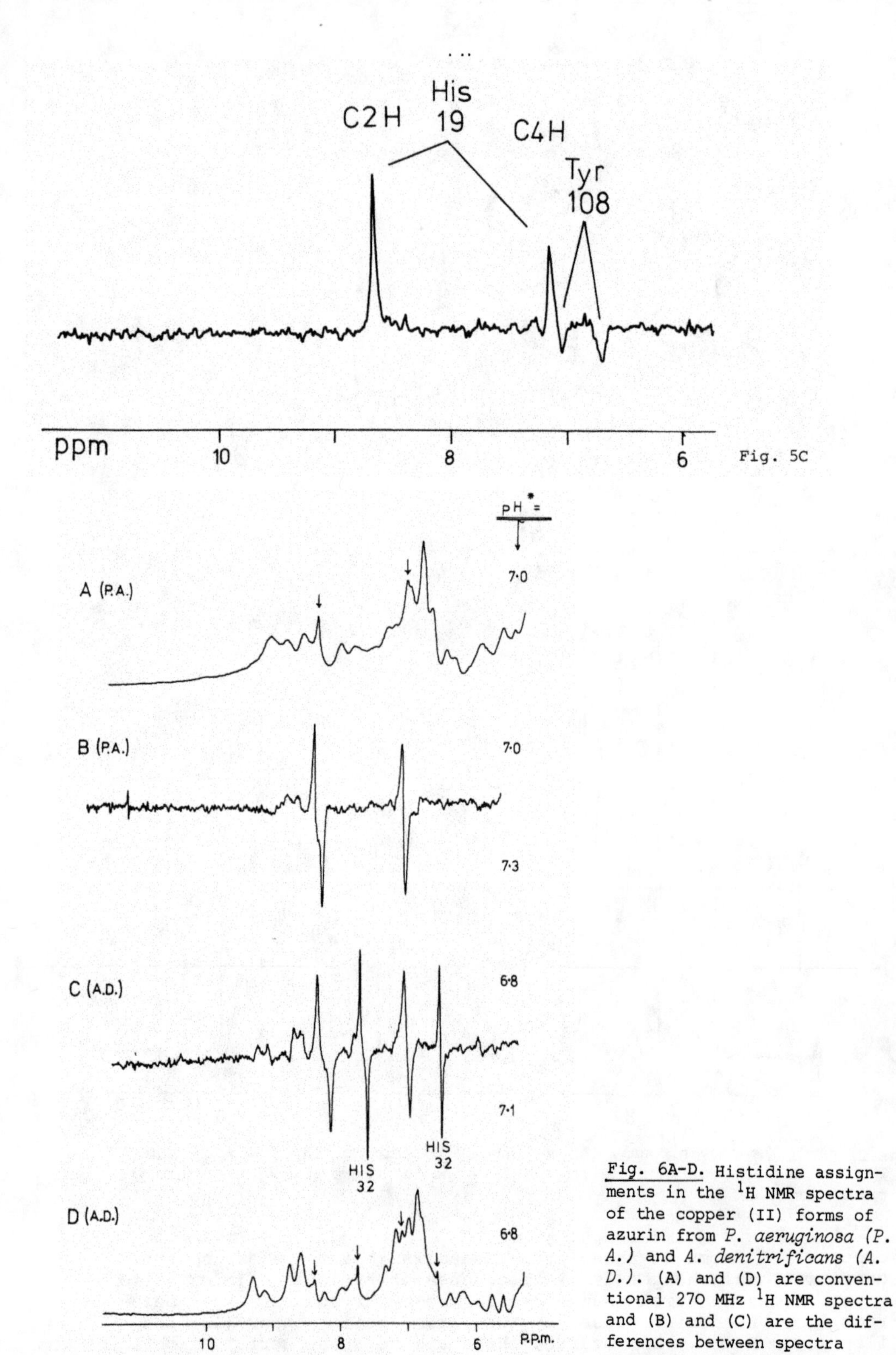

Fig. 5C

Fig. 6A-D. Histidine assignments in the ^{1}H NMR spectra of the copper (II) forms of azurin from *P. aeruginosa (P. A.)* and *A. denitrificans (A. D.)*. (A) and (D) are conventional 270 MHz ^{1}H NMR spectra and (B) and (C) are the differences between spectra recorded at different pH

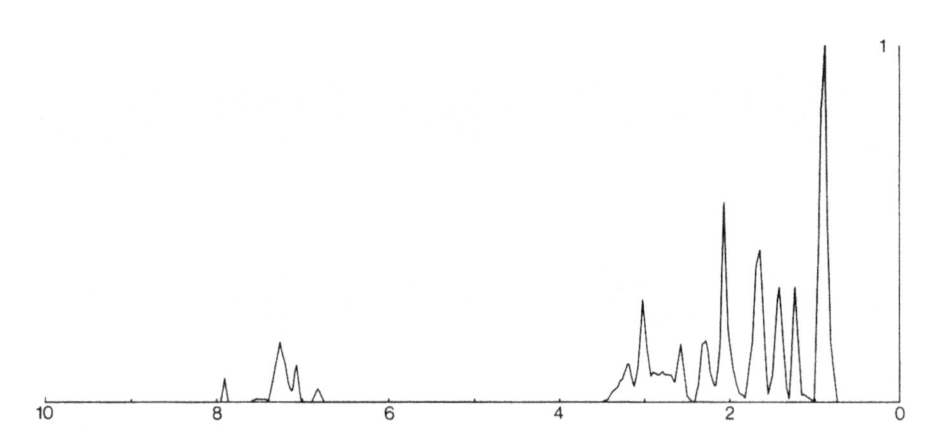

Fig. 7. The simulated ^{1}H NMR spectra of the random coil form azurin from *P. aeruginosa* using the method of McDonald and Phillips (1969)

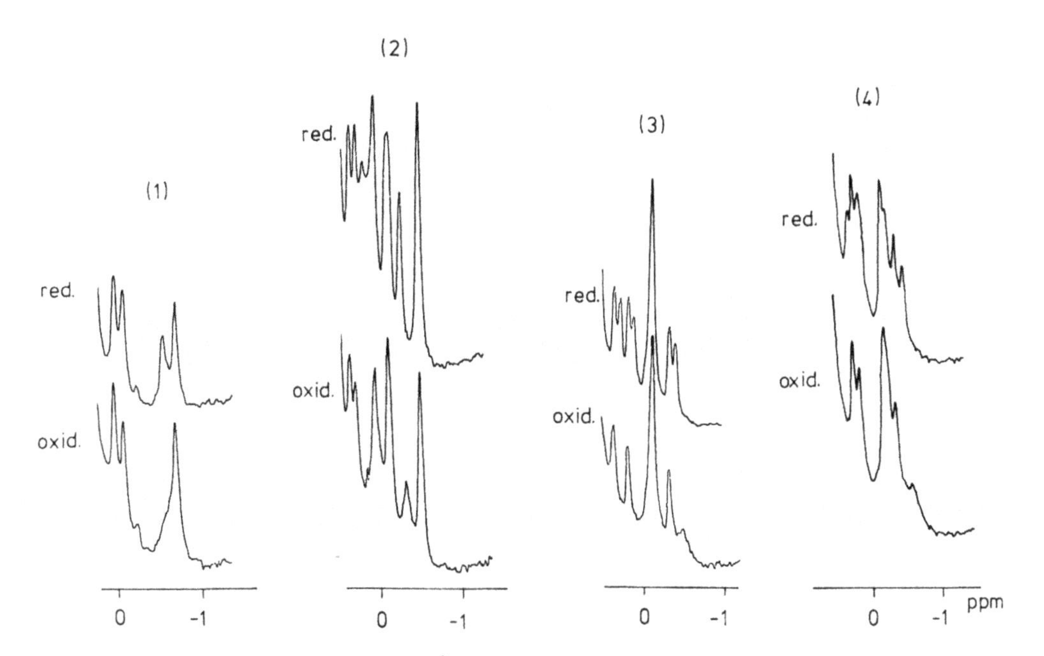

Fig. 8. This shows the high field ^{1}H NMR spectrum of plastocyanins from *1*: *Chlorella fusca*; *2*: *Anabaena variabilis*; *3*: spinach; *4*: french bean

as a result of change in oxidation state, replacement of copper (II) by mercury (II) (Hill et al., unpublished) or even complete removal of the metal ion. The evolved tertiary structure is such that the metal-binding site is preserved intact even in the absence of the metal ion. The spectra of many blue copper proteins indicate that there are a number of peptide NH resonances which are not readily exchanged (Fig. 1). It appears, therefore, that there is a region, or regions, of the protein which is inaccessible to the solvent and its attendant protons.

The investigation of the pH-dependence of the histidine C-2H and C-4H provides more quantitative information, in particular concerning the

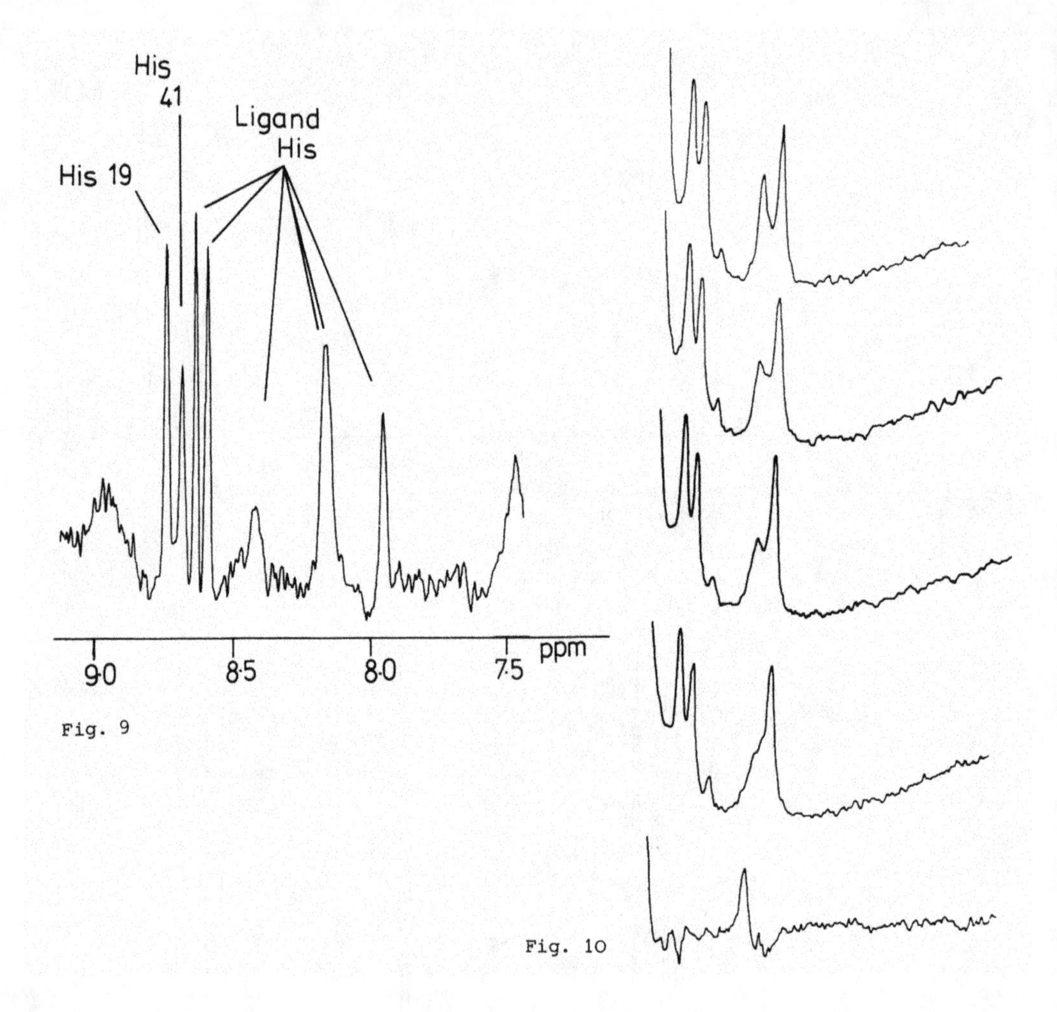

Fig. 9. The histidine C2H region of the 270 MHz convolution difference ^{1}H NMR spectrum of the apo form of superoxide dismutase from bovine erythrocytes

Fig. 10. The effect on the ^{1}H NMR spectrum of the ring-current shifted methyls of the copper (I) form of the plastocyanin from *Chlorella fusca* (*top*) of the addition of incremental amounts of ferricyanide. *Bottom:* difference between the spectra of the copper (I) and copper (II) forms

coordination sphere of the metal ion. For example, superoxide dismutase contains eight histidines per subunit and all of these are apparent in the histidine C-2H region of the spectrum (Cass et al., unpublished) of the apo-protein, (Fig. 9). The chemical shifts of seven of the eight resonances are markedly pH dependent in the pH^* region 6-8. In the copper (I) - zinc (II) protein only one of these seven resonances titrates normally, which is consistent with six histidines being involved in coordination to the two metal ions. With a knowledge of the crystal structure (Richardson et al., 1975) it is possible to assign histidine 19 and histidine 41.

The perturbations so far discussed, whether intrinsic or extrinsic, have been diamagnetic. The most dramatic perturbations occur when a paramagnetic centre is introduced and for copper proteins this is inherent in the copper (II) form. The large fluctuating magnetic fields generated by the unparied electron shortens the relaxation times of neighbouring protons and, except for ligand protons, this is essentially due to a dipolar interaction which has (Solomon, 1955) a dependence on r^{-6}, where r is the distance between the copper (II) and the proton in question. Obviously the effect is most marked on protons close to the copper (II), especially the ligand protons (whose relaxation times are shortened in addition by a scalar mechanism). The decreased relaxation times lead to increased line widths. The selectivity of the broadening is illustrated, in Figure 10, by the effect on the ring current shifted methyl region of the spectrum of *Chlorella* plastocyanin of oxidation of the copper (I) form under 'fast' exchange conditions. The distance within which resonances in the spectra of the copper (II) forms are broadened beyond detection varies from protein to protein but is generally within 10-15 Å. Obviously the ligand resonances are unobservable in the copper (II) forms, the effect extending in superoxide dismutase to the nearby histidines which coordinate to the zinc (II). In cases where the dipolar mechanism is dominant it will be possible to estimate quantitatively the distances between copper and residues corresponding to assigned resonances. It is possible that integration of the data derived from this approach with others from spectroscopic and theoretical methods will allow the proposal of three-dimensional structures for these copper proteins in solution. Such structures will lead to predictions of further assignments of the ^{1}H NMR spectra and hence an iterative approach to the determination of structures of proteins in solution will have been developed.

Acknowledgments. We thank the Science Research Council for a Fellowship to B.E.S. and a Studentship to A.E.G.C. and the Oxford Enzyme Group, of which H.A.O. Hill is a member, for support.

References

Beattie, J.K., Fenson, D.J., Freeman, H.C., Woodcock, E., Hill, K.A.O., Stokes, A.M.: An NMR investigation of electron transfer in the copper-protein, plastocyanin. Biochim. Biophys. Acta 405, 109-114 (1975)

Campbell, I.D., Dobson, C.M., Williams, R.J.P., Xavier, A.J.: Resolution enhancement of protein PMR spectra using the difference between a broadened and a normal spectrum. J. Magnet. Res. 11, 172-181 (1973)

Campbell, I.D., Dobson, C.M.: Spin echo double resonance: a novel method for detecting decoupling in Fourier transform nuclear magnetic resonance. Chem. Commun. 750-751 (1975)

Hill, H.A.O., Leer, J.C., Smith, B.E., Storm, C.B., Ambler, R.P.: A possible approach to the investigation of copper proteins: ^{1}H NMR spectra of azurins. Biochem. Biophys. Res. Commun. 10, 331-338 (1976)

McDonald, C.C., Phillips, W.D.: Proton magnetic resonance spectra of proteins in random-coil configurations. J. Am. Chem. 91, 1513-1521 (1969)

Richardson, J.S., Thomas, K.A., Rubin, B.H., Richardson, D.C.: Crystal structure of bovine Cu, Zn superoxide dismutase at 3 Å resolution: chain tracing and metal ligands Proc. Natl. Acad. Sci. 72, 1349-1353 (1975)

Solomon, I.: Relaxation processes in a system of two spins. Phys. Rev. 99, 559-565 (1955)

Flourescence and Absorption Studies of *Limulus* Hemocyanin and Its Components

J. K. H. MA, L. A. LUZZI, J. Y. C. MA, AND N. C. LI

Introduction

The absorption spectrum of a deoxygenated hemocyanin is in general a typical protein spectrum. Upon binding oxygen, hemocyanin exhibits a profound change in the absorption spectrum. Shaklai and Daniel (1970), in studying the fluorescence of *Levantina hierosolima*, reported that when the hemocyanin is deoxygenated, or when copper is removed, the fluorescence yield at pH 6.6 increases four fold in each case. The quenching on oxygen binding is traced to radiationless energy transfer from the tryptophanyl residues to the Cu...O groups. Bannister and co-workers (Bannister and Wood, 1971; Bannister et al., 1973) reported that for *Murex trunculus* the fluorescence is attributable to tryptophan, and the fluorescence is enhanced by 513 per cent when the apohemocyanin is prepared. Since there are significant differences between hemocyanins from molluscs and arthropods, including differences in molecular architecture as revealed by electron microscopy, circular dichroism, and subunit dissociation properties (Sullivan et al., 1974), we have carried out fluorescence and absorption studies using *Limulus*, an arthropod hemocyanin, and the results are presented in this paper. The choice of this hemocyanin takes on medical significance because *Limulus* lysate is now extensively investigated for rapid endotoxin assay (Sullivan and Watson, 1974).

Recently, Sullivan et al. (1974) have reported that *Limulus* hemocyanin prepared by extensive dialysis against 0.052 M Tris-glycine, 0.01 M EDTA at pH 8.9 (referred to as stripped hemocyanin), has molecular weight of only 66,000-70,000 as compared to native protein of molecular weight of 3.7 million. Furthermore, the stripped hemocyanin can be fractionated into at least five distinct subunits of different oxygen affinities and different responses to chloride ions (Bonaventura et al., 1974; Sullivan et al., 1974). In particular, the P_{50} and the state of aggregation of subunit II are changed substantially by the presence of 3 M NaCl, whereas the salt has relatively little effect on subunit V. We considered it worthwile to determine the fluorescence characteristics of the stripped hemocyanin and its subunits II and V. The results are given in this paper.

Experimental

Limulus hemocyanin, obtained from the Marine Biological Laboratory, Woods Hole, Mass., was purified in the following way, according to Ghiretti-Magaldi et al. (1966), Ke et al. (1973). The hemolymph was centrifuged in the cold using a Spinco Model L ultracentrifuge at 120,000 g, the pellet was dissolved with a small amount of water, dialyzed against water at 4° for two days and ultracentrifuged again, and the pellet was dissolved in water. The purified protein was stored at 4°C. Dilutions of the solutions were made using pH 7 Trizma buffer. The protein concentration of hemocyanin was determined from the absorbance at 280 nm ($E_{1\,cm}^{1\%}$ = 15.71), and using a value of 74,900 for the molar weight carrying one oxygen binding site in hemocyanin (Ghiretti-Magaldi et al., 1966). The purified protein did not display a detectable Cu(II) ESR signal measured at 77°K with a Varian E-4 ESR spectrometer operating at 9.06 GHz.

Apohemocyanin was prepared by dialyzing a 3% solution of the purified oxyhemocyanin against a 0.1 M phosphate buffer solution containing 0.1 M KCN and 0.05 M $MgCl_2$ at pH 7.8 for two days. Excess cyanide was removed by dialyzing against water for two days. The apohemocyanin thus prepared showed no absorption at 340 nm.

Stripped *Limulus* hemocyanin and its subunits II and V were prepared as described (Sullivan et al., 1974). The molecular weight of each species is taken to be 70,000 and $E_{1\,cm}^{1\%} = 13.9$ at 280 nm.

Fluorescence measurements were made with an Aminco-Bowman spectrophotofluorometer equipped with a 150 W xenon lamp and a 1P21 photomultiplier tube. The relative fluorescence intensity was recorded directly from fluorometer readings. The absorption spectra were recorded using a Cary 118 Model C spectrophotometer. Matched sets of 1 cm fluorescence and UV cells were used. The temperature of all experiments was controlled at 25 ± 0.5°.

Energy Transfer Calculations. According to the Förster theory (Förster, 1959), it is possible to calculate R_o, the characteristic distance between donor and acceptor where the probability of energy transfer is equal to the probability of fluorescence. In our system the donor and acceptor are the tryptophan residues and the Cu...O groups, respectively. Assuming a random distribution of the trytophans with respect to the Cu-O groups, the distance between the donor and acceptor, R, may be calculated by the following equation (Förster, 1959; Weber, 1960):

$$(F/F_o) - 1 = (R_o/R)^6 \tag{1}$$

where F_o and F are the fluorescence yields of the donor with and without energy transfer, respectively. According to the Förster theory,

$$R_o = \left[1.69 \times 10^{-33}\ \tau\ J/n^2 \left(\frac{\nu_a + \nu_e}{2}\right)^2\right]^{1/6} \text{cm} \tag{2}$$

where τ is the lifetime of the excited state, n is the refractive index of the solvent, ν_a and ν_e are the wave numbers of maximum absorption of the acceptor and of maximum emission of the donor, respectively. J is the overlap integral, which can be calculated by Equation (3) using Gaussian approximations[1]:

$$J = \frac{E_1 E_2 (2\pi)^{1/2}}{(\sigma_a^{-2} + \sigma_e^{-2})^{1/2}} \exp - 1/2 \left[(\nu_a - \nu_e)^2 / (\sigma_a^2 + \sigma_e^2)\right] \tag{3}$$

In Equation (3) E_1 and E_2 are the molar absorption coefficients at the maxima of the absorption bands of donor and acceptor, respectively, σ_a and σ_e are the standard deviations, in wave numbers, of the absorption and emission bands, respectively.

Results and Discussion

Native Limulus Hemocyanin. The fluorescence spectra of 1.36×10^{-6} M *Limulus* hemocyanin were measured in pH 7 Trizma buffer. The excitation and

[1] Equation (3) is slightly different from that used by Weber (1960) and Shaklai and Daniel (1970). Its derivation will be publsihed elsewhere.

emission maxima of oxyhemocyanin were found to be 280 and 345 nm, respectively. Deoxygenation of the protein, by adding sodium sulfite (final concentration 0.003 M) and allowed to stand in a covered cell for 25 min results in a six-fold increase of the fluorescence intensity, accompanied by a shift of the emission maximum from 345 to 350 nm. The shift indicates a change of the hydrophobic environment of the tryptophan residues in the protein upon oxygen binding. Apohemocyanin showed a ten-fold increase in intensity as compared to oxyhemocyanin, at 350 nm.

Table 1 (a). Fluorescence of *Limulus* hemocyanin in pH 7 Trizma buffer

	Fluorescence yield (%)
Oxyhemocyanin	1.2
Deoxyhemocyanin	10.5
Apohemocyanin	11.3

(b). Fluorescence of *Limulus*-stripped hemocyanin, II and V in pH 7- 0.05 M Tris-glycine buffer

	Fluorescence yield (%)
Stripped oxygemocyanin	2.0
Stripped deoxyhemocyanin	6.9
Oxyhemocyanin II	1.2
Deoxyhemocyanin II	4.7
Oxyhemocyanin V	2.1
Deoxyhemocyanin V	6.9

Table 1 (a) summarizes the emission maxima (excited at 280 nm) and fluorescence yields of the emission bands at pH 7.0 of *Limulus* oxy-, deoxy-, and apohemocyanin (1-5 × 10^{-6} M). The fluorescence yield was calculated by comparison of area under the emission peak with that of a bovine serum albumin solution of equal absorbance at 280 nm, and taking 15.2% for the fluorescence yield of bovine serum albumin (Teale, 1960). Complete deoxygenation, which was obtained by treating the oxygenated protein with 0.005 M sodium sulfite and 10^{-5} M $CuCl_2$ (final concentration), results in a nine-fold enhancement of the fluorescence yield. As comparison, Shaklai and Daniel (1970) reported only a four-fold enhancement of the fluorescence yield for deoxy- and apohemocyanin of mollusc *L. hierosolima* over that of the oxy- form at pH 6.6.

The fluorescenct changes of hemocyanin upon oxygen binding is due to the energy transfer from the tryptophan residues to the Cu...O groups. Using the following values: $E_1 = 0.59 \times 10^4$ 1/mol/cm (molar absorption of trytophan at 280 nm); $E_2 = 2.996 \times 10^4$ 1/mol/cm (molar absorption of the Cu-O group at 340 nm, calculated from the value of $E_{1\,cm}^{1\%}$ = 4.0 and using a value of 74,900 for the molar weight carrying one oxygen-binding site); $\nu_a = 2.94 \times 10^4$/cm ; $\nu_e = 2.86 \times 10^4$/cm; σ_a = 1850/cm; σ_e = 1710/cm, n = 1.33, τ = 1.26 ns (calculated from the natural lifetime of excitation of trytophan, which was taken to be 12 ns (Weber, 1961), and the fluorescence yield of deoxyhemocyanin), the energy tranfer parameters were calculated as follows: J = 5.26 × $10^{11} cm^3/mmol^2$, and R_o = 30.2 Å. The average distance, R, between the trytophan residues and the Cu...O groups was calculated, based on the fluorescence yield data of Table 1 and Equation (1), to be 21.4 Å. As comparison, for the mollusc *L. hierosolima*, Shaklai and Daniel

(1970) reported R_o and R to be 29.5 and 24.6 Å respectively, at pH 6.6. Using Equations (2) and (3), the values of R_o and R become 29.4 and 25.4 Å respectively.

The distance between the donor (trytophan residues) and the acceptor (Cu...O) groups is thus smaller in the arthropod *Limulus* compared to the mollusc *L. hierosolima* hemocyanin. One would expect, therefore, that quenching of fluorescence by Cu...O groups would be more pronounced for the arthropod than for the mollusc hemocyanin. The experimental data support this expectation: nine-fold enhancement of fluorescence yield of deoxy- over oxyhemocyanin for *Limulus* and only four-fold enhancement for *L. hierosolima*. For mollusc *Murex trunculus* hemocyanin, Bannister and Wood (1971) reported a five-fold enhancement. In both mollusc and arthropod hemocyanins, the energy transferred is related to the oxygenation function.

Limulus-Stripped Hemocyanin and Components II and V. The excitation and emission maxima for the proteins studied in this section, in 0.05 M Tris-glycine buffer, were found to be 290 and 350 nm, respectively. In the concentration range 0.5 to 3.5 ($\times$ 10^{-6}) M hemocyanin, the fluorescence intensity increases linearly with concentration increase.

Table 1 (b) gives the fluorescence yield data of stripped hemocyanin and components II and V. Deoxyhemocyanin was obtained by treating oxyhemocyanin solutions with 0.005 M Na_2SO_3 and 10^{-5} M $CuCl_2$ (final concentrations). Comparison of Table 1 (a) and (b) shows that in going from oxy- to deoxyhemocyanin, the native protein of molecular weight 3.7 million undergoes a greater enhancement of fluorescence yield (9-fold), as compared to the smaller proteins of molecular weight 70,000 (only 3-4-fold enhancement).

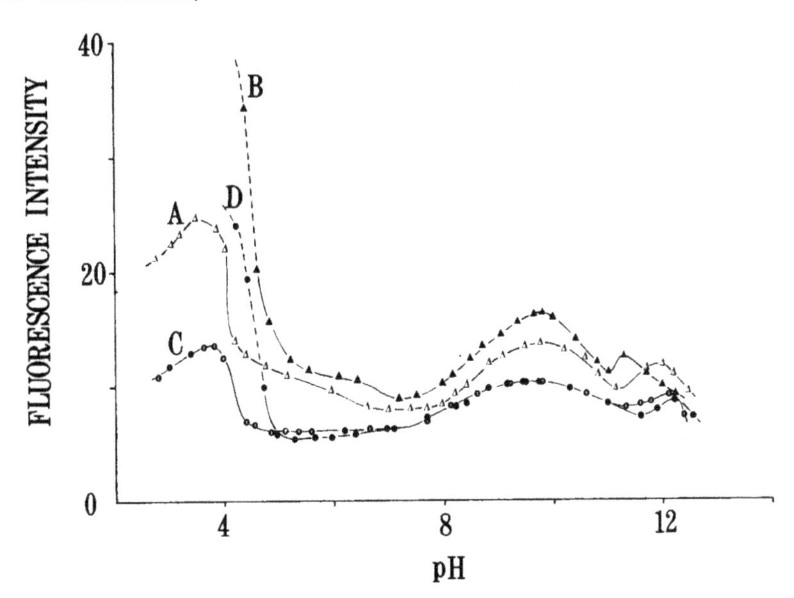

Fig. 1. Plots of fluorescence intensity of 1.55 $\times$ 10^{-6} M component V (*circles*) and 0.67 $\times$ 10^{-6} M component II (*triangles*) of *Limulus* hemocyanin in the absence (*unfilled circles or triangles*) and presence (*filled circles or triangles*) of 1 M NaCl. *Dashed curves*: cloudy suspension

Plots of fluorescence intensity vs. pH are given in Figure 1 for 1.55 × 10^{-6} M component II and 0.67 × 10^{-6} M component V, in the absence and presence of 1 M NaCl. The titrations were carried out by adding HCl or NaOH to the protein solutions prepared in pH 7, 0.05 M Trizma glycine buffer. It is seen that in the pH range 7-11, the fluorescence intensity of component V is unaffected by the presence of 1 M NaCl, while the intensity of component II does depend on the presence or absence of the salt.

The fluorescence intensity of 100% oxygenated hemocyanin, I_o, is taken to be the value at pH 7.0, since maximum oxygen binding occurs in neutral solution. The fluorescence intensity is linearly related to the area under the emission peak and hence to the fluorescence yield. I_o has the values 13.0, 8.0, and 6.4, respectively, for the stripped hemocyanin, components II and V. The corresponding intensities for the 100% deoxygenated hemocyanins, I_d, are 43.8, 30.0, and 22.5, respectively. The fluorescence of the completely deoxygenated hemocyanins is not affected by the presence of 1 M NaCl.

Plots of fluorescence ratio, X, vs. pH are given in Figure 2 for components II and V, in the absence and presence of 1 M NaCl. The fluorescence ratio is defined by the equation:

$$X = (I_d - I)/(I_d - I_o) \tag{4}$$

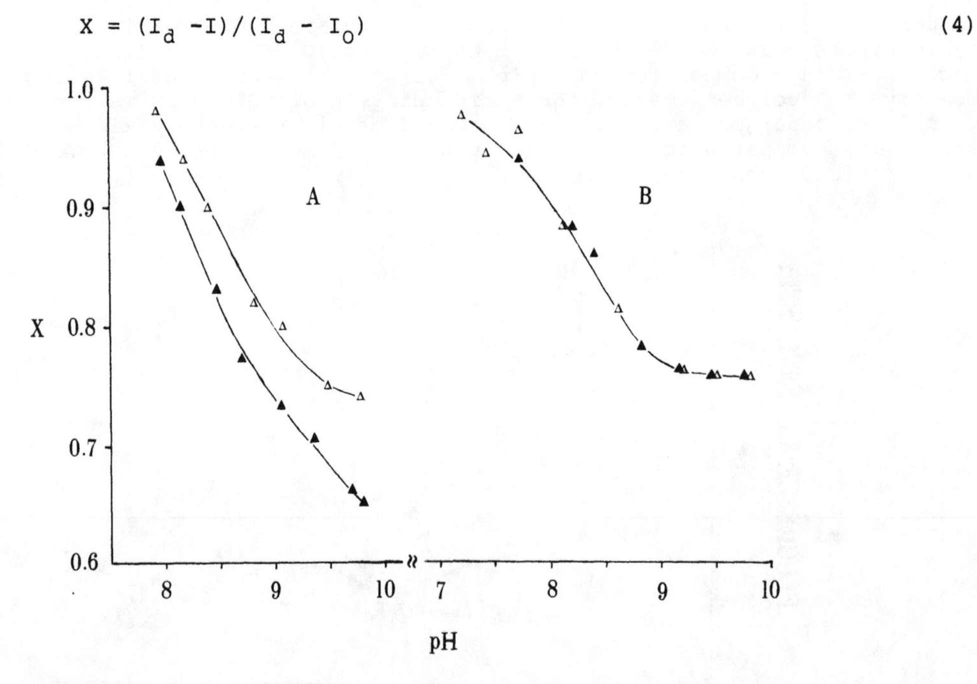

Fig. 2. Plots of fluorescence ratio, X, as functions of pH for (A) component II, (B) component V. *Filled triangles*: presence of 1 M NaCl

where I_d and I_o are as defined already, and I is the fluorescence intensity at a given pH, taken from Figure 1. It is seen that for component V the values of X are unaffected by the presence of NaCl, whereas for component II, the curve obtained in the presence of 1 M NaCl is below that obtained in the presence of the salt. The results of Figures 1 and 2 are in line with the findings of Sullivan et al. (1974) that for component II, oxygen affinity and state of aggregation are changed substantially by the presence of 3 M NaCl, whereas the salt has relatively little effect on component V. The different responses of components II and V to chloride ions is most likely due to difference in conformation of the two components.

Effect of Metal Ions on Deoxygenation by Sodium Sulfite. Oxyhemocyanin is slowly deoxygenated by sodium sulfite. Complete deoxygenation, however, can be rapidly achieved by adding traces of certain metal ions. In a solution containing 4.4×10^{-6} M oxyhemocyanin and 0.005 M sodium sulfite, the addition of 1×10^{-4} M Co(II), Cu(II), or Mn(II) resulted in an immediate diminution of the copper absorption band at 340 nm, as well as a large enhancement of the fluorescence of the protein. In the absence of these metal ions, the fluorescence or absorption bands did not change during the measuring time. Zn(II) ion, on the other hand, did not affect the fluorescence or absorption spectrum of the sulfite-treated protein. The plots of the absorbance of the copper band at 340 nm as a function of time for solutions containing 4.26×10^{-6} M oxyhemocyanin and 0.003 M sodium sulfite, in the presence and absence of Co(II), Mn(II) and Cu(II) ions, are shown in Figure 3. The rate of

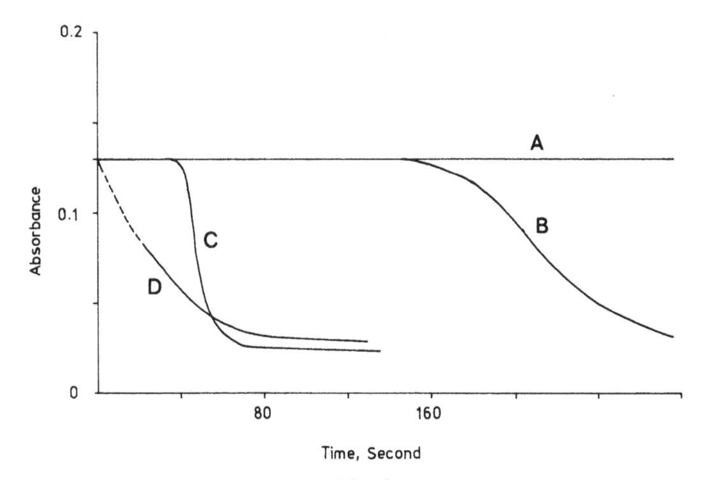

Fig. 3. Metal ion-catalyzed deoxygenation of native *Limulus* hemocyanin by sodium sulfite. The absorbance of 4.26×10^{-6} M oxyhemocyanin at 340 nm is plotted as function of time for solutions containing 0.002 M sodium sulfite and *A*: no metal ion added; *B*: 6.67×10^{-5} M $MnCl_2$; *C*: 3.33×10^{-5} M $CuCl_2$; *D*: 2×10^{-5} M $CoCl_2$

reaction of oxygen with sodium sulfite is significantly increased by the addition of these ions. The catalytic effect was found to be in the following decreasing order: Co, Cu, and Mn. The diminution of the copper band was too fast to measure when the concentration of $CoCl_2$ was increased to 3.33×10^{-5} M, the same as the $CuCl_2$ concentration. When the concentration of sodium sulfite was reduced to 0.001 M, the copper band of protein was not affected by $MnCl_2$ during the time interval measured, even when the concentration of $MnCl_2$ was increased to 1×10^{-4} M.

Veprek-Siska and Lunak (1974) have discussed the role of copper ion in copper-catalyzed oxidation of sulfite

$$Cu^{2+} + SO_3{}^{2-} = Cu^{+} + SO_3{}^{-}$$

and have proposed the formulation of a sulfito-cuprous complex $Cu(SO_3)_n{}^{-n+1}$. Oxygen would react with the sulfito-cuprous complex to form intermediate products of the type $[O_2Cu(SO_3)_n{}^{-n+1}]$. It may well be that the same mechanism holds for the copper-catalyzed reaction of oxygen in hemocyanin with sodium sulfite in the formation of deoxyhemocyanin.

Summary

Limulus hemocyanin, a weakly fluorescent copper protein when oxygenated, is highly fluorescent in the deoxygenated form. A large enhancement of the fluorescence yield, 8.8-fold at neutral pH, is observed on complete deoxygenation, and a 9.4-fold increase is observed on removal of copper from oxyhemocyanin. The weak fluorescence of oxyhemocyanin is attributed to radiationless energy transfer from the tryptophan residues to the Cu-O groups. The average distance between these two moieties is found to be 21.4 A. From the fluorescence results, it is found that the Cu-O group in *Limulus*, an arthropod hemocyanin, is nearer and interacts more strongly with the tryptophan residues than those in the mollusc (*L. hierosolima* and *M. trunculus*) hemocyanins. As for the *Limulus* subunits, the fluorescence properties of component II are affected by the presence of 1 M NaCl, whereas the fluorescence properties of component V are unaffected. The different responses of components II and V to chloride ions have been noted previously by Sullivan et al. (1974) with regard to oxygen affinity and state of aggregation. Deoxygenation of oxyhemocyanin by sodium sulfite is found to be catalyzed by trace amounts of certain metal ions. The reaction rate is significantly increased by Co(II), Cu(II), and Mn(II), in the decreasing order, but not by Zn(II).

Acknowledgments. The authors thank Irene Lee and Nan-sing Kan for sypplying apohemocyanin and the subunits. We are indebted to Professor R. Lontie for a helpful discussion on part of the paper. The work at Duquesne University was supported by the National Science Foundation grant PCM 76-11744.

References

Bannister, W.H., Cammilleri, P., Chantler, E.N.: Relation of trytophan to the copper-oxygen complex in oxyhemocyanin. Comp. Biochem. Physiol. 45B, 325-333 (1973)

Bannister, W.H., Wood, E.J.: Ultraviolet fluorescence of *Murex trunculus* hemocyanin in relation to the binding of copper and oxygen. Comp. Biochem. Physiol. 40B, 7-18 (1971)

Bonaventura, C., Sullivan, B., Bonaventura, J., Bourne, S.: CO binding by hemocyanins of *Limulus polyphemus*, *Busycon carica*, and *Callinectes sapidus*. Biochemistry 13, 4784-4789 (1974)

Förster, T.: Transfer mechanisms of electronic excitation. Disc. Faraday Soc. 27, 7-17 (1959)

Ghiretti-Magaldi, A., Nuzzolo, C., Ghiretti, F.: Chemical studies on hemocyanins. I. Amino acid composition. Biochemistry 5, 1943-1951 (1966)

Ke, C.H., Schubert, J., Lin, C.I., Li, N.C.: Nuclear magnetic resonance study of the binding of glycine derivatives to hemocyanin. J. Am. Chem. Soc. 95, 3375-3379 (1973)

Shaklai, N., Daniel, E.: Fluorescence properties of hemocyanin from *Levantina hierosolima*. Biochemistry 9, 564-568 (1970)

Sullivan, B., Bonaventura, J., Bonaventura, C.: Functional differences in the multiple hemocyanins of the horseshoe crab, *Limulus polyphemus* L. Proc. Natl. Acad. Sci. 71, 2558-2562 (1974)

Sullivan, J.D., Jr., Watson, S.W.: Factors affecting the sensitivity of *Limulus* lysate. Appl. Microbiol. 28, 1023-1026 (1974)

Teale, F.W.J.: The ultraviolet fluorescence of proteins in neutral solutions. Biochem. J. 76, 381-388 (1960)

Veprek-Siska, J., Lunak, S.: Role of copper ions in copper catalyzed autoxidation of sulfite. Z. Naturforsch. 28B, 689-690 (1974)

Weber, C.: Fluorescence-polarization spectra and electronic energy transfer in tryosine, trytophan and related compounds. Biochem. J. 75, 335-345 (1960)

Weber, G.: Light and Life. McElroy, W.D., Glass, B. (eds.). Baltimore: John Hopkins, 1961, p. 82

Limulus polyphemus Hemocyanin. A Nuclear Magnetic Resonance Study of Its Subunits

S. C. Chiang, J. Bonaventura, C. Bonaventura, B. Sullivan, F. K. Schweighardt, and N. C. Li

Summary

Hemocyanin from *Limulus polyphemus* is a 3.3 million dalton oligomeric protein made up of about eight different kinds of polypeptide chains. The subunit mixture obtained after dissociation of the oligomer can be fractionated into five distinct chromatographic zones. While homogeneous in molecular weight, the subunits in these zones appear to have unique functional properties. This uniqueness is also manifest in their NMR spectra. A simulated NMR spectrum, based on a random coil configuration, has been compared with the spectrum of the unfractionated subunit mixture. Notable differences between the observed and simulated spectra are apparent in the region of the spectrum, corresponding to aromatic amino acid side chains. These differences may provide information about the conformation of the native subunits and the environment of the copper atoms in the active site. Chloride ions are known to alter the oxygen affinities of some of the *Limulus* hemocyanin subunits. This functional differentiation is supported by the marked effects of chloride ions on the NMR spectra of these particular subunits.

Introduction

Hemocyanins (Hcy) are high molecular weight copper proteins which occur in molluscs and arthropods. In the blood of molluscs they are decamers or dodecamers with molecular weights from 4.5×10^6 to 9×10^6, while in the arthropods they vary from hexamers to 48-mers with molecular weights from 4.5×10^5 to 3.3×10^6 (van Holde and van Bruggen, 1971; Bonaventura et al., 1976). The high molecular weight of these aggregates precludes NMR spectroscopy at present. The molecular weight of molluscan hemocyanin subunits appears to be around 365,000, while that of arthropod hemocyanin subunits is 66 to 80,000 (Bonaventura et al., 1976). The relatively low molecular weight of arthropod hemocyanin subunits makes them attractive for nuclear magnetic resonance (NMR). The NMR studies reported here add to our knowledge of the hemocyanin system of the horseshoe crab, *Limulus polyphemus*, which has been the subject of considerable investigations. The subunits of *Limulus* hemocyanin comprise a family of charge-isomers, all having molecular weights in the range of 66 to 70,000. They can be separated into zones, Hcy I-V, by ion exchange chromatography. The native subunits in these zones differ in their oxygen affinities and in the effect of chloride on oxygen binding. Chloride ions lower the oxygen affinities of Hcy II and Hcy III but have little or no effect on the oxygen affinities of the other zones. NMR spectra have now been obtained with the subunit mixture (called whole-stripped *Limulus* hemocyanin) and the isolated zones in the presence and absence of chloride. All of these preparations, in D_2O, show rather distinct spectra. Knowledge of the amino acid composition of these subunits (Sullivan et al., 1976) and their relative proportions (Sullivan et al., 1974) in the subunit mixture has permitted us to prepare a simulated NMR spectrum based on a random-coil configuration of the protein.

Materials and Methods

L. polyphemus was collected in the waters around Beaufort, North Carolina. Stripped *Limulus* hemocyanin and purified chromatographic zones were prepared as previously described (Sullivan et al., 1974, 1976). The hemocyanin preparations were concentrated to about 50 mg/ml by ultrafiltration and dialyzed vs. 0.05 M Tris-glycine, 0.01 M EDTA pH 8.9. Immediately prior to the NMR studies, samples in aqueous media were exchanged with D_2O through Amicon ultrafiltration. Proton NMR spectra were obtained using the MPC-HF 250-MHz super-conducting spectrometer (Dadok et al., 1970) using the NMR correlation technique (Dadok and Sprecher, 1974). The spectrometer was operated in the linear frequency sweep mode with an internal homonuclear lock on HDO. The sweep was under program control using a Xerox Data Systems Sigma 5 computer with a 15-bit digital to analog converter. The data were collected at the output of a Princeton Applied Research Model 121 synchronous detector using a 12db per octave low-pass filter in that device, and a Xerox Data Systems Model MD 51 analog to digital converter (Dadok et al., 1970). Proton chemical shifts were referenced to the solvent-HDO resonance in each of the figures and assigned negative values when they were down field from HDO. Simulated NMR spectra (McDonald and Phillips, 1969) of whole stripped *Limulus* hemocyanin were calculated, based upon a random-coil configuration of the polypeptides using the known amino acid composition of the chromatographic zones (Sullivan et al., 1976) and their relative proportions in the unfractionated mixture (Sullivan et al., 1974). Since tryptophan is destroyed during acid hydrolysis prior to amino acid composition analysis, it is not usually reported. We know, however, that tryptophan is present in *Limulus* hemocyanin, as evidenced by the characteristic tryptophan notch observed in its UV absorption spectrum (Bonaventura et al., 1974). The loss of quantitative data on tryptophan prevented inclusion of contributions from this side chain in the simulated spectra.

Results and Discussion

NMR spectra, showing resonances of aromatic amino acid side chains of *Limulus* hemocyanin chromatographic zones, are shown in Figure 1. Comparable spectra showing aliphatic side chain resonances are shown in Figure 2. The aromatic region clearly shows differences among the spectra of the chromatographic fractions. In contrast, the spectra of the aliphatic region are much more similar.

At this point it is necessary to point out that the chromatographic zones do not all represent pure subunits. Inspection of the elution patterns obtained from DEAE-Sephadex chromatography of the mixture reveals that some of the zones are asymmetric. Regular disc gel electrophoresis also shows that some of the zones contain more than one band (Sullivan et al., 1976). Improved methodology has partially improved resolution of Zone III but not enough for subfractionation. Zone V, however, is now subdivided into Va and Vb.

Table 1 gives the mol % of the various amino acids present in Hcy I through Hcy V. The zones have very similar amino acid compositions, which may account for the similarity of their NMR spectra in the aliphatic region. The aromatic region includes contributions from histidine, phenylalanine, tryptophan, and tyrosine; residues which may be in the active site. The uniqueness of the NMR spectra in the aromatic region suggest differences in the active sites. These differences might explain the differences in oxygen affinity observed for each component.

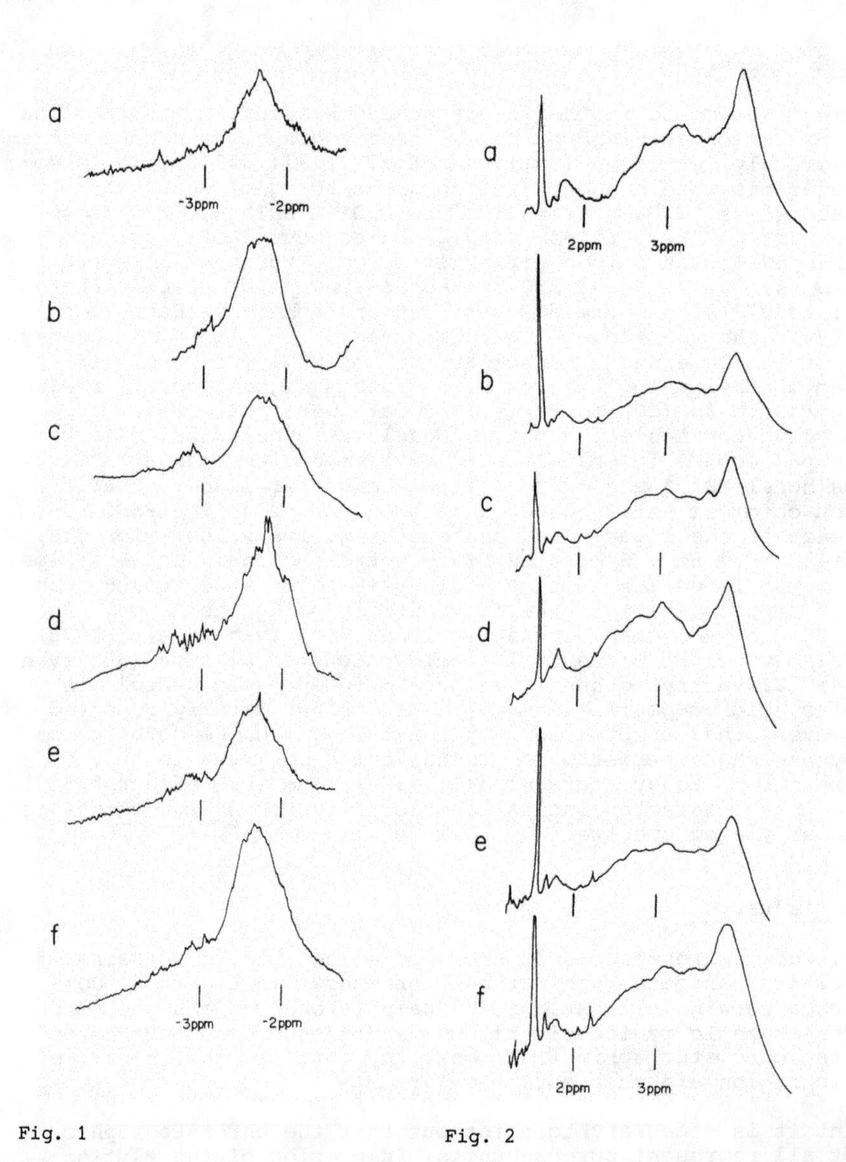

Fig. 1 Fig. 2

Fig. 1a-f. NMR spectra of the aromatic region of *Limulus* hemocyanins: (a) component I, 2.7% Hcy, pD 7.92; (b) component II, 3.8% Hcy, pD 8.25; (c) component III, 1.8% Hcy, pD 8.0; (d) component IV, 2.9% Hcy, pD 8.35; (e) component Va, 2.6% Hcy, pD 8.07; (f) component Vb, 4.5% Hcy, pD 8.75

Fig. 2a-f. NMR spectra of the aliphatic region. Experimental conditions (a) to (f) are the same as Figure 1

The weight percentages of components I, II, III, IV, and V in stripped *Limulus* hemocyanin are 13, 25, 37, 16, and 9 respectively (Sullivan et al., 1974). Using the amino acid compositions of these components (Sullivan et al., 1976) we can calculate the molar percentage of the various amino acids in whole stripped *Limulus* hemocyanin. This is done by making weighted averages of the component's molar percentage, resulting in the listed values in the last column of Table 1. With the aid of Table 1, we have prepared simulated NMR spectra of *Limulus* hemocyanin and its

Table 1. Mol % of amino acids in stripped *Limulus* hemocyanin and its components

Amino acid	I	II	III	IV	Va + Vb	Stripped
Lysine	6.3	5.8	5.6	5.2	5.2	5.7
Histidine	6.6	6.4	7.2	7.7	6.6	6.9
Arginine	4.4	5.2	5.4	5.9	5.3	4.7
Aspartic acid	11.6	13.0	13.7	10.9	11.8	12.7
Threonine	5.5	5.2	5.5	5.1	4.4	5.3
Serine	5.5	6.5	5.9	6.2	5.5	6.0
Glutamic acid	10.4	9.6	10.0	12.3	10.6	10.4
Proline	4.4	4.8	4.5	4.0	5.9	4.6
Glycine	7.8	6.6	7.6	6.6	6.6	7.1
Alanine	5.8	5.4	5.3	4.5	4.8	5.2
Half cystine	1.6	2.0	1.9	1.9	1.5	1.9
Valine	6.3	5.9	6.2	5.8	6.2	6.1
Methionine	2.1	1.8	2.6	2.1	2.7	2.3
Isoleucine	5.7	4.9	4.1	5.3	4.8	4.8
Leucine	8.2	8.1	9.7	8.3	9.2	8.9
Tyrosine	2.9	3.9	3.5	3.2	3.8	3.5
Phenylalanine	5.1	5.0	4.4	5.1	5.1	3.7
Tryptophan	Not determined					
Total	100.2	100.1	100.1	100.1	100.0	99.8

The values are calculated from the amino acid composition reported in Sullivan et al. (1976)

five components by the method of McDonald and Phillips (1969), which is based on a random-coil protein configuration in neutral D_2O medium. It should be noted that such simulated spectra are random-coil representations, which assume all side chains to be in a solvent environment, and can provide no direct structural information about the nature of the native protein or its components. However, the departures of an observed spectrum from a spectrum based on a random-coil configuration may provide information about the conformation of the native protein. Figure 2a shows the observed NMR spectrum of unfractionated stripped *Limulus* hemocyanin at a pD value of approximately 8. In Figure 3b the simulated spectrum of unfractionated whole stripped *Limulus* hemocyanin is shown. The aliphatic and aromatic resonances of the observed spectrum (3a) are clearly in the same frequency region as in the simulated spectrum (3b). The observed spectrum lacks the sharp peaks of the simulated spectrum. It should be noted that tryptophan was not determined in the amino acid compositions reported in Table 1, and therefore this residue was not included in the simulated spectrum of Figure 3b. The shoulders in the region of -2 ppm in Figure 1 are indicative of tryptophan residues (McDonald and Phillips, 1969).

Figure 4 shows the NMR spectra of the aliphatic region of Hcy II and Hcy Vb in the presence of 2 M NaCl. Comparison of these spectra with those shown in Figures 2b and 2f shows that NaCl has a greater effect on the NMR spectrum of Hcy II than on that of Hcy Vb. It is known that NaCl causes Hcy II to aggregate, from high O_2 affinity molecules whose molecular weight is 70,000 to low affinity forms whose molecular weight is about 400,000. In contrast, Hcy V remains a low affinity monomer after the addition of NaCl (Bonaventura et al., 1975, 1976). This difference in NaCl sensitivity is thus manifest in the NMR spectra presented here.

Moss et al. (1973) have reported an NMR study of *Cancer magister* hemocyanin at pD 11.2, a condition where it is dissociated into 80,000

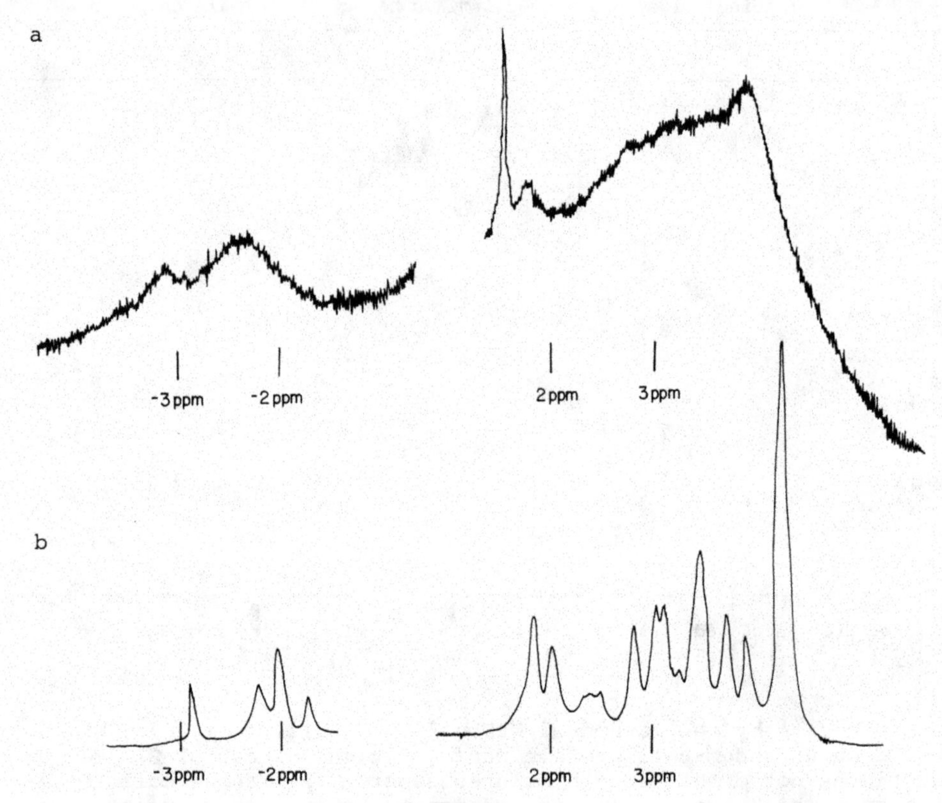

Fig. 3. (a) NMR spectrum of stripped *Limulus* hemocyanin, 15% Hcy, pD 7.98; (b) simulated NMR spectrum of stripped *Limulus* hemocyanin

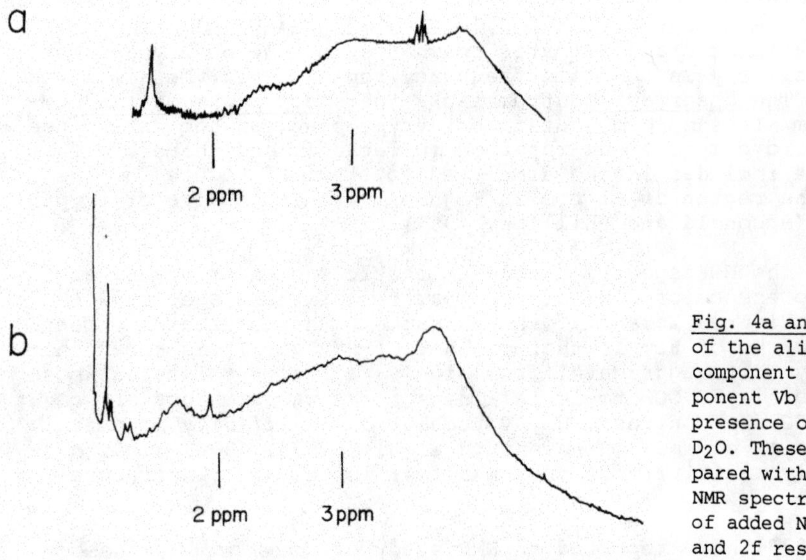

Fig. 4a and b. NMR spectra of the aliphatic region of component II (a) and component Vb (b), both in the presence of 2 M NaCl in D_2O. These are to be compared with the components' NMR spectra in the absence of added NaCl (Figs. 2b and 2f respectively)

dalton subunits. Their spectra, however, are not well structured "because of the relatively large subunit size and the possible subunit associative equilibria". We have thus obtained, for the first time, structured NMR spectra of arthropod hemocyanin subunits. We plan to continue these investigations in an attempt not only to identify the peaks but to elucidate the active sites and the mechanism of the hemocyanin-chloride interaction.

Acknowledgments. This investigation was supported by NSF Grants to Duquesne (PCM 76-11744) and Duke Universities and an NIH grant to Duke University; the NMR biomedical facilities supported by Public Health Service Grant Number RR-00292, were used. The authors are indebted to Gerald Godette for help in the preparation of the hemocyanin components and to K. Seman for assistance in preparing the simulated NMR spectra. Joseph Bonaventura is an established investigator of the American Heart Association. This paper represents paper number 2 on structure-function relationships in *Limulus* hemocyanin.

References

Bonaventura, C., Sullivan, B., Bonaventura, J., Bourne, S.: CO binding by hemocyanins of *Limulus polyphemus, Busycon carica* and *Callinectus sapidus*. Biochemistry 13, 4784-4789 (1974)

Bonaventura, J., Bonaventura, C., Sullivan, B.: Hemoglobins and hemocyanins. Aspects of structure and function. J. Exptl. Zool. 194, 155-174 (1975)

Bonaventura, J., Bonaventura, C, Sullivan, B.: Oxygen and Physiological Function. Jobsis, F. (ed.). Professional Information Library, Dallas, Texas (In press, 1976)

Dadok, J., Sprecher, R.F.: Correlation NMR spectroscopy. J. Magnetic Resonance 13, 243-248 (1974)

Dadok, J., Sprecher, R.F., Bothner-By, A.A., Link, T.: Correlation NMR spectroscopy. Abstr. 11th Exptl. NMR Conf., Pittsburgh, PA (1970)

McDonald, C.C., Phillips, W.D.: Proton magnetic resonance spectra of proteins in randon-coil configuration. J. Am. Chem. Soc. 91, 1513-1521 (1969)

Moss, T.H., Gould, D.C., Ehrenberg, A., Loehr, J.S., Mason, H.S.: Magnetic properties of *Cancer magister* hemocyanin. Biochemistry 12, 2444-2449 (1973)

Sullivan, B., Bonaventura, C., Bonaventura, J.: Functional differences in the multiple hemocyanins of the horseshoe crab, *Limulus polyphemus* (L.). Proc. Natl. Acad. Sci. 71, 2558-2562 (1974)

Sullivan, B., Bonaventura, J., Bonaventura, C., Godette, G.: Hemocyanin of the horseshoe crab, *Limulus polyphemus*. I. Structural differentiation of the isolated components. J. Biol. Chem. (In press, 1976)

Van Holde, K.E., van Bruggen, E.F.J.: Subunits in Biological Systems. Timasheff, S.N., Fasman, G.D. (eds.). New York: Marcel Dekker, 1971, pp. 1-53

Hemocyanin as a Copper Complex

On the Active Site of Molluscan Haemocyanin and of Tyrosinases. Opening Address

R. LONTIE

Introduction

In the next Haemocyanin Workshop we ought to celebrate the centennial of the terms "haemocyanin" and "oxyhaemocyanin", coined in 1878 by Leon Fredericq (1851-1935), who later became professor of physiology at the State University of Liege. Préparateur at the time at the State University of Ghent, he spent the second half of July and the month of August 1878 at the newly established laboratory in Roscoff, Brittany, working on the physiology of the octopus (Florkin, 1943). He realized that the colourless blood substance, a protein, became blue on binding oxygen reversibly and acted as an oxygen carrier. He could not detect any iron, but found instead the presence of copper (Fredericq, 1878).

The blue colour of the blood of a snail had probably been described for the first time in 1669 by the Dutch naturalist Jan Swammerdam (1637-1680) (Swammerdam, 1669). The blue blood had also been noted in snails by Erman in 1817, in river crayfish by Carus in 1824, in crabs by Wharton Jones in 1846, in Cephalopods by Harless in 1847, and in cuttlefish by Bert in 1867, who had also recognised the respiratory function of the pigment (Kobert, 1903).

Harless (1847) had not only described the blue colour taken on by the winter blood of the Roman snail in the air, but found that carbon dioxide destroyed the blue colour, which returned completely with oxygen. He demonstrated, moreover, the presence of copper in the blood of the Roman snail and of Cephalopods.

Spectra of the Haemocyanins and Stoicheiometry of the Oxygen Binding

Rabuteau and Papillon (1873), who had also observed that the blood of the octopus took on a blueish colour in the air which was lost reversibly under carbon dioxide, could not find any absorption band with a visual spectroscope. With a photographic method, Dhéré discovered in 1908 a typical band for oxyhaemocyanin at 346 nm besides the protein band at 278 nm (Dhéré, 1928). In the visible an absorption band was also observed with a maximum at 579 nm. The spectra were further investigated by Quagliariello (1923), Redfield (1934), and Roche and Dubouloz (1933) to cite only a few. It was also established that the typical bands disappeared reversibly on deoxygenation.

The recent progress can best be evaluated by comparing the wealth of the circular dichroism spectra (Bannister and Wood, 1972; Lontie and Vanquickenborne, 1974). With β-haemocyanin of *Helix pomatia*, proteolytic fragments were obtained, which showed the presence of two classes of copper groups, characterized by a positive maximum at 455 nm and at 500 nm respectively (Gielens et al., 1975).

Dhéré had already found that 1 g of Cu binds on the average 135 cm^3 of oxygen for molluscan and 224 cm^3 for arthropodan haemocyanins. The mean corresponded to 1 O_2/2 Cu (Quagliariello, 1923). This ratio was accurately determined by Redfield et al. (1928) and for *H. pomatia* haemocyanin by Guillemet and Gosselin (1932).

The Active Site of Molluscan Haemocyanins

From the ratio of 1 O_2/2 Cu, as from 1 O_2/2 Fe for haemerythrin, followed the traditional nonplanar μ-dioxygen bridge:

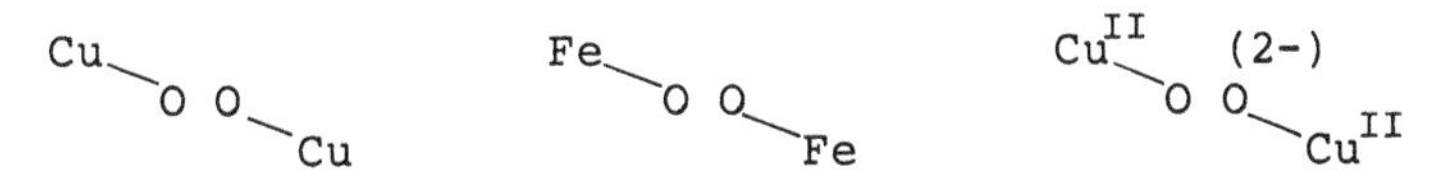

An interpretation of the optical spectra led to a μ-peroxo bridge (Bannister and Wood, 1969; Lontie and Vanquickenborne, 1974).

The presence of a peroxide group was demonstrated by resonance Raman by Freedman et al. (1976); the method has, however, a qualitative aspect. An O-O stretching vibration at 749/cm ($^{16}O_2$) was observed for the oxyhaemocyanin of *Busycon canaliculatum*. It shifted to 708/cm with 91 atom % $^{18}O_2$. This value is rather low compared to the average of 807 (range 790-844)/cm for μ-peroxo complexes (Vaska, 1976). This low figure was attributed by the authors to a very hydrophobic environment. Similarly, for oxyhaemerythrin values of 844/cm ($^{16}O_2$) and 798/cm ($^{18}O_2$) were observed (Klotz et al., 1976). Freedman et al. (1976) identified, moreover, in oxyhaemocyanin a frequency at 267/cm, which they assigned to a Cu-N(imidazole) vibration. Thus, two ligands of copper in oxyhaemocyanin were identified:

$$\text{im} - \text{Cu} - \text{O} - \text{O}^{(2-)} \qquad \text{Cu}^{I} - \text{L} - \text{Cu}^{I} - \text{O} = \text{C} \qquad \text{Cu}^{II} - \text{L} - \text{Cu}^{II} - \text{O} - \text{O}^{(2-)}$$

A study by Fourier transform infrared spectroscopy of the carbon monoxide derivative of the haemocyanins of squid and limpet yielded CO stretching vibrations near 2063/cm (Fager and Alben, 1972). These data were interpreted by a lateral binding of CO to a single copper atom, most likely through the oxygen. This lateral binding of carbon monoxide, albeit to Cu(I), suggests the presence of a bridging ligand L and a lateral binding of dioxygen as a peroxide ion.

In accordance with this model is our inability to insert potential bridging ligands, like imidazole and pyrazine, able to yield binuclear complexes with Cu-Cu distances of 6.0 and 6.7 Å respectively. The ready obtention of methaemocyanin on treating oxyhaemocyanin of *H. pomatia* with HN_3 or HF near pH 5 for 2 days at 37°C also suggested a lateral binding of dioxygen, and a displacement of peroxide by azide or fluoride, similar to that of superoxide in myoglobin or haemoglobin (Witters and Lontie, 1975), the catalase activity of molluscan haemocyanin (Ghiretti, 1956) and the very fast reduction of methaemocyanin of *H. pomatia* with traces of H_2S (De Ley et al., 1975).

The active site of superoxide dismutase of bovine erythrocytes (Fig. 1) provides a model for an imidazole anion as a bridging ligand (Richardson et al., 1975). One axial direction on the Cu is accessible to the solvent.

The proposal of Salvato et al. (1973) that each copper atom is surrounded by four histidines apparently does not allow a binding of carbon monoxide. But a cluster of histidines with an imidazole anion as a bridging ligand between the copper atoms would not be inconsistent with their titration data.

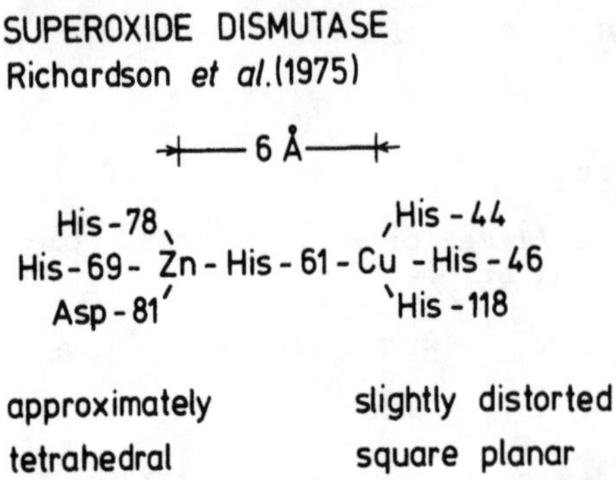

SUPEROXIDE DISMUTASE
Richardson *et al.*(1975)

→|— 6 Å —|←

His-78╲ ,His-44
His-69- Zn - His-61 - Cu - His-46
Asp-81╱ ╲His-118

approximately tetrahedral   slightly distorted square planar

Fig. 1. Structure of the active site of bovine Cu,Zn superoxide dismutase

A similar situation prevails with haemerythrin. Gray (1975) postulated an oxobridge Fe(III) dimer with a laterally coordinated peroxide: Fe^{III}—O—Fe^{III}-OOH. This structure, which apparently can not be reconciled with recent X-ray data on methaemerythrin of *Themiste dyscritum* at 2.8 Å resolution (Klotz et al., 1976), seems able to resolve the problems arising from Mössbauer spectroscopy. The 2 Fe atoms show a similar environment in deoxy- and in methaemerythrin, but two different ones in oxyhaemerythrin.

The magnetic properties of keyhole limpet oxyhaemocyanin (linear volume susceptibilities versus 1/T) call for a very strong coupling of the presumed Cu(II) atoms (Solomon et al., 1976). Very fresh oxyhaemocyanin of *H. pomatia* shows a total absence of electron paramagnetic resonance signals at g = 4 and at g = 2 (De Ley et al., unpublished results). Very weak irreversible signals, that are not removed by traces of hydrogen peroxide, appear on freezing. Moreover, the electron paramagnetic signals at g = 4 and at g = 2 of methaemocyanin of *H. pomatia* vanish completely on addition of fluoride (Witters et al., 1977).

All these observations could be explained simply by the rather extreme formulation of a binuclear Cu(I)-Cu(III) complex, where Cu(III) is stabilized by the "hard" ligands peroxide and fluoride:

$$Cu^{I}\text{-}L\text{-}Cu^{III}\text{-}O\diagdown O^{(2-)} \qquad Cu^{I}\text{-}L\text{-}Cu^{III}\text{-}F^{-}$$

Rather stable and simple Cu(III) complexes have been described recently, like the tetraglycine complex with deprotonated peptide nitrogens (Margerum et al., 1975). A low spin Cu(III) (d^8) becomes plausible with a square planar or nearly square planar geometry (Freeman, 1967). It might also be worthwhile to consider the crystal structure of bis-imidazolato-copper(II), where the copper atoms show two alternating environments: square planar and square planar distorted toward tetrahedral (Freeman, 1967).

A model involving Cu(I) and Cu(III) is excluded for type 3 copper in laccase, that seems to be an antiferromagentically coupled Cu(II) dimer (Solomon et al., 1976). Similarly, methaemocyanin of *H. pomatia* with ligands softer than F^- seems to contain a Cu(II) dimer, as shown by the electron paramagnetic resonance signals at g = 4 (Witters et al., Chap. 30, this vol.), which could be confirmed by magnetic susceptibilities.

Although the pitfalls of a chemical determination of oxidation numbers does not have to be stressed, could the proposed model explain the observation by Klotz and Klotz (1955)? On treating oxyhaemocyanin with a Cu(I) specific reagent and glacial acetic acid only half of the copper reacts. Cox and Elliott (1974) observed also an exchange of only half of the haemocyanin copper with copper(I) complexes. The reconstitution of haemocyanin followed, however, a quadratic relationship with the copper content, indicating equivalent copper atoms (Konings, 1969). The same relationship was observed on removing copper with cyanide (Lontie and Vanquickenborne, 1974).

Any model for the active site must, moreover, provide a basis for the cooperativity of the oxygen binding, which is optimal with *H. pomatia* haemocyanin near pH 8 in the presence of 10 mM Ca^{2+}. The following changes on oxygenation could be considered: a modification of the geometry, an alteration of bond lengths, and/or an expansion of the coordination sphere. An example of the latter possibility is given by the structures of bis(imidazole)copper(I) perchlorate and bis(imidazole)copper(II) diacetate, where the oxidation of copper is accompanied by the coordination of two carboxylic groups (Henriksson et al., 1976).

The Active Site of Tyrosinases

Kubowitz (1938) had already stressed the similarity of polyphenoloxidase and haemocyanins. Jolley et al. (1974) made the important observation that resting tyrosinase of mushrooms developed a haemocyanin spectrum by the addition of traces of hydrogen peroxide.

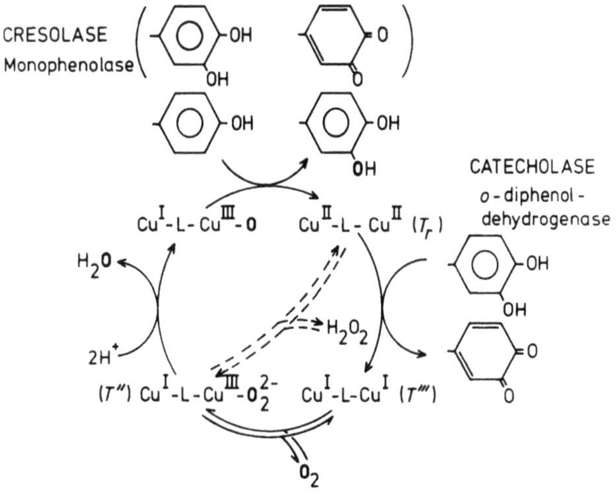

Fig. 2. Model of the active site and of the mechanism of mushroom tyrosinase. The symbols T_r, T'', T''' are taken from Jolley et al. (1974)

The model of the active site of molluscan haemocyanin with a bridging ligand, like an imidazole anion, and a lateral binding of dioxygen can immediately be transposed to the tyrosinase of mushrooms (Fig. 2). The oxene formulation for the hydroxylating intermediate is borrowed from the model for cytochrome P450 (Lichtenberger et al., 1976). The active copper must also be able to coordinate a phenolate, the substrate.

Brief Comparison of Arthropodan and Molluscan Haemocyanins

The active sites of arthropodan and molluscan haemocyanins show very similar properties on the whole: the optical spectra (Witters et al.,

1974), the presence of peroxide as shown by resonance Raman (Freedman et al., 1976), the diamagnetism of oxyhaemocyanin (Moss et al., 1973). There are, however, striking differences: the low catalase activity (Ghiretti, 1956) and the formation of methaemocyanin with traces of hydrogen peroxide with arthropodan haemocyanins (Felsenfeld and Printz, 1959), the much lower affinity for carbon monoxide of arthropodan haemocyanins (Bonaventura et al., 1974), and the lack of methaemocyanin formation with HN_3 or HF, as shown for *Limulus polyphemus* haemocyanin (Witters and Lontie, 1975).

Outlook

After one century we might be dismayed by our ignorance, and often by the absence of irrefutable and unequivocal proofs. We turn our hopes to high resolution nuclear magnetic resonance and X-ray diffraction. With molluscan haemocyanins homogeneous functional fragments can be expected soon, but the yields will remain a serious problem. The unravelling of the differences near the active site of deoxy-, oxy-, and methaemocyanins, and the understanding of the cooperativity of the oxygenation, will still require much effort and dedication. These data will help to clarify the mechanism of tyrosinases and to understand type 3 copper.

At present we can only express our admiration for the intricacies of nature and look forward to the proceedings of this Workshop and of many to come.

References

Bannister, W.H., Wood, E.J.: Free electron model for the absorption of oxygen-bridged binuclear complexes in the near ultraviolet. Nature (London) 223, 53-54 (1969)

Bannister, W.H., Wood, E.J.: Gaussian analysis of the visible and near-ultraviolet absorption and circular dichroism spectra of haemocyanin from *Murex trunculus*. Comp. Biochem. Physiol. 43B, 1033-1037 (1972)

Bonaventura, C., Sullivan, B., Bonaventura, J., Bourne, S.: CO binding by hemocyanins of *Limulus polyphemus*, *Busycon carica*, and *Callinectes sapidus*. Biochemistry 13, 4784-4789 (1974)

Cox, J.A., Elliott, F.G.: Isotopic copper exchange in *Pila* haemocyanin with three radioactive cuprous complexes. Biochim. Biophys. Acta 371, 392-401 (1974)

De Ley, M., Candreva, F., Witters, R., Lontie, R.: The fast reduction of *Helix pomatia* methaemocyanin with hydrogen sulphide. FEBS Lett. 57, 234-236 (1975)

Dhéré, Ch.: Sur quelques pigments respiratoires des Invertebres. Revue suisse Zool. 35, 277-288 (1928)

Fager, L.Y., Alben, J.O.: Structure of the carbon monoxide binding site of hemocyanins studied by Fourier transform infrared spectroscopy. Biochemistry 11, 4786-4792 (1972)

Felsenfeld, G., Printz, M.P.: Specific reactions of hydrogen peroxide with the active site of hemocyanin. The formation of "methemocyanin". J. Am. Chem. Soc. 81, 6259-6264 (1959)

Florkin, M.: Leon Fredericq et les Debuts de la Physiologie en Belgique. Brussels: Office de Publicite, 1943, p. 32

Fredericq, L.: Sur l'hemocyanine, substance nouvelle due sang de Poupe (*Octopus vulgaris*). C. R. Acad. Sci. (Paris) 87, 996-998 (1878)

Freedman, T.B., Loehr, J.S., Loehr, T.M.: A resonance Raman study of the copper protein, hemocyanin. New evidence for the structure of the oxygen-binding site. J. Am. Chem. Soc. 98, 2809-2815 (1976)

Freeman, H.C.: Crystal structures of metal-peptide complexes. Advan. Protein Chem. 22, 257-424 (1967)

Ghiretti, F.: The decomposition of hydrogen peroxide by *Octopus* hemocyanin. Bull. Soc. Chim. Belg. 65, 103-106 (1956)

Gielens, C., Preaux, G., Lontie, R.: Limited trypsinolysis of β-haemocyanin of *Helix pomatia*. Characterization of the fragments and heterogeneity of the copper groups by circular dichroism. Europ. J. Biochem. 60, 271-280 (1975)
Gray, H.B.: Polynuclear iron (III) complexes. In: Proteins of Iron Storage and Transport in Biochemistry and Medicine. Crichton, R.R. (ed.). Amsterdam-Oxford: North-Holland, 1975, p. 7
Guillemet, R., Gosselin, G.: Sur le rapport entre le cuivre et la capacite' respiratoire dans les sangs hemocyaniques. C. R. Soc. Biol. 111, 733-735 (1932)
Henriksson, H.-Å., Sjoberg, B., Osterberg, R.: Model structures for a copper(I)-copper(II) redox couple in copper proteins: X-ray powder structure of bis(imidazole)copper(I) perchlorate and crystal structure of bis(imidazole)copper(II) diacetate. J. Chem. Soc. Chem. Comm. 130-131 (1976)
Jolley, R.L., Evans, L.H., Makino, N., Mason, H.S.: Oxytyrosinase. J. Biol. Chem. 249, 335-345 (1974)
Klotz, I.M., Klippenstein, G.L., Hendrickson, W.A.: Hemerythrin: Alternative oxygen carrier. Science 192, 335-344 (1976)
Klotz, I.M., Klotz, T.A.: Oxygen-carrying proteins: A comparison of the oxygenation reaction in hemocyanin and hemerythrin with that in hemoglobin. Science 121, 477-480 (1955)
Kobert, R.: Über Hemocyanin nebst einigen Notizen über Hamerythrin. Pflüger's Arch. Physiol. 98, 411-427 (1903)
Kubowitz, F.: Spaltung und Resynthese der Polyphenoloxydase und des Hamocyanins. Biochem. Z. 299, 32-57 (1938)
Lichtenberger, F., Nastainczyk, W., Ullrich, V.: Cytochrome P450 as an oxene transferase. Biochem. Biophys. Res. Comm. 70, 939-946 (1976)
Lontie, R., Vanquickenborne, L.: The role of copper in hemocyanins. In: Metal Ions in Biological Systems. Vol. III. High Molecular Complexes. Sigel, H. (ed.). New York: Marcel Dekker, 1974, pp. 183-200
Margerum, D.W., Chellappa, K.L., Bossu, F.P., Burce, G.L.: Characterization of a readily accessible copper(III)-peptide complex. J. Am. Chem. Soc. 97, 6894-6896 (1975)
Moss, T.H., Gould, D.C., Ehrenberg, A., Loehr, J.S., Mason, H.S.: Magnetic properties of *Cancer magister* hemocyanin. Biochemistry 12, 2444-2449 (1973)
Quagliariello, G.: Das Hamocyanin. Naturwissenschaften 11, 261-268 (1923)
Rabuteau, Papillon, F.: Observations sur quelques liquides de l'organisme des Poissons, des Crustaces et des Cephalopodes. C. R. Acad. Sci. (Paris) 77, 135-138 (1873)
Redfield, A.C.: The haemocyanins. Biol. Rev. 9, 175-212 (1934)
Redfield, A.C., Coolidge, T., Montgomery, H.: The respiratory proteins of the blood. II. The copper-combining ratio of oxygen and copper in some bloods containing hemocyanin. J. Biol. Chem. 76, 197-205 (1928)
Richardson, J.S., Thomas, K.A., Rubin, B.H., Richardson, D.C.: Crystal structure of bovine Cu,Zn superoxide dismutase at 3 Å resolution: Chain tracing and metal ligands. Proc. Natl. Acad. Sci. 72, 1349-1353 (1975)
Roche, J., Dubouloz, P.: Etude de la constitution des hemocyanines et des hemerythrines au moyen de leur spectre ultraviolet. C. R. Acad. Sci. (Paris) 196, 646 (1933)
Salvato, B., Zatta, P., Ghiretti-Magaldi, A., Ghiretti, F.: On the active site of hemocyanin. FEBS Lett. 32, 35-36 (1973)
Solomon, E.I., Dooley, D.M., Wang, R.H., Gray, H.B., Cerdonio, M., Mogno, F., Romani, G.L.: Susceptibility studies of laccase and oxyhemocyanin using an ultrasensitive magnetometer. Antiferrognetic behaviour of the type 3 copper in *Rhus* laccase. J. Am. Chem. Soc. 98, 1029-1031 (1976)
Swammerdam, J.: Historia Insectorum Generalis ofte algemene verhandeling van de bloedeloose dierkens. Utrecht: van Dreunen, 1669
Vaska, L.: Dioxygen-metal complexes: Toward a unified view. Accounts Chem. Res. 9, 175-183 (1976)
Witters, R., Goyffon, M., Lontie, R.: Etude de l'hemocyanine de Scorpion par dichroisme circulaire. C. R. Acad. Sci. (Paris) 278, 1277-1280 (1974)
Witters, R., Lontie, R.: The formation of *Helix pomatia* methaemocyanin accelerated by azide and fluoride. FEBS Lett. 60, 400-403 (1975)

Photooxidative and Spectral Studies of *Octopus vulgaris* Hemocyanin

G. JORI, B. SALVATO, AND L. TALLANDINI

Introduction

Although hemocyanins represent a family of proteins which are the subject of intensive investigations, very little information is presently available as regards those sites of the protein molecule which are critical for the conformational stability and the biological function. Such a scarcity of information largely arises from the structural complexity of these proteins, which hinders the attainment of refined knowledge about their main physical and chemical properties, including the subunit structure, the association-dissociation equilibria, the primary and tertiary structure.

The use of photochemical techniques to elucidate some structure-function relationships in hemocyanins was pioneered by Wood and Bannister (1968), who studied the methylene blue-sensitized photooxidation of *Murex trunculus* hemocyanin. A more specific approach was described by Tallandini et al. (1975), using the copper-oxygen absorption bands as a built-in photochemical probe of the amino acid residues at the active site of *Octopus vulgaris* hemocyanin. In this paper, we show how the irradiation of *O. vulgaris* hemocyanin in the presence of proflavine as an external photosensitizer, under conditions allowing the preferential or even selective modification of the tryptophyl side chains, yields information on the conformational and functional role of this amino acid, as well as on some more general conformational features of the protein. The photooxidation data have been complemented by fluorescence emission studies, since this kind of spectroscopy can provide a detailed description of the protein environment of the tryptophyl residues (Longworth, 1971).

Experimental Section

Preparation of Hemocyanin Solutions. Hemocyanin from *Octopus vulgaris* was prepared according to the standard method detailed elsewhere (Ghiretti-Magaldi et al., 1966). The native or copper-free protein was dissolved at a concentration of 1 mg/ml in the following solvent systems: (1) 0.1 M Tris-acetate buffer, pH 6.0, plus 3 M urea; (2) 0.1 M Tris-acetate buffer, pH 6.0, plus 0.02 M $CaCl_2$.

Sedimentation Velocity Measurements. The aggregation state of hemocyanin under the aforesaid experimental conditions was controlled in the ultracentrifuge. The single experiments were carried out at 40,000 r.p.m. and at 20°C by means of a Beckman Model E analytical ultracentrifuge.

Amino Acid Analyses. Samples of native or photooxidized hemocyanin were dialyzed exhaustively against twice-distilled water. The protein was precipitated by means of acid acetone and exhaustively washed. The dye- and buffer-freed protein samples were hydrolyzed in deoxygenated sealed tubes in 6 M HCl at 110°C for 22 h; norleucine was added as an internal standard. Alkaline hydrolyses were performed in 3.75 M NaOH according to Jori et al. (1968). Analyses were carried out on a Carlo Erba 3A27 amino acid analyzer following the method of Moore and Stein (1963). The content of tryptophan was quantitated on the intact protein by the spectrophotofluorimetric method described by Genov and Jori (1973). To avoid artifacts arising from the effect of the protein conformation on the emission properties of the tryptophyl residues (Longworth, 1971), the protein samples were previously denatured by diluting 0.5 ml of the irradiated solution with 2.5 ml of double-crystallized 7 M guanidinium chloride. Control experiments showed that there was a linear relationship between protein concentration and the intensity of the 290 nm-excited emission.

Fluorescence Measurements. The fluorescence emission of the native or photooxidized hemocyanin was monitored by means of a Perkin Elmer 44 spectrophotofluorimeter, operated in the ratio mode, adopting the usual precautions to avoid optical artifacts (Longworth, 1971; Jori et al., 1974). The analyzed solutions were contained in 1 cm quartz cells, which were kept at 17.5°C by thermostated water circulating through the cell holder. The excitation monochromator was set at 290 nm, where the light absorption is exclusively due to the indole chromophores; at this wavelength, the optical density of the protein solutions was lower than 0.1 so that inner filter effects were negligible. In the denaturation experiments involving addition of urea, appropriate corrections were applied for volume changes and solvent emission. Quantum yields were evaluated according to Parker and Rees (1960), using a pH 7 aqueous solution of L-tryptophan (Q = 0.20 upon excitation at 290 nm) as a standard (Longworth, 1971).

Photooxidation Experiments. In a typical experiment, 5 ml of a pH 6 - buffered protein solution at a concentration of 1 mg/ml were added in the dark with 0.5 mg of proflavine and exposed to the light of four 300 watt tungsten bulbs. The experimental arrangement and the irradiation apparatus have been detailed elsewhere (Jori et al., 1974). During irradiation, a stream of purified oxygen from a capillary tube was slowly flushed through the protein solutions. The temperature was maintained at 18° ± 1°C by circulating water connected with a thermostated bath. At fixed intervals, suitable aliquots were taken for the spectrophotofluorimetric determination of tryptophan or other analyses.

Results

Homogeneity of 11S and 50S Hemocyanin. As pointed out by Salvato and Tallandini (Chap. 28, this vol.), in the presence of 3 M urea and at pH 8.0, *O. vulgaris* hemocyanin exhibits a single symmetrical sedimenting peak of about 8S. This value, when reported to the standard conditions, gives a value of $S^{o}_{20,w}$ = 11.5. In the presence of Ca^{2+} ions, *O. vulgaris* hemocyanin is fully aggregated into 50S components, with only very small peaks appearing at lower sedimentation coefficients. These conclusions were fully reproduced by us upon sedimentation measurements at pH 6.0.

Fluorescence Studies. The fluorescence emission spectra of native and copper-free hemocyanin, both in the 11S and in the 50S aggregation states, are shown in Figure 1. In all cases, the exciting wavelength was 290 nm, so that the observed emission can be safely ascribed to the tryptophyl fluorophores (Longworth, 1971). In this connection, the localization of the fluorescence maximum around 330 nm points out that the emission of *O. vulgaris* hemocyanin, predominantly occurs from tryptophyl side chains deeply buried in hydrophobic regions, although the presence of a shoulder located at about 340 nm suggests that a group of partially buried tryptophyl residues appreciably contributes to the overall emission (Burstein et al., 1973). As shown in Table 1, upon removal of copper from both the nonassociated and the associated

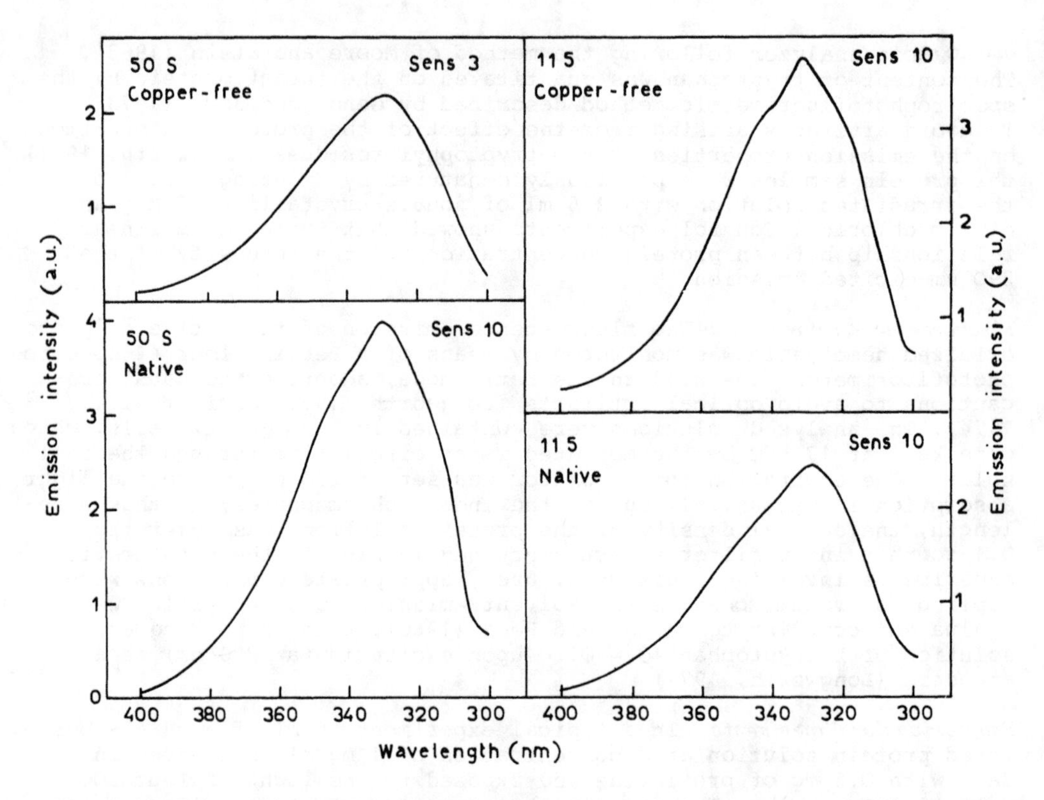

Fig. 1. Fluorescence emission spectra of native and copper-free *O. vulgaris* hemocyanin in the 11S and 50S aggregation state. The spectra were recorded at 17.5°C, in pH 6 buffered solutions (0.1 M Tris-acetate, added with 3 M urea for the 11S proteins and with 0.02 M $CaCl_2$ for the 50S proteins), using quartz cells of 1 cm optical path. The excitation monochromator was set at 290 nm and the emitted fluorescence was observed at 90° with respect to the incident beam. The spectra are uncorrected

Table 1. Some fluorescence properties of *Octopus vulgaris* hemocyanin

Sample	Irradiation	Emission λ maximum (nm)	Emission quantum yield
11S native	unirradiated	330	0.020
	photooxidized	332	0.018
11S copper-free	unirradiated	331	0.094
	photooxidized	332	0.047
50S native	unirradiated	328	0.016
	photooxidated	331	0.014
50S copper-free	unirradiated	331	0.063
	photoóxidized	334	0.044

All measurements were made at 17.5°C by exciting the protein solutions with 290 nm radiation. The quantum yields were calculated using L-tryptophan as a standard. The photooxidized products contained one modified tryptophyl residue.

protein, the emission quantum yield displayed about a four-fold increase, in good agreement with the findings of other authors with different hemocyanins (Shaklai and Daniel, 1970; Bannister and Wood, 1971). This enhancement reflects the quenching of tryptophan emission by the copper-oxygen system. On the other hand, the emission yield slightly increases on going from the associated to the nonassociated state.

Photooxidation Studies. The proflavine-sensitized photooxidation of *O. vulgaris* hemocyanin under different conditions (see Experimental section) provoked the selective modification of the tryptophyl side chains, as shown by amino acid analysis of extensively irradiated products. This result was expected, since at pH values below neutrality some highly photosensitive amino acid residues, such as histidine and tyrosine, are generally protected from photooxidative attack (Spikes and Macknight, 1970), while methionine, which is photoreactive throughout the entire pH range, is usually affected at a much lower rate than tryptophan (Benassi et al., 1967).

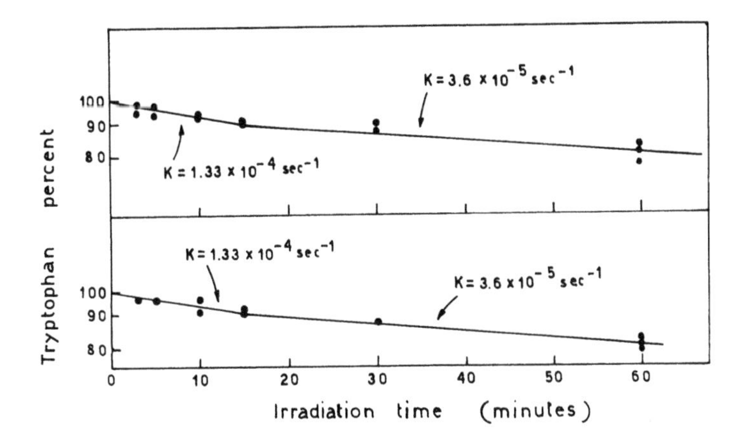

Fig. 2. Time-course of tryptophan photooxidation upon irradiation of 11S native hemocyanin (*upper plot*) and of 11S copper-free hemocyanin (*lower plot*) at 18°C with visible light in the presence of proflavine as a photosensitizer. The irradiated solutions were constituted by 5 ml of a pH 6 0.1 M Tris-acetate buffer, added with 3 M urea, containing 5 mg of protein and 0.5 mg of dye. The tryptophan content was estimated by diluting 0.5 ml of irradiated solution with 2.5 ml of 7 M guanidinium chloride and by estimating the fluorescence emitted by the resulting solution in the 300-420 nm range upon excitation at 290 nm

As usual for dye-sensitized photooxidation of proteins (Spikes and Macknight, 1970), the photoprocess followed a first-order kinetics with respect to tryptophan concentration. The semilogarithmic plots describing the time-course of tryptophan degradation are shown in Figures 2 and 3 for the 11S and, respectively, for the 50S proteins. With the exception of the native 50S hemocyanin, the plots are clearly biphasic as a consequence of the photooxidative destruction of a group of fast-reacting and a group of slow-reacting tryptophyl residues (Galiazzo et al., 1972). The pseudo-first order rate constants, as calculated from the slopes of the experimental plots, can be compared with the value $K = 0.83 \times 10^{-3}$/s, which was found for the photosensitized oxidation of tryptophan in oligopeptides and of tryptophans exposed at the surface of protein molecules (Genov and Jori, 1973). It appears that 0.5 tryptophyl residues per 50,000 daltons are preferentially affected in both the 11S native and copper-free protein samples, while 1.0 tryptophan is fast-reacting in the copper-free 50S protein. On the other hand, no preferential attack is apparent from the experi-

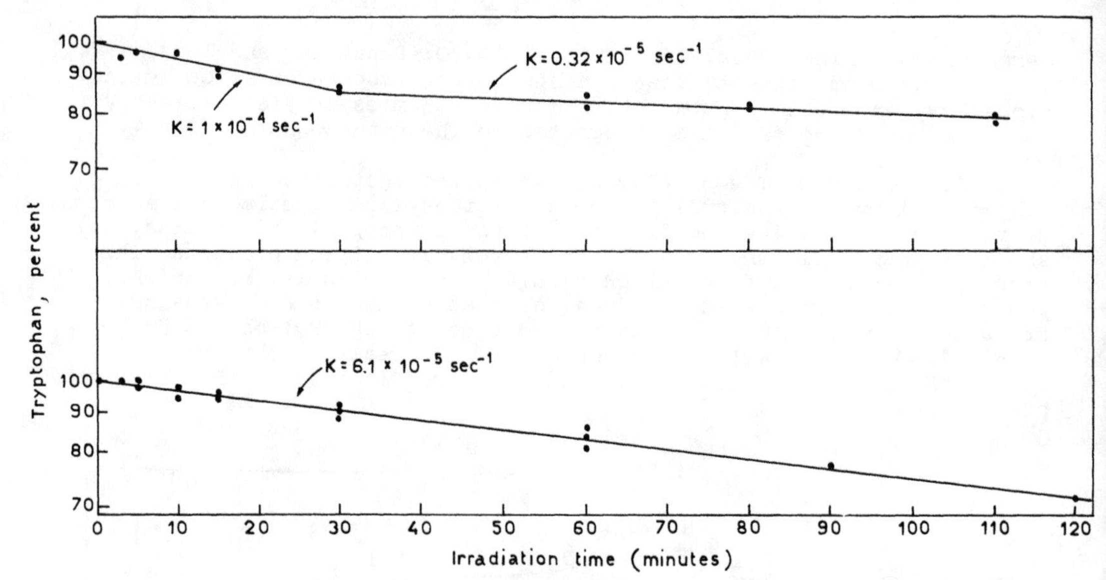

Fig. 3. Time-course of tryptophan photooxidation upon irradiation of 50S copper-free hemocyanin (*upper plot*) and of 50S native hemocyanin (*lower plot*) in pH 6 0.1 M Tris-acetate solution added with 0.02 M $CaCl_2$. Other conditions and procedures were the same as described for Figure 2

mental plot obtained for associated native hemocyanin, at least up to 120 min of irradiation, corresponding with the modification of 2.3 tryptophyl residues per 50,000 daltons. However, the photooxidation rate is significantly lowered.

Conformational Studies. As shown by ultracentrifugation studies the aggregation state of the photooxidized derivatives and of the unirradiated hemocyanins was the same.

Some fluorescence parameters of the various unirradiated and photooxidized products, containing one modified tryptophan per 50,000 daltons, are summarized in Table 1. There were only minor changes, if any, in the position of the emission maxima, whereas appreciable decreases of the emission quantum yield were observed after irradiation in the case of the copper-free proteins.

The effect of increasing urea concentration on the fluorescence emission energy yields of the unirradiated and photooxidized hemocyanins is shown in Figure 4 for the 11S products and in Figure 5 for the 50S products. It is worthwhile to remark that the emission maximum in 8 M urea of the unirradiated proteins, as well as of the photooxidized derivatives of the copper-containing 11S and 50S hemocyanins, was located between 330 nm and 340 nm, i.e. at wavelengths significantly lower than that typical of fully exposed tryptophyl residues, i.e. 355 nm (Longworth, 1971).

Discussion

The fluorescence emission data, obtained in the present investigation, suggest that the three-dimensional organization of *O. vulgaris* hemocyanin undergoes a limited destabilization both by dissociation and/or by removal of the copper ions. Actually, both the native and the copper-free 11S proteins are significantly more sensitive to the effect of

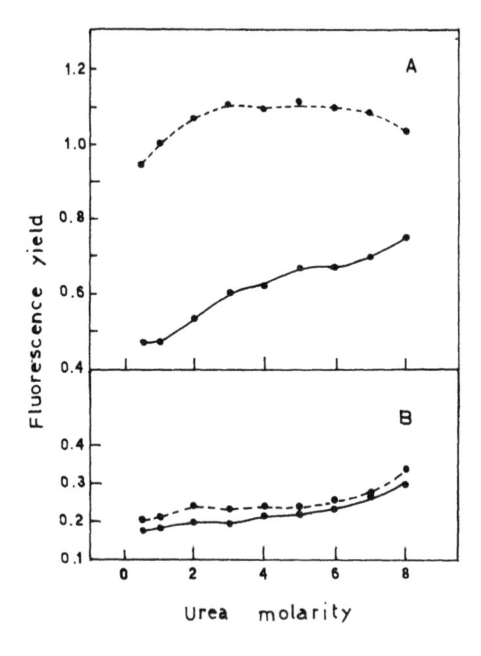

Fig. 4

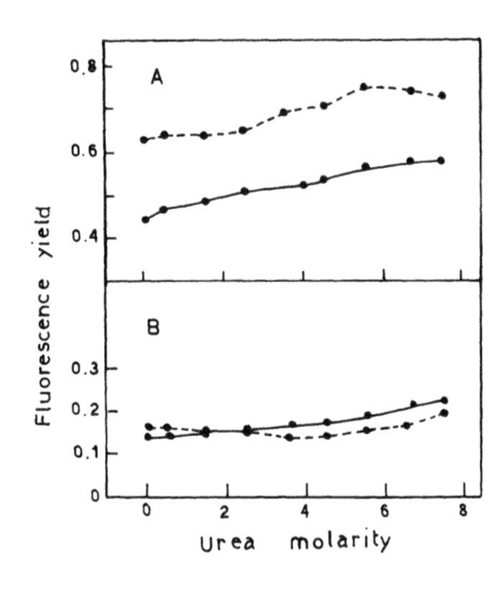

Fig. 5

Fig. 4A and B. The effect of urea concentration on the fluorescence quantum yield of copper-free hemocyanin (A) unirradiated .---.; 60 min - irradiated .——.; and of native hemocyanin (B) unirradiated .---.; 60 min - irradiated .——.

Fig. 5A and B. The effect of urea concentration on the fluorescence quantum yield of copper-free hemocyanin (A) unirradiated .---.; 60 min - irradiated .——.; and of native hemocyanin (B) unirradiated .---.; 60 min - irradiated .——.

urea concentration than the corresponding 50S proteins. Moreover, the influence of urea on the emission quantum yield is more pronounced for both 11S and 50S apohemocyanin. In any case, the presence of a large number of deeply buried tryptophyl moieties, which are not completely exposed to the aqueous solvent even by urea concentrations as high as 8 M, would demonstrate that the overall conformation of *O. vulgaris* hemocyanin is remarkably tight. The phenomenon is more evident for the native proteins; hence, the copper ions appear to exert an important structural role, including an higher stability to denaturing agents.

The photooxidation kinetic experiments provide the same line of evidence. Thus, in all the cases investigated, only a very limited number of tryptophyl side chains was preferentially attacked by the photooxidizing agent, the photooxidation rate constant being noticeably lower than that typical of readily accessible indole side chains. The identity of the photoreaction rate, observed for native and copper-free 11S hemocyanin, demonstrates that the low reactivity of most tryptophyl residues is not to be ascribed to protection by paramagnetic copper, and probably arises from a screening effect performed by the protein spatial arrangement. In particular, the greatly reduced rate constant estimated for the photooxidation of native 50S hemocyanin, and the absence of any preferentially reactive tryptophan, support the hypothesis that the overall conformation of this protein is significantly compact. Furthermore, the conformational differences between the associated and dissociated proteins appear to be spread over the entire molecular structure instead of being localized at specific sites.

It is likely that the preferentially modified tryptophyl residue plays an important role in stabilizing the tertiary structure of *O. vulgaris* hemocyanin. Actually, its photooxidation is paralleled by an enhanced sensitivity of the protein to the denaturing action of urea. Now, the phenomenon is less pronounced for the associated and for the copper-containing derivatives, which suggests that the conformational perturbation caused by the chemical modification of the tryptophyl side chain is somewhat counteracted by aggregation of the subunits, as well as by the presence of copper.

Interestingly, the photooxidation of one tryptophyl side chain results in a drastic depression of the emission quantum yield for both the copper-free proteins; hence, this residue should give a noticeable contribution to the overall fluorescence emission of *O. vulgaris* hemocyanin. Since no remarkable reduction of the fluorescence efficiency is observed for the photooxidized derivatives of the copper-containing protein, it appears reasonable to propose that the emission from the photosensitive residue is originally quenched in the native hemocyanin. Two hypotheses can account for this fact: (1) upon removal of copper, a subtle conformation change involves the molecular region including this residue, so that its distance and/or orientation from a quenching group is modified; (2) the tryptophyl residue is located near the copper-oxygen site. Shaklai and Daniel (1970) estimated a critical distance of 29.5 Å for energy transfer at the first excited singlet level from tryptophan to the copper-oxygen complex. In this connection, it would be of interest to investigate whether such a proximal tryptophyl residue has any functional role in the oxygenation process. Studies are planned in our laboratory in order to shed further light on this point.

References

Bannister, W.H., Wood, E.J.: Ultraviolet fluorescence of *Murex trunculus* haemocyanin in relation to the binding of copper and oxygen. Comp. Biochem. Physiol. 40B, 7-18 (1971)

Benassi, C.A., Scoffone, E., Galiazzo, G., Jori, G.: Proflavine-sensitized photooxidation of tryptophan and related peptides. Photochem. Photobiol. 6, 857 (1967)

Burstein, E.A., Vedenkina, N.S., Ivkova, M.N.: Fluorescence and the location of tryptophan residues in protein molecules. Photochem. Photobiol. 18, 263-279 (1973)

Galiazzo, G., Jori, G., Scoffone, E.: Dye-sensitized photooxidation as a tool for elucidating the topography of proteins in solution. In: Research Progress in Organic, Biological and Medicinal Chemistry. Gallo, L., Sanatamaria, L. (eds.). Amsterdam-London: North-Holland, 1972, Vol. III, part 1, pp. 137-154

Genov, N., Jori, G.: Conformational studies on the alkaline protease from *Bacillus mesentericus*. Intern. J. Protein. Res. 5, 127-133 (1973)

Ghiretti-Magaldi, A., Nuzzolo, C., Ghiretti, F.: Chemical studies on hemocyanins. I. Amino acid composition. Biochemistry 5, 1943-1951 (1966)

Jori, G., Folin, M., Gennari, G., Galiazzo, G., Buso, O.: Photosensitized oxidation of lanthanide ion-lysozyme complexes. A new approach to the evaluation of intramolecular distances in proteins. Photochem. Photobiol. 19, 419-434 (1974)

Jori, G., Galiazzo, G., Marzotto, A., Scoffone, E.: Dye-sensitized selective photooxidation of methionine. Biochem. Biophys. Acta 154, 1-9 (1968)

Longworth, J.: Luminescence of polypeptides and proteins. In: Excited States of Proteins and Nucleic Acids. Steiner, R.F., Weinry, I. (eds.). New York: Plenum Press, 1971, pp. 319-484

Moore, S., Stein, W.H.: Chromatographic determination of amino acids by the use of automatic recording equipment. Methods Enzymol. 6, 819-831 (1963)

Parker, C.A., Rees, W.T.: Correction of fluorescence spectra and measurement of fluorescence quantum efficiency. The Analyst 85, 587-600 (1960)

Shaklai, N., Daniel, E.: Fluorescence properties of haemocyanin from *Levantina hierosolima*. Biochemistry 9, 564-568 (1970)

Spikes, J.D., Macknight, M.L.: Dye-sensitized photooxidation of proteins. Ann. N. Y. Acad. Sci. 171, 149-162 (1970)
Tallandini, L., Salvato, B., Jori, G.: Photochemical effects associated with the copper absorption bands of the native hemocyanin from *Octopus vulgaris*. FEBS Lett. 54, 283-285 (1975)
Wood, E.J., Bannister, W.H.: Methylene blue-sensitized photooxidation of *Murex trunculus* hemocyanin. Biochem. Biophys. Acta 154, 10-16 (1968)

The Photooxidation of *Helix pomatia* Haemocyanin

M. De Ley and R. Lontie

Introduction

Photooxidation with or without a sensitizer has been widely used in order to gather structural information on the active site in enzymes. Methylene blue-sensitized photooxidation of *Murex trunculus* (Wood and Bannister, 1967, 1968) and *Helix pomatia* haemocyanin (Engelborghs et al., 1968) resulted in the parallel lowering of histidine content, oxygen-binding capacity, and absorbance at 346 nm, the latter being attributed to a charge transfer transition in the copper-oxygen complex. Simultaneously other amino acids were degraded. Using the copper-oxygen chromophore as a built-in photosensitizer of *Octopus vulgaris* haemocyanin the photodegradation was limited to histidine, albeit in a ratio of six residues per two copper atoms (Tallandini et al., 1975). The present paper describes the selective photooxidation of *H. pomatia* haemocyanin by irradiation in the 346 nm chromophore. It will be shown that even in the case of selective photooxidations caution should be applied in drawing conclusions.

Materials and Methods

Haemocyanin. H. pomatia haemocyanin was isolated according to Heirwegh et al. (1961). Its concentration was determined spectrophotometrically, A (0.1%, 1 cm, 278 nm) = 1.416 at pH 9.2. Prior to the experiments it was regenerated with hydrogen peroxide (R = 10) as described by Heirwegh et al. (1965). Copper-free haemocyanin was prepared by cyanide treatment. A solution of 1 M KCN (rendered 3 mM in $NH_2OH.HCl$ and 0.8 M in HOAc) was added drop-wise to a 3% haemocyanin solution in 0.1 M sodium acetate buffer, pH 5.6, until the blue colour completely vanished and finally an additional ml was added. This mixture was dialyzed against two 500 ml changes of 50 mM KCN rendered 10 mM in $Ca(OAc)_2$, 3 mM in $NH_2OH.HCl$ and 40 mM in HOAc. Cyanide and hydroxylamine were removed by extensive dialysis against 0.1 M sodium acetate buffer, pH 5.6. Copper-free haemocyanin was reconstituted by adding $K_3Cu(CN)_4$ (R_{Cu} = 1.5) in 0.1 M sodium acetate buffer, pH 5.6, under nitrogen atmosphere.

Absorption and Circular-Dichroic Spectra. Absorption spectra were measured with a Beckman DU (Munich, Germany) and a Beckman DB-G spectrophotometer (Fullerton, CA, USA). Circular-dichroic spectra were recorded with a Cary 61 spectropolarimeter (Monrovia, CA, USA).

Electron Paramagnetic Resonance. EPR spectra were recorded with an E-109 spectrometer (Varian, Palo Alto, CA, USA) at -170°C, microwave frequency 9.12 GHz with a power of 10 mW at g = 2 and 195 mW at g = 4, field modulation was 1 mT at a frequency of 100 kHz. The total scan time was 8 min for 0.2 T with a time constant of 0.5 s.

Haemocyanin samples for EPR measurements were concentrated in a preparative ultracentrifuge (Spinco model L, rotor 30, 3 h at 27,500 r.p.m. at 4°C) under conditions where whole molecules occur, i.e. in 0.1 M sodium acetate buffer, pH 5.6, or in 0.1 M borate-HCl buffer, pH 8.2, rendered 10 mM in $CaCl_2$. The sedimented haemocyanin was redissolved in a minimal amount of supernatant buffer.

Amino-Acid Analyses. Amino-acid analyses were carried out as described by Penke et al. (1974). About 3 mg haemocyanin in 200 µl 0.1 M sodium acetate buffer, pH 5.6, were hydrolyzed with 1 ml 3 N mercaptoethane-sulphonic acid (Pierce, Rockford, IL, USA) in an evacuated and sealed pyrex tube at 110°C for 24 h. After hydrolysis 2 ml 1 N NaOH were added to the contents of the tube and the volume brought to 5 ml with distilled water. Amino-acid analyses were carried out with a Beckman model 120 C amino-acid analyzer (Palo Alto, CA, USA) on 1 ml samples.

Copper Determinations. The copper content of photooxidized samples was determined spectrophotometrically with 2:9-dimethyl-1:10-phenanthroline (neocuproin) (UCB, Brussels, Belgium) according to McCurdy and Smith (1952). Two milliliters of a haemocyanin solution were mixed with 6 ml of the reagent solution containing 0.1 g neocuproin hydrochloride and 0.1 g $NH_2OH.HCl$ per 100 ml glacial acetic acid. After 2 h the absorbance at 454 nm was read against distilled water. The copper content was calculated from a calibration curve after subtraction of the suitable blanks.

Photooxidations. The photooxidations were carried out with a water-cooled super-high-pressure mercury-vapour lamp SP500W (Philips, Eindhoven, The Netherlands) at 8°C. In order to isolate the 365 nm spectral region, the light passes successively through an infrared filter, an ultraviolet filter (cooled with, and separated by a 1 cm layer of circulating water at 8°C), and a 1 cm layer of 90% saturated KNO_3 in water, held between two pyrex windows (each of them 3 mm thick). The haemocyanin solution (4 g/l) to be photooxidized was contained in a layer with 1 cm optical path, resulting in a transmission of less than 10% at 365 nm. In order to use the 546 and 578 nm spectral lines, the ultraviolet filter was replaced by a 475 nm cut-off filter.

Quantum-Yield Determinations. The quantum yield of the photochemical degradation during irradiation with light at 365 nm was determined with a chemical actinometer consisting of uranyl sulphate (0.1 M) and oxalic acid (50 mM) in a 10 cm cuvette. After irradiation the remaining oxalic acid was titrated at 50°C with 0.1 N $KMnO_4$, using sodium oxalate (p.a. according to Sorensen, Merck, Darmstadt, Germany) as a primary standard. The potentiometric titrations were carried out with a platinum and a glass electrode, the potential was followed with a Vibron electrometer (Electronic Instruments Ltd., Richmond, Surrey, England). Photooxidation of haemocyanin (0.4 g/l in 0.1 M borate-HCl buffer, pH 8.2) was carried out in the same 10 cm cuvette, the amount of degraded active sites was determined from the absorbance change at 346 nm.

Results

Absorbance and Circular-Dichroic Spectra. Irradiation of haemocyanin solutions (4.7 g/l) in 0.1 M sodium acetate buffer, pH 5.6, and in 0.1 M borate-HCl buffer, pH 8.2, resulted in a gradual decrease of the absorbance at 346 nm as a function of time, the reaction rates being comparable as evident from Figure 1. A comparison of the circular-dichroic spectra of native and photooxidized haemocyanin (69% remaining copper-band at 346 nm) in 0.1 M sodium acetate buffer, pH 5.6, shows the disappearance of all circular-dichroic bands as a result of photodegrations (Fig. 2).

Although the combined light intensity of the spectral lines at 546 and 578 nm was about twice that at 365 nm (following the diagram supplied

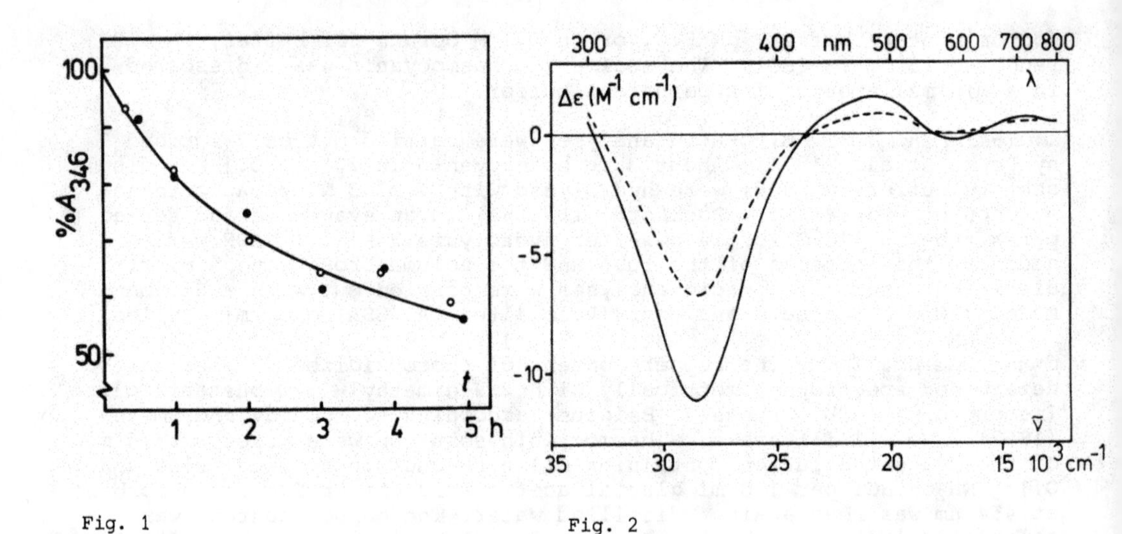

Fig. 1

Fig. 2

Fig. 1. Percentage remaining copper-band after photooxidation of *H. pomatia* haemocyanin (4.7 g/l) in 0.1 M sodium acetate buffer, pH 5.6 (•), and in 0.1 M borate-HCl buffer, pH 8.2 (o) as a function of the irradiation time, at $8^{o}C$ in a 1 cm cuvette

Fig. 2. Circular-dichroic spectra of native (——) and photooxidized (69% remaining copper-band at 346 nm) haemocyanin (3.71 g/l in 0.1 M sodium acetate buffer, pH 5.6) (---)

by the manufacturers), no photooxidation could be achieved by irradiation of a haemocyanin solution (17.6 g/l in 0.1 M borate-HCl buffer, pH 8.2) in a 2 cm optical path cuvette with visible light.

Quantum Yield. In order to determine the quantum yield of the above reaction at 365 nm, the following solutions were successively irradiated in a 10 cm cuvette: (a) a mixture containing 0.1 M uranyl sulphate and 0.05 M oxalic acid, (b) a haemocyanin solution (0.4 g/l) in 0.1 M borate-HCl buffer, pH 8.2, and (c) the same as (a). The degree of conversion was determined by potentiometric titration with $KMnO_4$ for (a) and (c), and by absorbance measurement at 346 nm for (b). From the known quantum yield of the chemical actinometer at 365 nm, i.e. 0.49 at $25^{o}C$ (Borell, 1973), the quantum yield of the photooxidation of haemocyanin could be determined as 1.6×10^{-4}.

Regeneration. A series of haemocyanin solutions (3.7 g/l in 0.1 M borate-HCl buffer, pH 8.2) was photooxidized to respectively 89, 82 and 65% of their original copper band. Reaction with hydrogen peroxide (R = 10) did not result in any regeneration of the copper bands. As the regeneration with hydrogen sulphide proceeds much faster (De Ley et al., 1975), the latter was also tried, with the same negative result however.

Amino-Acid Analyses. A 2% solution of haemocyanin in 0.1 M borate-HCl buffer, pH 8.2, was subjected to photooxidation in 600 µl aliquots in a 2 mm cuvette for different periods of time. As it was shown in preliminary experiments that none of the neutral nor acidic amino acids were degraded, analyses were limited to the basic amino acids. The sum of the arginine and lysine present was used as the internal standard, as these two amino acids cannot be degraded photochemically. From each sample 100 µl were diluted with 2 ml half-saturated sodium borate, pH 9.2, in order to determine the protein concentration at 278 nm and the copper band at 346 nm. Figure 3a shows the percentage of remaining

Fig. 3. Percentage remaining histidine (•) and tryptophan (o) after photooxidation of haemocyanin in 0.1 M borate-HCl buffer, pH 8.2 (a) and in 0.1 M sodium acetate buffer, pH 5.6 (b) as a function of the percentage remaining copper-band at 346 nm

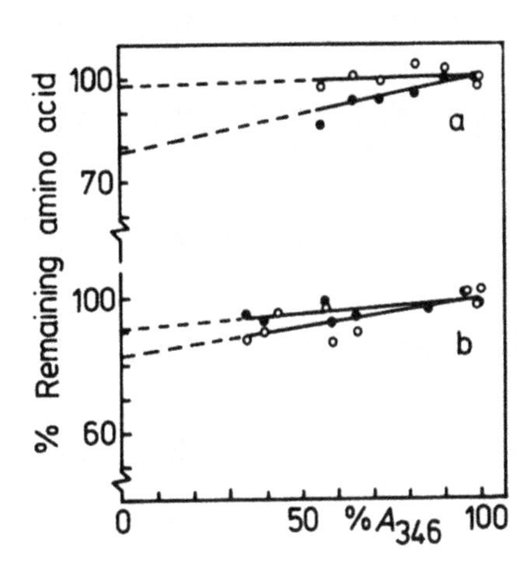

histidine and tryptophan plotted as a function of the percentage remaining copper band at 346 nm. By extrapolation (method of least squares) a value of 78% remaining histidines and 98% tryptophan residues upon complete photooxidation could be calculated, corresponding to a destruction of 4.8 (i.e. 5) histidine residues out of 22, and 0.2 tryptophan residues out of 11 per subunit of 55,000 daltons, i.e. per two copper atoms. This value was confirmed by the amino-acid analysis of a completely photooxidized haemocyanin sample in 0.1 M borate-HCl buffer, pH 8.2.

In a similar experiment haemocyanin (4 g/l in 0.1 M sodium acetate buffer, pH 5.6) was photooxidized in a 1 cm cuvette for different periods of time. The samples were concentrated by preparative ultracentrifugation; the sediment was redissolved in a minimal amount of the supernatant buffer. Amino-acid analyses were carried out on 2 mg samples. Figure 3b shows again the percentage remaining histidine and tryptophan as a function of percentage remaining copper-band at 346 nm, measured at pH 9.2. Extrapolation to zero percent copper band resulted in a value of 91% remaining histidine, 83% tryptophan, which points to the destruction of 2.1 (i.e. 2) histidine and 1.9 (i.e. 2) tryptophan residues.

Electron Paramagnetic Resonance. Haemocyanin solutions (4 g/l) in either 0.1 M sodium acetate buffer, pH 5.6, or 0.1 M borate-HCl buffer, pH 8.2, the latter rendered 10 mM in $CaCl_2$, were photooxidized for different periods of time in a 1 cm cuvette. The copper bands were determined after dilution with half-saturated sodium borate, pH 9.2 (rendered 20 mM in EDTA in order to remove Ca^{2+} for the measurements at pH 8.2). The haemocyanin solutions were then concentrated by preparative ultracentrifugation, the sediment was redissolved in a minimal amount of supernatant buffer, reaching a copper concentration of 2.9 mM. At both pH values a signal for Cu^{2+} was observed, the intensity of which increased with decreasing copper band (Fig. 4), the EPR parameters being $g_{\parallel} = 2.287$, $g_{\perp} = 2.073$ and $A_{\parallel} = 17.1$ mK for photooxidation carried out in 0.1 M sodium acetate buffer, pH 5.6, and $g_{\parallel} = 2.315$, $g_{\perp} = 2.071$ and $A_{\parallel} = 19.0$ mK in 0.1 M borate-HCl buffer, pH 8.2. A small signal at g = 4 was observed, slightly increasing with decreasing copper. The blank solution also showed a small EPR signal, to a much lesser extent however.

Photooxidation of Copper-Free Haemocyanin. Solutions of copper-free (0.03% of Cu w/w) haemocyanin (2.4 g/l in 0.1 M sodium acetate buffer, pH 5.6)

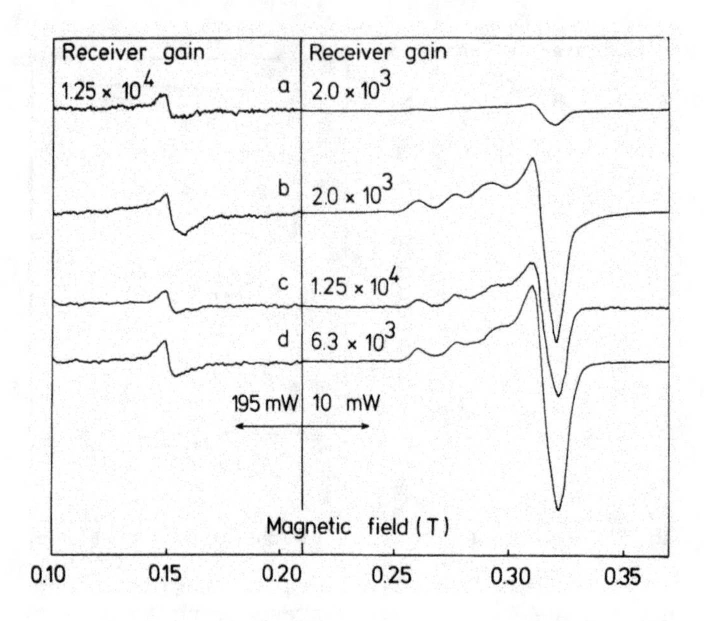

Fig. 4. Electron paramagnetic resonance spectra of native (*a*) and photooxidized (*b*, 55% remaining copper-band at 346 nm) haemocyanin in 0.1 M sodium acetate buffer, pH 5.6, and in 0.1 M borate-HCl buffer, pH 8.2, native (*c*) and photooxidized (*d*, 61% remaining copper-band at 346 nm). Time constant 0.5 s, scan time 8 min per 0.2 T, modulation amplitude of 1 mT, modulation frequency 100 kHz, temperature -170°C, microwave frequency 9.124 GHz, total copper concentration 2.9 mM

were photooxidized for respectively 4 and 7 h at 8°C in a 1 cm cuvette at 365 nm. They were then divided in two parts: the first one was concentrated by preparative ultracentrifugation and used for amino-acid analyses, the second one was reconstituted with $K_3Cu(CN)_4$ (R_{Cu} = 1.5) under a nitrogen atmosphere. The extent of reconstitution was determined by measuring the absorption spectrum of the solutions in equilibrium with the air after dilution with half-saturated sodium borate, pH 9.2. From the results summarized in Table 1, it can be concluded that both histidine and tryptophan residues were photooxidized and the maximum extent of reconstitution was lowered.

Table 1. Reconstitution of photooxidized copper-free haemocyanin as a function of remaining histidine and tryptophan residues

Duration of photooxidation	Percentage reconstitution	Number of histidines per 55,000 daltons	Number of tryptophans per 55,000 daltons
0	(100)	(22)	(11)
4 h	85	21.4	10.2
7 h	70	19.8	9.9

Copper Determinations. After photooxidation of haemocyanin (4 g/l in 0.1 M sodium acetate buffer, pH 5.6) at 8°C for different periods of time, the samples were treated with Chelex 100 (Bio Rad, Richmond, CA, USA). The copper content was determined as described in Materials and Methods. After dilution with half-saturated sodium borate, pH 9.2, the protein concentration was determined at 278 nm and the copper band at 346 nm.

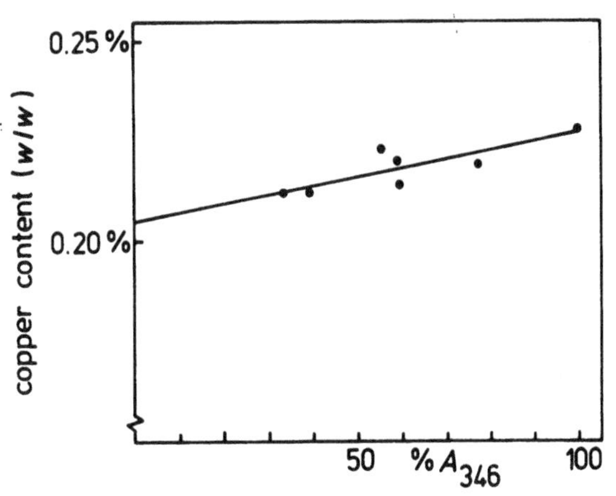

Fig. 5. Copper content of photooxidized haemocyanin (4 g/l in 0.1 M sodium acetate buffer, pH 5.6) after treatment with Chelex 100, as a function of percentage remaining copper-band at 346 nm. Extrapolation to zero copper-band yields a value of 0.205% Cu (w/w)

Figure 5 shows the copper content (w/w) of photooxidized haemocyanin plotted as a function of percentage remaining copper band at 346 nm. Extrapolation to zero copper band yields a value of 0.205% Cu (w/w) pointing to the loss of 10% of the bound copper upon complete photooxidation.

Photooxidation under Physiological Conditions. Fresh haemocyanin solutions (4 g/l in 0.1 M borate-HCl buffer, pH 8.2, rendered 6 mM in $CaCl_2$) were photooxidized to a different degree, the remaining copper band was determined after dilution with half-saturated sodium borate, pH 9.2 (rendered 10 mM in EDTA in order to remove Ca^{2+}). The original solutions were diluted with 0.1 M borate-HCl buffer, pH 8.2 (rendered 6 mM in $CaCl_2$), the oxygen dissociation curves were determined with a tonometer according to Pantin and Hogben (1925) with a Beckman DU spectrophotometer at 25°C. Figure 6 shows the Hill coefficient (n_H) and

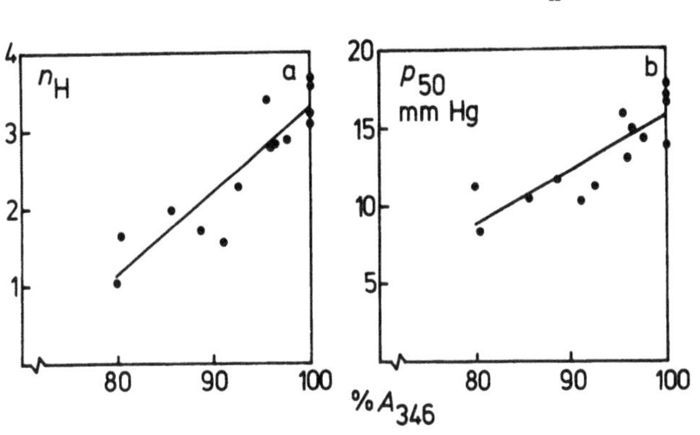

Fig. 6a and b. Hill coefficient n_H (a) and half-saturation pressure P_{50} (b) of haemocyanin photooxidized under physiological conditions (4 g/l in 0.1 M borate-HCl buffer, pH 8.2, rendered 6 mM in $CaCl_2$) as a function of percentage remaining copper-band at 346 nm

the half-saturation pressure (P 50) for the different haemocyanin samples plotted as a function of the percentage remaining copper band at 346 nm. During photooxidation the Hill coefficient dropped from 3.4 at 100% copper band to 1.0 at 80%, corresponding to a decrease in half-saturation pressure from 16 to 8 mm Hg.

Discussion

Irradiation of *H. pomatia* haemocyanin with light at 365 nm results in the destruction of the oxygen-binding sites, as shown by a decrease of the absorbance at 346 nm, and a disappearance of all the bands in the circular-dichroic spectrum. Although the photooxidation rates at pH 5.6 and 8.2 are equal, the mechanisms are distinctly different, two histidine and two tryptophan residues being destroyed at pH 5.6, five histidine residues at pH 8.2, per domain of 55,000 daltons, i.e. per two copper atoms. Simultaneously an EPR signal for Cu^{2+} arises, demonstrating the oxidation and partial or complete decoupling of the copper pairs of native haemocyanin. Only 10% of the copper atoms in completely photooxidized haemocyanin become available to Chelex 100.

Although the photochemical irradiation is carried out directly in the absorption band at 346 nm, the activation of oxygen in the active site itself seems highly improbable for several reasons: (1) the quantum yield for the photochemical degradation is extremely low, viz. 1.6×10^{-4}. For its calculation the absorbance due to copper-oxygen complexes was taken into account. If, however, the photochemically active complex absorbs to a much lesser extent (e.g. 10^3 times) and its absorption band is thus hidden under the copper band, the quantum yield could be multiplied by this factor, bringing it to a more realistic value. (2) For oxygen to be excited in the active site, the stoicheiometry of the destruction of histidine is unusually high. It is improbable that after the oxidation of one or two histidine residues the characteristic absorbance at 346 nm, necessary for the further activation of oxygen, would still exist. (3) Although the light intensity in the visible region was about twice that at 365 nm, and the energy per quantum of light at 546 and 578 nm is still sufficiently high to produce singlet oxygen, no photooxidation could be achieved by irradiation with visible light. (4) Copper-free haemocyanin, though completely missing the copper-oxygen absorption band at 346 nm, could be photooxidized by irradiation at 365 nm, resulting in the destruction of histidine and tryptophan residues and in a decrease of the extent of reconstitution with $K_3Cu(CN)_4$.

As a possible mechanism for the photooxidation of haemocyanin, we propose an activation of oxygen by a chromophore near the active site, which absorbs light around 365 nm and is not degraded during the reaction. With respect to this, a charge-transfer complex between oxygen and tryptophan, absorbing between 350 and 400 nm, was described (Slifkin, 1971). The mechanism would then be:

$$\mathrm{Trp} + O_2 \rightleftharpoons \mathrm{Trp.}\ O_2$$

$$\mathrm{Trp.}\ O_2 \xrightarrow{h\nu} \mathrm{Trp} + O_2^*$$

$$\mathrm{His} + O_2^* \longrightarrow \mathrm{His}_{ox}$$

$$\mathrm{Trp} + O_2^* \longrightarrow \mathrm{Trp}_{ox} \text{ (only at pH 5.6)}$$

Haemocyanin was photooxidized under physiological conditions resulting in a loss of cooperativity, the latter being complete when still 80% of the oxygen binding sites were intact.

Acknowledgments. We are grateful to the National Fonds voor Wetenschappelijk Onderzoek for research grants and for a fellowship (M.D.L.). This work was supported by the Fonds voor Collectief Fundamenteel Onderzoek (Contract Nr. 2.0016.76) and the Fonds Derde Cyclus, Katholieke Universiteit te Leuven. We are grateful to Mr. J. Pauwels and Mr. G. Goemans for the construction of the photooxidation apparatus and to M.J. Rycke-boer for carrying out the amino-acid analyses.

References

Borrell, P.: Photochemistry: A Primer. In: Studies in Chemistry. Stokes, B.J., Malpas, A.J. (eds.). London: Arnold, 1973, p. 30

De Ley, M., Candreva, F., Witters, R., Lontie, R.: The fast reduction of *Helix pomatia* methaemocyanin with hydrogen sulphide. FEBS Lett. 57, 234-236 (1975)

Engelborghs, Y., Witters, R., Lontie, R.: Photooxidation of *Helix pomatia* haemocyanin. Arch. Intern. Physiol. Biochim. 76, 372-373 (1968)

Heirwegh, K., Blaton, V., Lontie, R.: The regeneration of the copper bands of aged *Helix pomatia* haemocyanin with hydrogen peroxide or cysteine. Arch. Intern. Physiol. Biochim. 73, 149-150 (1965)

Heirwegh, K., Borginon, H., Lontie, R.: Separation and absorption spectra of α- and β-haemocyanin of *Helix pomatia*. Biochim. Biophys. Acta 48, 517-526 (1961)

McCurdy, W.H., Smith, G.F.: 1:10 Phenanthroline and mono-, di-, tri-, and tetramethyl-1:10-phenanthrolines as chelated copper complex cations. The Analyst 77, 846-857 (1952)

Pantin, C.F.A., Hogben, L.T.: A colorimetric method for studying the dissociation of oxyhaemocyanin suitable for class work. J. Mar. Biol. Assoc. U.K. 13, 970-980 (1925)

Penke, B., Ferenczi, R., Kovacs, K.: A new acid hydrolysis method for determining tryptophan in peptides and proteins. Anal. Biochem. 60, 45-50 (1974)

Slifkin, M.A.: Charge Transfer Interactions in Biomolecules. London-New York: Academic Press, 1971, p. 54

Tallandini, L., Salvato, B., Jori, G.: Photochemical effects associated with the copper absorption bands of the native haemocyanin from *Octopus vulgaris*. FEBS Lett. 54, 283-285 (1975)

Wood, E.J., Bannister, W.H.: Photooxidation of haemocyanin in the presence of methylene blue. Biochem. J. 104, 42P (1967)

Wood, E.J., Bannister, W.H.: The effect of photooxidation and histidine reagents on *Murex trunculus* haemocyanin. Biochim. Biophys. Acta 154, 10-16 (1968)

X-Ray Photoelectron Spectroscopic Studies of Hemocyanin and Superoxide Dismutase

H. VAN DER DEEN, R. VAN DRIEL, A. H. JONKMAN-BEUKER, G. A. SAWATZKY, AND R. WEVER

Summary

X-ray photoelectron spectra of the $2p_{3/2}$ and 3p levels of Cu and Zn in superoxide dismutase and Cu in hemocyanin have been recorded and compared. From the intensity ratios of the $2p_{3/2}$ and $3p_{3/2}$ peaks in the spectra, it is concluded that Cu is located at the surface of superoxide dismutase, whereas Zn must be bound more inside the protein. This confirms conclusions of earlier X-ray photoelectron spectroscopic studies of the Cu and Zn $2p_{3/2}$ levels in this metallo-protein. Because the $2p_{3/2}$ and $3p_{3/2}$ binding energies of Cu in hemocyanin and superoxide dismutase are nearly equal, the experiments suggest that the state of copper in oxyhemocyanin is cupric.

Introduction

X-ray photoelectron spectroscopy involves the determination of the kinetic energy distribution of electrons emitted from X-ray-irradiated compounds. The binding energies of inner shell electrons can be calculated from the difference between the kinetic energy of the emitted photoelectrons and the energy of the exciting radiation. From these binding energies, conclusions can be drawn concerning the charge, oxidation state and the chemical environment of an atom. The books by Siegbahn et al. (1967, 1969) give a general introduction in this field.

Applications of X-ray photoelectron spectoscopy (XPS) are described and discussed in recent reviews (Jolly, 1974; Hercules and Carver, 1974 and Siegbahn, 1974). A growing number of biochemical applications of this spectroscopic technique has started to appear. Most of these applications were studies of metal ions bound to proteins or amino acids. For example XPS was used in the study of non-heme iron proteins (Kramer and Klein, 1972), superoxide dismutase (Jung et al., 1973) and metallothionein (Sokolowski et al., 1974).

XPS is a convenient method for the investigation of surfaces. Photoelectrons which are liberated at a depth of more than about 100 Å are scattered and absorbed by the sample. The mean escape depths of photoelectrons for various materials indicate that these depths are very sensitive to the surface composition and the kinetic energy of these electrons (Jolly, 1974). $3p_{3/2}$ level electrons have a much higher kinetic energy after their liberation than the $2p_{3/2}$ level electrons, since their binding energy is much smaller. For this reason $3p_{3/2}$ electrons, when compared to $2p_{3/2}$ electrons, are less sensitive to scattering and absorbing by the sample. Therefore, in this study $2p_{3/2}$ / $3p_{3/2}$ intensity ratios of metal ions in proteins are used as an indication of the distance of the ions to the surface of the protein. For the present study the metalloproteins superoxide dismutase and hemocyanin were used for several reasons.

Superoxide dismutase from cytosol of eukaryotic organisms is an enzyme, which contains 2 mol Cu and 2 mol Zn per MW of 32,600. The protein has a high enzymatic activity, in which Cu is involved while Zn only plays a structural role (Weser, 1973 and Fridovich, 1974). The crystal structure shows at 3 Å resolution that Cu and Zn have a common histidine ligand and are about 6 Å apart (Richardson et al., 1975). From the intensities of the X-ray photoelectron spectra of the Cu and Zn $2p_{3/2}$ levels of superoxide dismutase, Jung et al. (1973) concluded that Cu is located at the surface of this metalloprotein, whereas Zn must be bound more inside the molecule. We have extended the measurements to the Cu and Zn 3p level electrons, since these core electrons, as mentioned before, are less sensitive to the depth at which metal ions are buried in the protein.

Direct and indirect experimental evidence suggests that in deoxyhemocyanin the copper is completely in the cuprous state, while in oxyhemocyanin a pair of closely linked Cu(II) ions is present in the active site (Moss et al., 1973; Lontie and Vanquickenborne, 1974 and Freedman et al., 1976). The aim of our X-ray photoelectron spectroscopic study is to gather more direct information about the active site and the valence state of copper in hemocyanin by comparing the spectoscopic data of hemocyanin with data of superoxide dismutase.

Materials and Methods

Chemicals. All chemicals were reagent grade and were used without further purification. Dried bovine serum albumin (demineralized) was bought from Povite (Amsterdam, Holland).

Hemocyanin. α-hemocyanin was isolated from Roman snails (*Helix pomatia*) according to Heirwegh et al. (1961), as modified by Konings et al. (1969) and Siezen and van Driel (1973). The protein was lyophilized in the presence of sucrose and stored at -20°C. Before the protein was used for our experiments, the sucrose was removed by dialysis against deionized water, to which a small amount of inorganic phosphate was added in order to keep the protein in solution. After this procedure the protein was lyophilized.

Superoxide Dismutase. Superoxide dismutase was isolated from bovine erythrocytes according to McCord and Fridovich (1969). The chromatographic purification steps during isolation of the enzyme were carried out in potassium phosphate buffer (pH 7.2). Cu and Zn contents were measured by atomic absorption spectroscopy. The data obtained with a Perkin Elmer 403 Atomic absorption spectrophotometer gave a Cu/Zn ratio of 0.99. After dialysis against distilled water (16 h, 4°C) the enzyme was lyophilized.

Recording of X-Ray Photoelectron Spectra. Photoelectron spectra were collected using an AEI-ES 200 spectrometer using Mg-K_α or Al-K_α radiation. The energy of the exciting X-rays was 1253.6 and 1486.6 eV, respectively and a X-ray source of 12 kV and 15 mA was used. The samples were fixed on cellotape. The samples were maintained at a temperature of about -25°C and the vacuum in the sample chamber was around 10^{-8} torr.

Only the most intense and characteristic signals, e.g. 2p and 3p for Cu and Zn and 1s for C and O were selected for comparisons. For superoxide dismutase, data of about 70 scans, with a sweep time of 20 min, were collected with a PDP-8 computer. For hemocyanin 100 scans of 10 min each were taken, while for the calibration experiments with Ag powder fewer scans were sufficient.

When a sample in contact with the spectrometer is irradiated with X-rays, the ejection of electrons can cause the sample to have excess positive charge if the conductivity of the sample is inadequate to permit electrical equilibration. The binding energies were corrected for these "charging up effects" by using the C 1s line of the aliphatic carbon atoms of the samples. These lines were calibrated with the $3d_{5/2}$ at 367.8 eV of Ag powder, which was mixed with the sample and fixed with layers of different thickness on cellotape. The C 1s line of cellotape was found at 284.0 eV when measured with Ag powder attached to it, while the C 1s line of the proteins, including bovine serum albumin was found at 283.3 eV.

Analysis of X-Ray Photoelectron Spectra. Most of the spectra were corrected for the energy dependent transmission of the analyzer, scattered electrons and X-ray satellites using a computer program developed for this purpose (Antonides, E., and G.A. Sawatzky, unpublished). The same computer program was used for a fit of the spectrum with Lorentzian and Gaussian line components and was able to calculate peak intensities, positions, etc.

The $2p_{1/2}$ and $2p_{3/2}$ peaks of Cu and Zn are well separated. The binding energy differences (B.E. $2p_{1/2}$ -B.E. $2p_{3/2}$) were about 20 eV for Cu as well as for Zn (E. Antonides, unpublished). On the other hand the $3p_{3/2}$ and $3p_{1/2}$ bands of both Cu and Zn have been found to overlap somewhat. The binding energy differences (B.E. $3p_{1/2}$ -B.E. $3p_{3/2}$) were 1.8 eV and 2.5 eV for Cu and Zn, respectively (E. Antonides, unpublished). Therefore, during the best fit simulation of the 3p peak the position of the $3p_{1/2}$ was coupled with the above mentioned energy difference to the $3p_{3/2}$ position. In addition the intensity of the $3p_{3/2}$ band was taken twice as large as the $3p_{1/2}$ band intensity.

Results and Discussion

The X-ray photoelectron spectra of the Cu and Zn 3p and $2p_{3/2}$ levels of superoxide dismutase are shown in Figures 1 and 2, respectively. About 70 scans were taken in order to get a good signal-to-noise ratio. However no significant changes in peak positions are observed between the 10th and 70th scan.

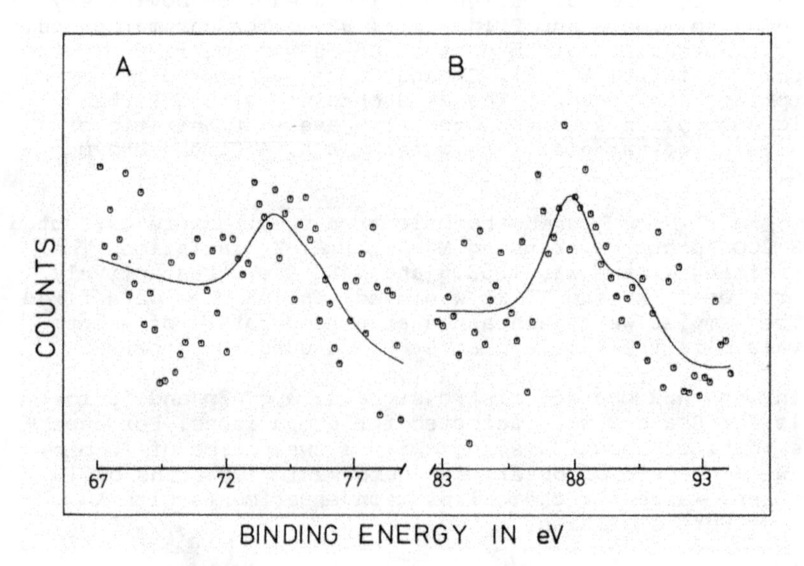

Fig. 1. X-ray photoelectron spectrum of the Cu (*A*) and Zn (*B*) 3p levels of superoxide dismutase

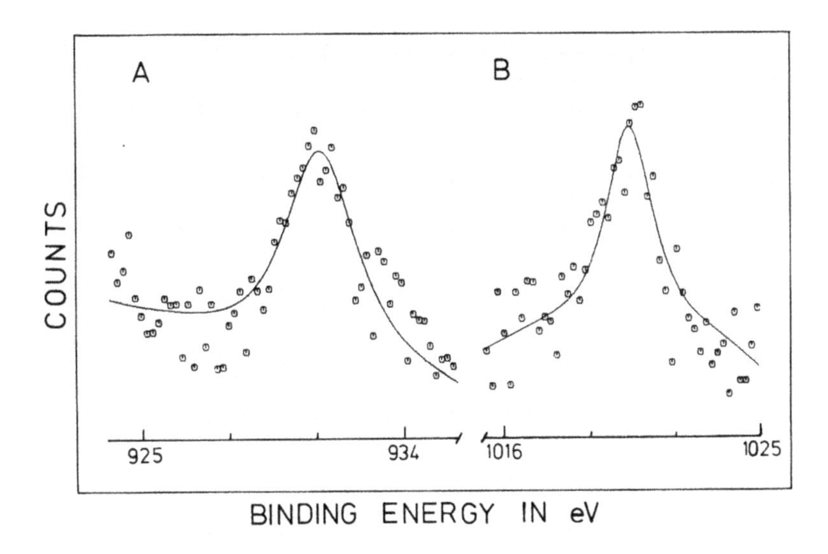

Fig. 2. X-ray photoelectron spectrum of the Cu (*A*) and Zn (*B*) $2p_{3/2}$ levels of superoxide dismutase

Table 1. Binding energies of the $2p_{3/2}$ and $3p_{3/2}$ levels of Cu and Zn in superoxide dismutase and Cu in *Helix pomatia* hemocyanin

Proteins	Metal	Binding energies in eV	
		$2p_{3/2}$	$3p_{3/2}$ [a]
Superoxide dismutase	Cu	931.1 ± 0.1	73.7 ± 0.2
	Zn	1020.6 ± 0.1	87.6 ± 0.2
Hemocyanin	Cu	931.0 ± 0.1	73.4 ± 0.2

[a] These positions have been calculated as described in the section Materials and Methods.

The peak positions which are calculated, using the C 1s line at 283.3 eV as reference, are summarized in Table 1. The binding energy of the Cu $2p_{3/2}$ level electrons is equivalent to the value found by Jung et al. (1973) at 931.9 eV, using the C 1s line of superoxide dismutase at 284 eV as internal standard. However, the binding energy of the Zn $2p_{3/2}$ level electrons is found almost 1 eV higher than reported by Jung et al. (1973) for superoxide dismutase, but is equivalent to their values on several Zn(II) complexes. As pointed out before, information about the relative location of the two metals in the protein can be obtained from the ratio of intensities of the $2p_{3/2}$ signals of Cu and Zn. Jung et al. (1973) found that in superoxide dismutase the $2p_{3/2}$ peak intensity of Cu was 2.8 times larger than $2p_{3/2}$ peak intensity of Zn, whereas the intensities of the Cu $2p_{3/2}$ signals of amino acid complexes were comparable with the intensities of the Zn $2p_{3/2}$ signals of the corresponding complexes. They concluded, therefore, that Cu should be more at the surface of the protein, whereas Zn should be more buried. The ratio of intensities of the $2p_{3/2}$ peaks of Cu and Zn found in this study is 1.8, a value somewhat lower than the value reported by Jung et al. (1973). Experimental errors are not sufficient to explain this difference completely. Part of the difference can be explained by assuming that in this study the surface of superoxide dismutase is covered with more layers of water than in their study.

A value of 0.9 was found for the ratio of intensities of the $3p_{3/2}$ signals of Cu and Zn in superoxide dismutase, which is close to the Cu/Zn ratio (0.99), as established with atomic absorption spectroscopy. If the difference in the $2p_{3/2}$ signal intensities of Cu and Zn is not only caused by different scattering behaviors of the two metal-binding sites, this would indeed have been expected.

In this study we will use, as described in the introduction, the $2p_{3/2}$/$3p_{3/2}$ intensity ratios of metal ions in proteins as an indication of the distance of this ion to the surface of the protein. The intensity ratio of the $2p_{3/2}$ and $3p_{3/2}$ peaks of Cu is twice that of Zn, as can be seen in Table 2. Thus, it can be concluded in line with the tentative con-

Table 2. Intensity[a] ratios of the $2p_{3/2}$ and $3p_{3/2}$ levels of Cu and Zn in superoxide dismutase and Cu in *Helix pomatia* hemocyanin

Protein	Metal	Intensity $2p_{3/2}$ / Intensity $3p_{3/2}$
Superoxide dismutase	Cu	12 ± 2
	Zn	6 ± 1
Hemocyanin	Cu	8 ± 2

[a]The intensities have been calculated from the computer fits of the spectra.

clusion of Jung et al. (1973) that Cu is situated closer to the surface than Zn. This conclusion is also supported by the observation of the high enzymatic activity of the protein in which Cu is only involved, whereas Zn plays a structural role (Weser, 1973 and Fridovich, 1974). The high enzymatic activity suggests a highly accessible copper-binding site.

Since the crystal structure of bovine superoxide dismutase is resolved at 3 Å resolution (Richardson et al., 1975) a comparison can be made with the XPS data. Although the amino acid side groups positions are not yet exactly identified in the crystal structure, the electron density maps at 3 Å resolution show that Cu has accessibility, through a large hole from the outside, whereas Zn, although close to Cu, is much more difficult to reach from the outside of the protein (Richardson, D.C., personal communication). Therefore, the XPS data are in good agreement with the crystal structure. However, before an exact comparison can be made calculations of the accessibility of Cu and Zn for water molecules should be made. The envelope of the protein, which is defined in this way, (van der Waal's surface) can be calculated according to, for example, Lee and Richards (1971) and should be compared with the XPS data.

The hemocyanin sample was prepared by lyophilizing oxyhemocyanin and was after this procedure still slightly blue in color. The deoxy hemocyanin sample prepared at the same time stayed uncoloured during this procedure. Although there could be different explanations for this blue color, we think that during the lyophilizing procedure oxyhemocyanin is fixated in its conformation and does not lose its oxygen. The possibility that the protein through this confirmation fixation stays oxyhemocyanin, even during the low pressures of the XPS measurements, seems therefore reasonable.

In Figure 3 the Cu $2p_{1/2}$ and $2p_{3/2}$ X-ray photoelectron spectra of *Helix pomatia* hemocyanin are depicted, while Figure 4 shows the spectrum of

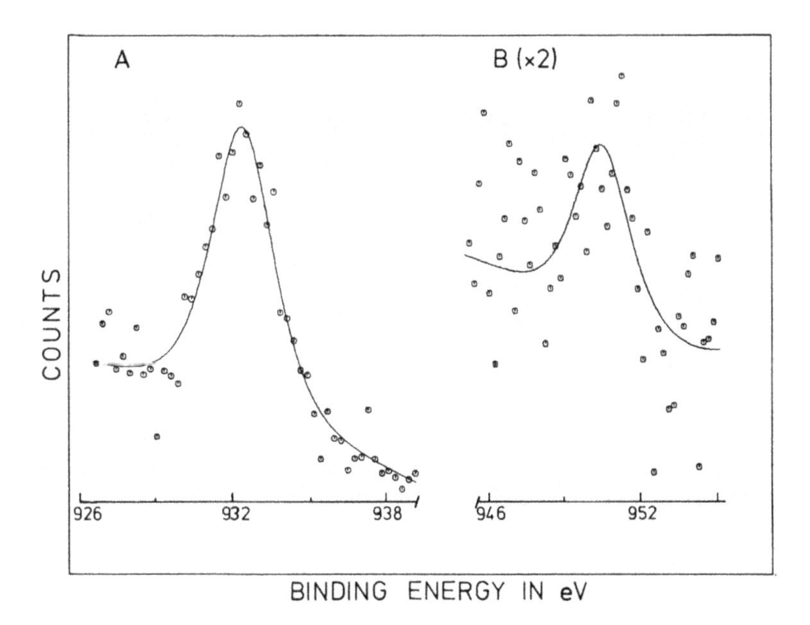

Fig. 3. X-ray photoelectron spectrum of the Cu $2p_{3/2}$ (*A*) and $2p_{1/2}$ (*B*) level of *Helix pomatia*

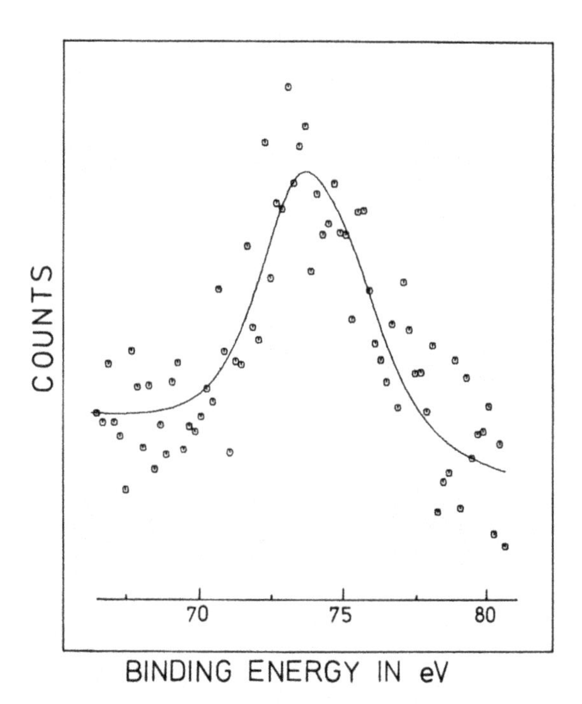

Fig. 4. X-ray photoelectron spectrum of the Cu 3p level of *Helix pomatia* hemocyanin

the Cu 3p level. The peak positions, using the C 1s line at 283.3 eV as internal standard, are summarized in Table 1. The peak positions of the $2p_{1/2}$ is found at 950.5 eV, which means that this level is separated by 20.5 eV from the $2p_{3/2}$ level. The values for the $2p_{3/2}$ (931.0 eV) and $3p_{3/2}$ (73.4 eV) Cu peak positions in hemocyanin are very similar to the values for Cu in superoxide dismutase, namely 931.1 and 73.7 eV, respectively.

Before one concludes from the similarity in these data that copper in oxyhemocyanin is in the cupric state, one has to be aware of some problems. First, one has to consider that Cu $2p_{3/2}$ peak positions in Cu_2O and CuO powders are found at 931.9 and 932.9 eV, respectively, by Robert et al. (1972). This difference in binding energy is much smaller than the one found in a row of Cu(II) amino acid complexes by Jung et al. (1973). Most of the binding energies of the $2p_{3/2}$ levels of these Cu(II) compounds were found at 934 eV, whereas in $Cu(Asp)_2$ two binding energies at 934.1 and 932.0 eV were found.

It is, therefore, impossible to state at this point, without knowledge of the binding energies of the Cu $2p_{3/2}$ levels in dimeric cuprous and cupric complexes, that the copper in our hemocyanin sample could only be in the cupric state. Different forms of this protein, as for example nitrosyl-hemocyanin, which contains half of its copper in the cupric form and the other half in the cuprous form (Schoot Uiterkamp, 1972), should also be studied and compared with the results of dimeric copper complexes before a final statement can be made.

Table 2 tabulates the intensity ratio of the Cu $2p_{3/2}$ and $3p_{3/2}$ peaks in hemocyanin. This value is close to the intensity ratio of Zn in superoxide dismutase. However, this gives us no indication about the depth of Cu in hemocyanin compared to Zn in superoxide dismutase. As discussed before, the number of water layers on the surface of the lyophilized proteins may differ and this can greatly influence the intensity ratios. What is clearly needed are XPS studies in which molecules of water or of a different compound are attached in monolayers, bilayers, etc. on the surface of a protein. In this way more indications can be obtained for the distance of a metal ion to the envelope of the protein.

Acknowledgments. We would like to thank Dr. D.C. Richardson for stimulating discussions and making available to us data of the crystal structure of superoxide dismutase prior to publication. We are also indebted to Drs. E. Antonides for his help with the computations and Ir. A. Heeres for running many of the spectra.

This investigation was supported by the Netherlands Foundation for Chemical Research (S.O.N.) and the Netherlands Foundation for Biophysics, both with financial aid from the Netherlands Organization for the Advancement of Pure Research (Z.W.O.).

References

Freedman, T.B., Loehr, J.S., Loehr, T.M.: A resonance Raman study of the copper protein, hemocyanin. New evidence for the structure of the oxygen-binding site. J. Am. Chem. Soc. 98, 2809-2815 (1976)

Fridovich, I.: Superoxide dismutases. Advan. Enzymol. 41, 35-97 (1974)

Hercules, D.M., Carver, J.C.: Electron spectroscopy: X-ray and electron excitation. Anal. Chem. 46, 133R-150R (1974)

Heirwegh, K., Borginon, H., Lontie, R.: Separation and absorption spectra of α- and β-hemocyanin of *Helix pomatia*. Biochim. Biophys. Acta 48, 517-526 (1961)

Jolly, W.L.: The application of X-ray photoelectron spectroscopy to inorganic chemistry. Coord. Chem. Rev. 12, 47-81 (1974)

Jung, G., Ottnad, M., Bohnenkamp, W., Bremser, W., Weser, U.: X-ray photoelectron spectroscopic studies of copper- and zinc- amino acid complexes and superoxide dismutase. Biochim. Biophys. Acta 295, 77-86 (1973)

Konings, W.N., Van Driel, R., Van Bruggen, E.F.J., Gruber, M.: Structure and properties of hemocyanins V. Binding of oxygen and copper in *Helix pomatia* hemocyanin. Biochim. Biophys. Acta 194, 55-66 (1969)

Kramer, L.N., Klein, M.P.: XPS of non-heme iron proteins. In: Electron Spectroscopy. Shirley, D.A. (ed.). Amsterdam: North-Holland, 1972, pp. 733-751

Lee, B., Richards, F.M.: The interpretation of protein structure: Estimation of static accessibility. J. Mol. Biol. 55, 379-400 (1971)

Lontie, R., Vanquickenborne, L.: The Role of Copper in Hemocyanins. In: Metal Ions in Biological Systems. Sigel, H. (ed.). New York: Marcel Dekker, 1974, pp. 183-200

McCord, J.M., Fridovich, I.: Superoxide dismutase. An enzymic function for erythrocuprein (hemocuprein). J. Biol. Chem. 244, 6049-6055 (1969)

Moss, T.H., Gould, D.C., Ehrenberg, A., Loehr, J.S., Mason, H.S.: Magnetic properties of *Cancer magister* hemocyanin. Biochemistry 12, 2444-2448 (1973)

Richardson, J.S., Thomas, K.A., Rubin, B.H., Richardson, D.C.: Crystal structure of bovine Cu, Zn superoxide dismutase at 3 Å resolution: Chain tracing and metal ligands. Proc. Natl. Acad. Sci. 72, 1349-1353 (1975)

Robert, T., Bartel, M., Offergeld, G.: Characterization of oxygen species absorbed on copper and nickel oxides by X-ray photoelectron spectroscopy. Surf. Sci. 33, 123-130 (1972)

Schoot Uiterkamp, A.J.M.: Monomer and magnetic dipole-coupled Cu^{2+} EPR signals in nitrosylhemocyanin. FEBS Lett. 20, 93-96 (1972)

Siegbahn, K.: Electron spectroscopy - an outlook. In: Electron Spectroscopy, Progress in Research and Applications. Caudano, R., Verbist, J. (eds.). Amsterdam: Elsevier, 1974, pp. 3-97

Siegbahn, K., Nordling, C., Fahlman, A., Nordberg, R., Hamrin, K., Hedman, J., Johansson, G., Bergmark, T., Karlsson, S.E., Lindgren, I., Lindberg, B.: ESCA-Atomic, Molecular and Solid State Structure Studied by Means of Electron Spectroscopy. Uppsala: Almquist and Wisells, 1967

Siegbahn, K., Nordling, C., Johansson, G., Hedman, J., Heden, P., Hamrin, K., Gelius, U., Bergmark, T., Werne, L.O., Manne, R., Baer, Y.: Esca Applied to Free Molecules. Amsterdam: North Holland, 1969

Siezen, R.J., Van Driel, R.: Structure and properties of hemocyanins: VIII. Microheterogenity of α-hemocyanin of *Helix pomatia*. Biochim. Biophys. Acta 295, 131-139 (1973)

Sokolowski, G., Pilz, W., Weser, U.: X-ray photoelectron spectroscopic properties of Hg-thionein. FEBS Lett. 48, 222-225 (1974)

Weser, U.: Structural aspects and biochemical function of erythrocuprein. Structure and Bonding 17, 1-65 (1973)

Electron Paramagnetic Resonance in the Study of Binuclear Centres of Copper Proteins

L. CALABRESE AND J. ROTILIO

Metalloproteins often use binuclear metal centres to perform important biological functions. This is the case of electron transfer proteins, such as ferredoxins, oxygen carriers, such as hemerythrin and hemocyanin, and oxygenases, such as tyrosinase. Hemocyanins contain copper pairs as the functional unit (Lontie and Witters, 1973) and are known since long time to show no electron paramagnetic resonance (EPR) signal in the native state, either oxy or deoxy (Nakamura and Mason, 1960). Though EPR can only detect paramagnetic copper and has therefore been extensively used in the study of copper proteins (Vanngard, 1972), the absence of EPR signals in molecules containing more than one copper, does not mean that the copper is in the diamagnetic cuprous state. In fact two paramagnetic ions, which would give EPR signals in mononuclear centres, may become EPR silent if they are very close to each other or are bridged by a common ligand. In the case, as for Cu(II), that they have each total electron spin S = 1/2, they can couple to each other according to the scheme shown in Figure 1. The coupling will result in

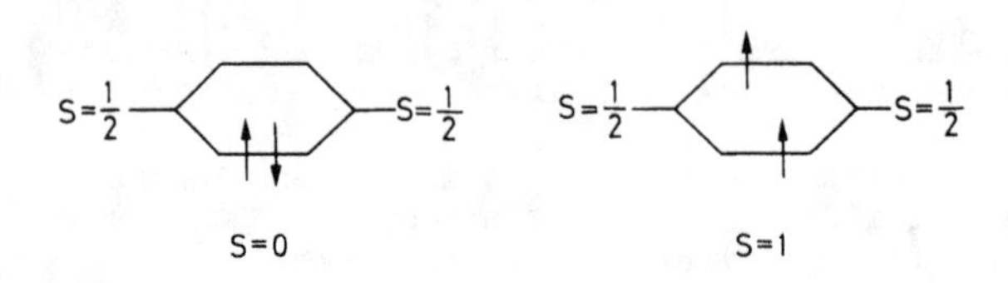

Fig. 1. Magnetic states produced as a result of coupling of two paramagnetic centres with total spin S = 1/2

a ground state with antiparallel spins, which will be diamagnetic (S = 0), and in an excited state, with parallel electron spins, which will behave as a triplet state (S = 1). The coupling between unpaired electrons will always have a dipolar component, that is an interaction between magnetic dipoles depending on their distance ($\alpha 1/r^3$) and on their orientation with respect to the magnetic field. This kind of interaction will produce zero field splitting of the triplet state (D), which will give rise to the magnetic field dependent separation of energy levels shown in Figure 2. It appears that at the frequency of a standard EPR experiment, transitions are possible only within the triplet state and can be either allowed $\Delta M_S = 1$ transitions in the region of g = 2 or $\Delta M_S = 2$ forbidden transitions at half field. $\Delta M_S = 1$ transitions between singlet and triplet are observable only for very low values of the coupling constant J.

The $\Delta M_S = 1$ transitions within the triplet, though allowed, are strongly orientation dependent. Therefore, in the powder samples used in low temperature EPR spectroscopy of metalloproteins (Beinert and Palmer, 1965), they are spread over a wide range of magnetic fields and could be so broad that they become undetectable. On the contrary the forbidden $\Delta M_S = 2$ transition is not so orientation-dependent and can be detected more easily. In other words the half-field signal ($g \simeq 4$) is diagnostic of the EPR of triplet states.

Fig. 2. Energy levels in a system of two interacting spins with S = 1/2

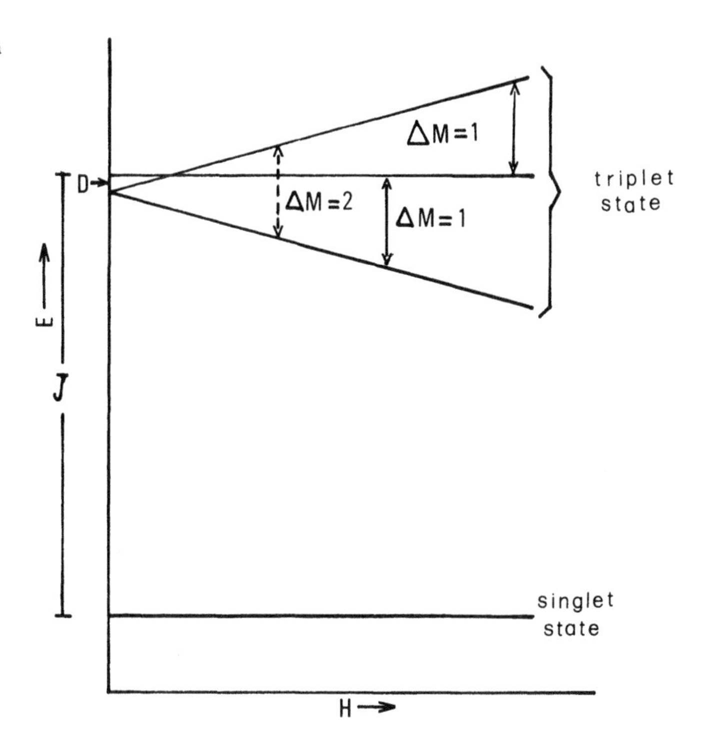

In the purely dipolar interaction the spectra can be used to calculate the distance between the two interacting centres (Schoot Uiterkamp et al., 1974), from the values of the dipolar splittings detectable in the $\Delta M_S = 1$ resonances. For instance, in the case of Cu(II) pairs, the dipolar splittings are of the order of 0.1/cm when r = 3 Å and one order of magnitude less when r = 7 Å.

In the presence of a bridging ligand, which is capable of transferring one electron to the unpaired orbital of each the two paramagnetic metals, the two centres can couple their spins in an exchange interaction. In Figure 3 the case of an oxo bridge between two cupric ions with square

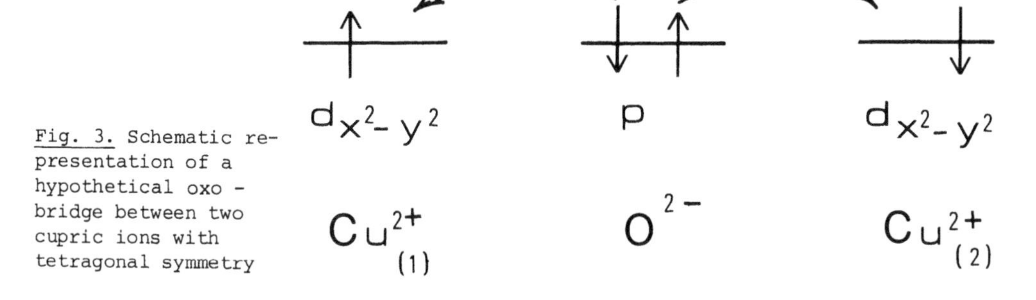

Fig. 3. Schematic representation of a hypothetical oxo - bridge between two cupric ions with tetragonal symmetry

planar geometry is reported as an example. In the case of exchange interaction the J value (Fig. 1) is much larger than for dipolar interaction (>> 0.1 K). When the S = 0 state lies lower in energy (antiferromagnetic exchange), magnetic susceptibility will show a reduction of the susceptibility value (χ) at the temperature corresponding to the splitting between singlet and triplet, that is when, at low enough temperature, the population of the diamagnetic ground state becomes higher. For EPR, the transitions are the same as described in Figure 2 but, again, temperature dependence of the signal intensity will allow the determination of J, since the signal intensity will increase with temperature until a value which marks the full population of the triplet state. Thus, deviation of the EPR signals of a metal pair from Curie behavior will differentiate an exchange interaction from a dipolar coupling.

While low molecular weight binuclear Cu(II) complexes give EPR spectra of the nature expected from the above considerations (Schoot Uiterkamp et al., 1974), no EPR signal has been observed, at any temperature, which could arise from binuclear centres existing in native copper proteins. Such centres have been suggested, independently of direct magnetic data, for three well distinct proteins: (a) *oxyhemocyanin;* the presence of a Cu(II)-Cu(II) pair at the active site is supported by the analysis of the optical transitions (Van Holde, 1967) and of the magnetic circular dichroism spectra (Mori et al., 1975); (b) *laccase;* a chromophore absorbing at 330 nm has been shown to behave as a two-electrons acceptor and identified as a Cu(II)-Cu(II) pair on the basis of the copper content of the protein (Malkin et al., 1969); (c) *tyrosinase;* redox titrations (Makino et al., 1974), have shown that it, but not apotyrosinase, contains a titratable group within n = 2, which appears to be a Cu(II)-Cu(II) pair from the metal content of the protein. For cases (a) and (b) however, a Cu(III)-Cu(I) pair would give the same results.

Recently, the potential application of EPR to the study of such centres in proteins has been expanded by the use of nitric oxide. Schoot Uiterkamp (1972) obtained clear evidence for dipolar interaction in EPR spectra induced in hemocyanin by treatment with nitric oxide. Seven hyperfine lines were observed on the half-field signal, showing that the triplet spectrum arises from two Cu(II) interacting with each other. It is likely that nitric oxide oxidizes the cuprous ions of the deoxy protein and binds to each of the resulting cupric ions, thus preventing an exchange interaction between them. In this pure dipolar situation a distance between the two centres of approximately 6 Å could be calculated from the EPR spectra, and it was considered to fit an exchange interaction of the two coppers through a peroxo bridge in the native oxy protein. The same experiment gave analogous results with tyrosinase (Schoot Uiterkamp and Mason, 1973), pointing to a close structural similarity between the two active sites. Lastly, Van Leeuwen et al. (1973) reacted with nitric oxide fully reduced ceruloplasmin, a blue copper protein which has similar spectroscopic properties and copper distribution to laccase. Figure 4 reports the EPR signal of the triplet

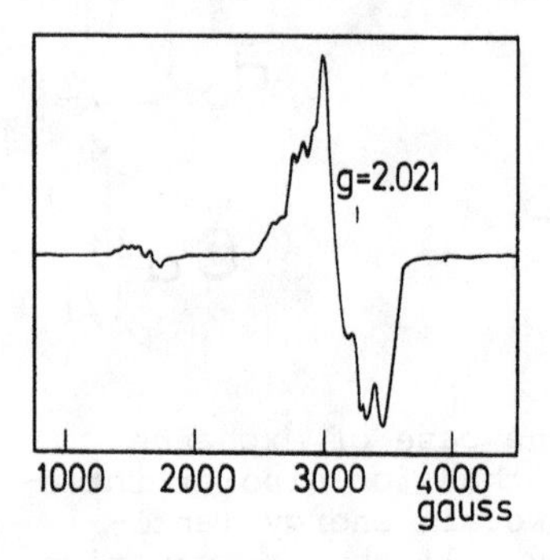

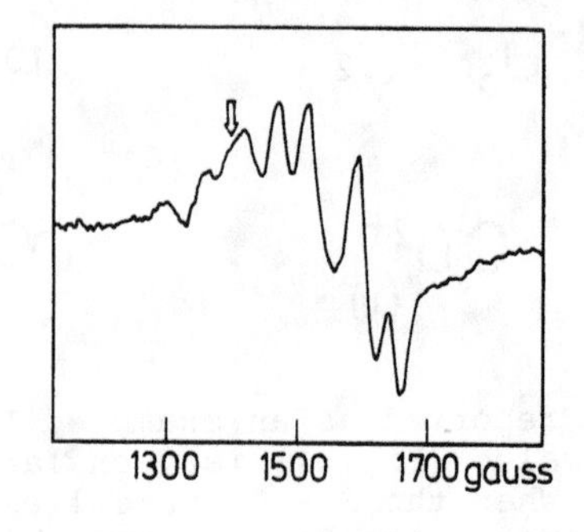

Fig. 4. EPR spectrum of reduced ceruloplasmin in the presence of nitric oxide. *On the right:* Magnification (× 12.4) of the $\Delta M_s = 2$ absorption (from Van Leeuwen et al., 1973)

type like those observed in nitric oxide-treated hemocyanin and tyrosinase, and confirms the existence of Cu(II) pairs in the blue copper oxidases (laccases, ceruloplasmin and ascorbate oxidase).

In the last year very sensitive magnetometers have become available and magnetic susceptibility measurements over a wide range of temperatures have been carried out on laccase and oxyhemocyanin (Solomon et al., 1976). For laccase the exchange coupling constant has been determined (J = 170/cm), thus showing the presence of an exchange coupling interaction. On the other hand, no deviation from linear χ vs. T^{-1} behavior was observed for oxyhemocyanin, and a lower limit of J near 625/cm was estimated.

A particular case of binuclear copper protein is erythrocuprein (superoxide dismutase). The native protein contains equivalent amounts of copper and zinc. Co(II) substitution for zinc (Calabrese et al., 1972) leads to disappearance of EPR signals of both metals (Rotilio et al., 1974), showing that cobalt and copper are magnetically coupled. X-ray crystallography (Richardson et al., 1975) later showed that the zinc and copper sites are bridged through an imidazolate residue as the common ligand. Thus, this protein is a new model for superexchange interactions in metalloproteins. Fee and Briggs (1975) have reported the preparation and properties of a derivative in which both the zinc and the copper sites contain copper. The EPR spectrum of such a derivative is typical of a triplet state, and from the temperature dependence of the signal intensity, J ≈ 52/cm was determined. A more precise evaluation of the coupling constant in this imidazolato-bridged pair has been attempted by magnetic susceptibility measurements of the cobalt-copper derivative. Moss and Fee (1975) could only conclude that the magnitude of the constant should be >> 5/cm, since the intrinsic nature of the measurement and the chemical composition of their samples did not permit a more precise assignment. More detailed measurements in prpgress in Rome have shown that J is near 20/cm, thus confirming the relatively low value determined in the Cu(II)-Cu(II) derivative.

In conclusion, EPR and magnetic susceptibility data on binuclear centres of copper proteins seem to indicate that the dioxygen bridge of oxyhemocyanin brings about the closest association of the two metal ions. The interaction is much weaker in laccase and considerations from model compounds (Jotham et al., 1972; Hodgson, 1975) suggest carboxylato- or hydroxo-bridges as possible candidates for the binuclear centre of laccase. The imidazolato-bridge, giving the weakest coupling, appears to be involved in a completely different functional requirement, such as that of the bimetal cluster of superoxide dismutase (Rotilio et al., 1976).

References

Beinert, H., Palmer, G.: Contribution of EPR spectroscopy to our knowledge of oxidative enzymes. Advan. Enzym. 27, 105-198 (1965)

Calabrese, L., Rotilio, G., Mondovi, B.: Cobalt superoxide dismutase: preparation and properties. Biochim. Biophys. Acta 263, 827-829 (1972)

Fee, J.A., Briggs, A.G.: Studies on the reconstitution of bovine erythrocyte superoxide dismutase. V. Preparation and properties of derivatives in which both zinc and copper sites contain copper. Biochim. Biophys. Acta 400, 439-450 (1975)

Hodgson, D.J.: Structural and magnetic properties of first-row transition metal dimers containing hydroxo, substituted hydroxo and halogen bridges. Prog. Inorg. Chem. 19, 173-177 (1975)

Jotham, R.W., Kettle, S.F.A., Marks, J.A.: Antiferromagnetism in transition metal complexes. IV. Low-lying excited states of binuclear copper (II)-carboxylate complexes. J. Chem. Soc., Dalton Trans. 428-438 (1972)

Lontie, R., Witters, R.: Hemocyanin. In: Inorganic Biochemistry. Eichhorn, G.L. (ed.). Amsterdam: Elsevier, 1973, Vol. I, pp. 344-358

Makino, N., McMahill, P., Mason, H.S.: The oxidation state of copper in resting tyrosinase. J. Biol. Chem. 249, 6062-6066 (1974)

Malkin, R., Malmstrom, B.G., Vanngard, T.: Spectroscopic differentiation of the electron accepting sites in fungal laccase. Europ. J. Biochem. 10, 324-329 (1969)

Moss, T.H., Fee, J.A.: On the magnetic properties of cobalt substituted bovine superoxide dismutase. Biochem. Biophys. Res. Commun. 66, 799-808 (1975)

Mori, W., Yamaguchi, O., Nakao, Y., Nakahara, A.: Spectroscopic studies on the active site of *Sepioteuthis lessoniana* hemocyanin. Biochem. Biophys. Res. Commun. 66, 725-730 (1975)

Nakamura, T., Mason, H.S.: An electron spin resonance study of copper valence in oxyhemocyanin. Biochem. Biophys. Res. Commun. 3, 297-299 (1960)

Richardson, J.S., Thomas, K.A., Rubin, B.H., Richardson, D.C.: Crystal structure of bovine Cu, Zn superoxide dismutase at 3 Å resolution. Chain tracing and metal ligands. Proc. Natl. Acad. Sci. 72, 1349-1353 (1975)

Rotilio, G., Calabrese, L., Mondovi, B., Blumberg, W.E.: Electron paramagnetic resonance studies of cobalt-copper bovine superoxide dismutase. J. Biol. Chem. 249, 3157-3160 (1974)

Rotilio, G., Morpurgo, L., Calabrese, L., Finazzi-Agro, A., Mondovi, B.: Metal ligand interaction in copper proteins. In: Metal Ligand Interactions in Organic Chemistry and Biochemistry. Pullman, B., Ginsburg, D. (eds.). Dordrecht: Reidel (in press, 1976)

Schoot Uiterkamp, A.J.M.: Monomer and magnetic dipole coupled Cu(II) EPR signals in nitrosyl hemocyanin. FEBS Lett. 20, 93-96 (1972)

Schoot Uiterkamp, A.J.M., Mason, H.S.: Magnetic dipole-dipole coupled Cu(II) pairs in nitric oxide treated tyrosinase: a structural relationship between the active sites of tyrosinase and hemocyanin. Proc. Natl. Acad. Sci. 70, 993-996 (1973)

Schoot Uiterkamp, A.J.M., Van der Deen, H., Berendsen, H., Boas, J.F.: Computer simulation of the EPR spectra of mononuclear and dipolar coupled Cu(II) ions in nitric oxide and nitrite treated hemocyanin and tyrosinase. Biochim. Biophys. Acta 372, 407-425 (1974)

Solomon, E.I., Dooley, D.M., Wang, R.H., Gray, H.B., Cerdonio, M., Mogno, F., Romani, G.L.: Susceptibility studies of laccase and oxyhemocyanin using an ultrasensitive magnetometer. Antiferromagnetic behaviour of type 3 copper. J. Am. Chem. Soc. 98, 1029-1031 (1976)

Van Holde, K.E.: Physical studies of hemocyanins. III. Circular dichroism and absorption spectra. Biochemistry 6, 93-96 (1967)

Van Leeuwen, F.X.R., Wever, R., Van Gelder, B.F.: EPR study of nitric oxide-treated reduced ceruloplasmin. Biochim. Biophys. Acta 315, 200-203 (1973)

Vanngard, T.: Copper proteins. In: Biological Applications of Electron Spin Resonance. Swartz, H.M., Bolton, J.R., Borg, D.C. (eds.). New York: Wiley-Interscience, 1972, pp. 411-447

Reactions of Hemocyanin

Some Reflections About Linked Phenomena in Macromolecules

J. WYMAN

The study of the structure and behavior of these often super-large molecules such as hemocyanin is a matter of real importance, providing as it does a stepping stone to the understanding of the regulation and assembly of still larger biological structures, such as ultimately, microtubules or ribosomes.

The behavior of the hemocyanins, like that of other proteins, may be studied at three levels: equilibrium, steady state, and transient state (relaxation). Equilibrium studies involve the binding of ligands and the associated ligand-linked conformational changes under conditions where the sure principles of thermodynamics are applicable. Thus they provide an insight into regulation and control at its simplest, where pure linkage theory governs. Such studies may be directed either towards the cooperative phenomena involving homotropic interactions between sites which bind the same ligand or towards the regulatory phenomena involving heterotropic interactions between sites which bind different ligands (Wyman, 1964). As is will known, in some of the larger hemocyanins there are more than 100 oxygen-binding sites, many of which, if not all, are proton-linked, and the oxygenation process may show such high cooperativity as to suggest something approaching a phase change (Colosimo et al., 1974). In the majority of proteins the observed interactions, whether heterotropic or homotropic, have their origin in ligand-linked conformational changes, and we may view these changes as the means by which the sites communicate with one another. Conformational changes may involve simply a change in the arrangement and contacts of the parts of the macromolecule, or, in extreme cases, an actual association or dissociation. In the former case we speak of the linkage effects as allosteric; in the latter, as polysteric (Colosimo et al., 1976). The reason for the distinction is that the analytical treatment applicable to the two kinds of linkage is different. All the thermodynamic linkage relations applicable to any system at equilibrium may be derived from a group of potentials derivable from the energy by a corresponding group of Legendre transformations. A principal member of this group is what has been called the binding potential (Wyman, 1965). In the case of allosteric systems the binding potential can be expressed as the logarithm of a polynomial in the activities of the various ligands; in the case of a polysteric system no such simple formulation is possible. Nevertheless both types of system respond to their ligands in much the same way. In both the conformational changes always give rise to cooperativity (which may of course be masked by site heterogeneity) and for both the Hill plots are alike, characterized by an asymptote of unit slope at each end.

When a macromolecule is not in equilibrium but only in a steady state, as in the case of a working enzyme present in a pool containing substrate and ligands maintained at constant concentrations, or in the case of a respiratory (or other) protein subjected to a constant photodissociating light, the situation is different (Giacometti et al., 1975). The amounts of the various liganded forms and conformations will be uniquely determined by the ligand activities and the system will be governed by a well-defined set of interactions, both homotropic and heterotropic. But it will no longer be possible to describe the system by a group of binding potentials. This results from the fact that it is no longer possible to write an expression for the energy in terms of S, V and the amounts of the various components. It reflects the fact that the microscopic kinetic constants are no longer subject to the restrictions of microscopic balance and with the result that the system enjoys a greater number of degrees of position.

It is interesting to compare the properties of a true allosteric system with those of a corresponding system in a steady state. As we have seen, both will show linkage phenomena, homotropic and heterotropic, which we may describe as true and steady-state linkage in the two cases respectively. Suppose we are concerned with the binding of just one ligand, the activities of all the other ligands being held constant: then it may be shown that in the steady state case the flux of free energy and matter through the system in the steady state can generate cooperativity (or anticooperativity, as the case may be). Thus even a one-site molecule can show a value of $n>1$ in a Hill plot. In other words, the steady state may give rise to homotropic effects which would be quite impossible under equilibrium conditions. Indeed the system will behave much like a true allosteric system with more than one site (Wyman, 1975). Indeed the system will closely resemble a true allosteric system with more than one site at equilibrium. In the absence of any knowledge of the actual number of sites present in the macromolecule one might, therefore, on the basis of binding data alone, mistake a steady-state system for one with a larger number of sites but at equilibrium. A possible way of distinguishing between the two interpretations would be to consider the asymptotes of a Hill plot, if they could be measured. In both cases the initial asymptote will have a slope of 1, but in the steady case the final asymptote will have a slope of 2 or more (except in the special case where there is only one binding site for the ligand being studied; in this case the final slope would be 1).

So far I have said nothing about relaxation. Relaxation studies of course take one into a new domain and add another dimension to the picture (Eigen, 1968). Both equilibrium and steady-state systems are asymptotically stable and each is characterized by a number of relaxation times 1 less than the number of different forms. This number will be the same for a given macromolecule in the presence of its ligands whether the system is at equilibrium or in a steady state, though of course the relative amounts of the various forms may vary widely between the two cases. (Thus forms which are negligible at equilibrium may in extreme cases become predominant in a steady state, or conversely. Possibly this is what is being observed in cytochrome oxidase under certain experimental conditions). When a steady state is replaced by an equilibrium (as when the supply of energy required to maintain the steady state is suddenly shut off), the system will relax. However, it is apparent from what has been said that no observation of the *number* of relaxation times can serve to distinguish between a steady state and equilibrium.

Much useful information can be obtained from a study of the heat of liganding of a protein under equilibrium conditions. The heat may either be measured directly in a suitable calorimeter or calculated on the basis of the Van't Hoff relation, from binding studies at several different temperatures. In interpreting the results it should be realized that temperature (as an intensive variable) is formally the exact equivalent of a chemical potential in linkage theory and that ΔH is the exact equivalent of the amount of ligand bound. Thus all the linkage relations developed for a set of ligands are applicable with this simple substitution. The same equivalence may of course be used in formulations of the binding potentials. Recently we in Rome have been fortunate in having the collaboration of Stanley Gill, of the University of Colorado, in a study of the point value of the heat of combination of hemoglobin Trout I with CO as a function of saturation (Wyman et al., 1976). It is striking that in this hemoglobin there is a change of sign of ΔH with saturation, which would seem to represent a profound difference between the properties of the unliganded and liganded conformations of this molecule. Nothing like this is encountered in the case of human hemoglobin.

Exactly the same considerations apply to studies involving the effect of pressure: p corresponds to a chemical potential and the corresponding variable ΔV to the amount of ligand bound in the linkage equations. I have just had the pleasure of hearing from Gregorio Weber, in the course of his EMBO lectures given in Rome, of his recent experiments on the effect of pressure in inducing conformational changes in a number of different proteins. These studies were largely made possible by the use of fluorescence techniques (Weber, 1976). Possibly these techniques could be applied to hemocyanins. In any case a study of the effects of temperature and pressure applied to hemocyanins should be illuminating.

It has now been established that many (probably all) proteins show spontaneous fluctuations of conformation (and liganding) under constant conditions. This has been amply demonstrated by both isotopic exchange methods and optical ones, particularly those involving fluorescence. Such fluctuations occur whenever the free energy changes involved are sufficiently small in relation to kT. They correspond to the imposed changes of conformation which result from changing of ligand activity. It is significant that they are characterized by a spectrum of relaxation times ranging from nanoseconds or less to hours or even more. The early measurements of dielectric dispersion, which provided values of the rotary diffusion constant of the protein molecule as a rigid body, represent but one aspect of this interesting field - the study of the dynamics of conformational changes, or as we might say, the dynamics of macromolecular breathing.

A somewhat simpler subject which might be worth looking into in the case of hemocyanins is ligand partition. Where two or more ligands bind at the same site or sites in a protein, the sites are shared by, or partitioned among, the ligands: a case in point is the binding of O_2 and CO by hemoglobin. Ligands which are subject to partition may be said to be *identically linked*, and it can be shown that if there is an observable partition at any one saturation there must be one at all, with the same value of the constant partition coefficient everywhere. From a formal point of view it is worth noting that in this case the activities of the two (or more) identically linked ligands will enter into the binding polynomial through an expression of the form (pO_2 + MpCO) (Wyman, 1964, 1965; Colosimo et al., 1974, 1976).

One final subject which I feel has been somewaht overlooked (possibly because of experimental difficulties) but which might be worth exploring is the solubility relations between the different crystalline forms (phases) of a protein, particularly those associated with the presence and absence of its ligands. It is well known, for instance, that oxy and deoxy human hemoglobins crystallize in different forms which have different solubilities. Studies of such phenomena, based on phase rule considerations, should bring out small differences in lattice free energy between various forms, liganded and unliganded (Wyman, 1964).

As stated earlier, conformational changes provide the means by which the various sites in a macromolecule communicate with one another. Thus they may be said to provide the vocabulary of a macromolecular language, and it is clear that the grammatical rules which govern this language are to be found in the linkage relations relating conformation and ligand binding. The rock on which the whole structure rests is thus the group of binding potentials derivable from the energy by the group of Legendre transformations from which these linakge relations are derived.

References

Colosimo, A., Brunori, M., Wyman, J.: Concerted changes in an allosteric macromolecule. Biophys. Chem. 2, 338-344 (1974)

Colosimo, A., Brunori, M., Wyman, J.: Polysteric linkage. J. Mol. Biol. 100, 47-57 (1976)

Eigen, M.: New looks and outlooks on physical enzymology. Quart. Rev. Biophys. 1, 3-33 (1968)

Giacometti, M.G., Focesi, A., Giardina, B., Brunori, M., Wyman, J.: Kinetics of binding of carbon monoxide to *Lumbricus* erythrocruorin: A possible model. Proc. Natl. Acad. Sci. 72, 4313-4316 (1975)

Weber, G.: EMBO lecture, Rome, Italy, 1976

Wyman, J.: Linked functions and reciprocal effects in hemoglobin: A second look. Advan. Prot. Chem. 19, 223-281 (1964)

Wyman, J.: The binding potential, a neglected linakge concept. J. Mol. Biol. 11, 631-644 (1965)

Wyman, J.: The turning wheel: A study in steady states. Proc. Natl. Acad. Sci. 72, 3983-3987 (1975)

Wyman, J., Gill, S.J., Noll, L., Giardina, B., Colosimo, A., Brunori, M.: The balance sheet of a hemoglobin. J. Mol. Biol. (In press, 1976)

Oxygen Binding to Haemocyanin: A Tentative Analysis in the Framework of a Concerted Model

A. Colosimo, M. Brunori, and J. Wyman

Introduction

In the last ten years several allosteric models (Monod et al., 1965; Koshland et al., 1966; Perutz, 1970), have been proposed in order to provide a description of the cooperative phenomena present in the ligand-binding curves of functioning proteins with complex quaternary structure. The principal distinction among the various models is the different relevance given to the conservation of symmetry at the level of quaternary structure in the low affinity (T) and high affinity (R) states: in one of the possible cases this principle is the most powerful constraint (concerted models), while in the other it is completely absent (sequential models). More recently a number of refinements as well as generalizations have been proposed, largely with the purpose to describe, within a comprehensive framework, the experimental data available for different systems (Wyman, 1969; Colosimo et al., 1976).

In a recent contribution (Colosimo et al., 1974) we have examined the relationship between (1) the parameters characteristic of cooperative ligand binding as obtainable from the geometrical features of the Hill plot, i.e. the total apparent interaction energy ΔF_I and the index of cooperativity n, and (2) the analytical parameters of the concerted model, i.e. the constants L and λ (see Table 1), as well as the total number of interacting sites, r.

Table 1. Analytical parameters associated with the theoretical curves of Figure 1

L	λ	ΔF_I	r	n_m
15×10^6	0.045	1800 cal/site	12	4.7

L, the allosteric constant, is defined as the relative abundance of the R and T forms in the absence of ligand, i.e. as $\frac{[R]}{[T]} = L$

λ is the ratio of affinities for the ligand of the two forms, i.e. $\frac{K_R}{K_T} = \lambda$

This work is aimed at the description in terms of a concerted model and at the interpretation of a feature common to the Hill plots of very large respiratory systems, such as chlorocruorin (Antonini et al., 1962) or hemocyanin (Hc) (Van Driel, 1973), i.e. a relatively high value of n coupled with a relatively small value of ΔF_I.

Results and Discussion

The symmetry principle, in fact, implies the transition from the low to the high affinity form of all the subunits together. This means that in the ligand equilibrium curve of a macromolecule with a large number of sites a very sudden increase of the slope in the middle part of the saturation, triggered by increasing the ligand concentration, should be observed.

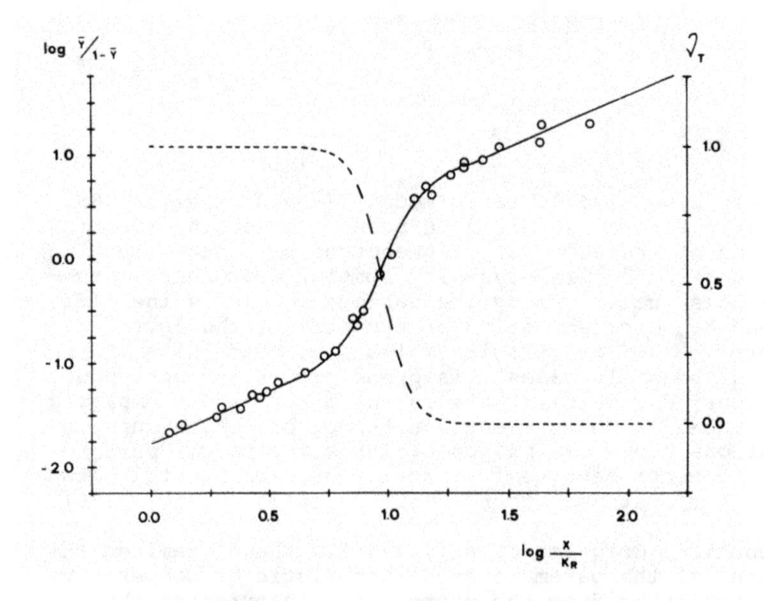

Fig. 1. Hill plot of O_2 binding by *Helix pomatia* α-Hc in Tris-HCl, pH 8.1, 10 mM $CaCl_2$ and T = 20°C. *Full line*: theoretical curve drawn on the basis of the M.W.C. concerted model with the parameters listed in Table 1. *Dotted line*: fraction of total molecules present in the low affinity form (ν_T), shown on the *right-hand coordinate*, as a function of ligand concentration

Figure 1 shows the O_2 binding data for H.P. α-Hc under conditions in which large cooperative effects are observed (Van Driel, 1973). The data have been fitted to a theoretical curve derived from the Monod-Wyman-Changeaux model, whose characteristic parameters are listed in Table 1. In the same figure the relative abundance of the T state, expressed as the fraction of total molecules present in the T form, is reported as a function of ligand concentration.

The fitting procedure was based on selecting sets of values for L, λ and r compatible with the value of $\Delta F_I \approx$ 1800 cal/site fixed from the observed asymptotes of the Hill plot. In doing so the quasi-symmetric shape of experimental data is taken into consideration and thus a "symmetry index"[1] not very far from unity applies. Following these constraints we have selected the values of L and listed them in Table 1. It was observed that the slope of theoretical curves in their middle part, namely $n_{1/2}$, is very sensitive to the number of interacting sites. Table 2 reports some of the theoretical parameters of the fit for three different values of r, i.e. the values of $n_{1/2}$, the symmetry index and an estimate of the goodness of matching between experimental points and theoretical curves based on a least squares method.

---

[1]The "symmetry index", defined as $L.\lambda^{r/2}$, gives a measure of the symmetry for a theoretical binding curves in the Monod-Wyman-Changeaux model. In the case of perfect symmetry its value must be exactly 1.

Table 2. Theoretical parameters obatined after fitting three different values of r by the Monod-Wyman-Changeaux Model

r	n_m [a]	Symmetry index	R [b]
10	4.4	2.77	0.445
12	4.7	0.12	0.082
15	4.1	0.002	0.607

[a] n_m is the medium slope of the theoretical curve calculated between saturations 0.3 and 0.7.
[b] R is an empirical parameter, defined as the sum of the squared distances, measured on the vertical axis, between experimental points and theoretical curves. The lowest value of R corresponds to the best matching.

The estimate of the minimum number of interacting sites emerging from this analysis as a suitable one to fit, within the assumptions of a concerted model, the shape of the O_2 binding curve for α-Hc is 12. This value should be compared with the total number of O_2 binding sites present in the entire molecule, i.e. ≈ 180. In this connection the following considerations are in order: since the possibility of simulating the characteristics of a physical system depends inversely on the complexity of the system and directly on the number of assumptions embodied in the model, it is quite surprising that the functional behaviour of a complex molecule like Hc can be fitted at such a degree by a model containing only two fundamental constraints (the existence of two conformations and the conservation of symmetry in the quaternary structure). In addition the analytical treatment used to draw the analytical curve is of the simplest type, taking no account of association-dissociation phenomena or of possible functional heterogeneity of various chains. Last, but not least, we should like to point out that the heuristic value which a theoretical model like the Monod-Wyman-Changeaux can have in understanding the function of a system as complex as Hc resides mainly in the general indications and stimulating suggestions it can give in addressing the experimental work.

As a conclusion it may be of particular interest to make a parallel between two ideas which have been developing in recent years starting from completely different contexts, i.e. the ideas of "functional constellations" and of "structural domains". The first one originated from general, statistical considerations on possible mixed conformations in a multisubunits, giant molecule (Wyman, 1969) and suggests that the very many ligand-binding sites present in the system could be distributed in a number of strongly interacting subgroups responsible for most of the cooperativity shown by the entire system. The second one, emerging from investigations of the structural properties of the Hc molecule, resides mainly on electron micrsocopy (Siezen et al., 1974; Van Breemen et al., Chap. 15, this vol.) and limited proteolysis studies (Brouwer, 1975; Gielens et al., Chap. 11, this vol.).

It is a challenging task for future studies to underscore the possible relationship, if any, between functional constellations and structural domains in giant functioning macromolecules such as haemocyanins.

References

Antonini, E., Rossi-Fanelli, A., Caputo, A.: Studies on chlorocruoin I. The oxygen equilibrium of *Spirographis* chlorocruorin. Arch. Biochem. Biophys. 97, 336-342 (1962)

Brouwer, M.: Structural domains in *Helix pomatia* α-hemocyanine. Doctoral dissertation, Rijksuniversiteit, Groningen (1975)

Colosimo, A., Brunori, M., Wyman, J.: Concerted changes in an allosteric macromolecule. Biophys. Chem. 2, 338-344 (1974)

Colosimo, A., Brunori, M., Wyman, J.: Polysteric linkage. J. Mol. Biol. 100, 45-47 (1976)

Driel, R. van: Oxygen binding and subunit interactions in *Helix pomatia* α-haemocyanin. Biochemistry 12, 2696-2699 (1973)

Koshland, D.E., Nemethy, G., Filmer, D.: Comparison of experimental binding data and theoretical models in proteins containing subunits. Biochemistry 5, 365-385 (1966)

Monod, J., Wyman, J., Changeaux, J.P.: On the nature of allosteric transitions: a plausible model. J. Mol. Biol. 12, 88-118 (1965)

Perutz, M.F.: Stereochemistry of cooperative effect in hemoglobin. Nature (London) 228, 726-739 (1970)

Siezen, R.J., Van Bruggen, E.F.J.: Structure and properties of hemocyanins XII. Electron microscopy of dissociation products of *Helix pomatia* α-hemocyanin. Quaternary structure. J. Mol. Biol. 90, 77-89 (1974)

Wyman, J.: Linked functions and reciprocal effects in hemoglobin: a second look. Advan. Protein Chem. 19, 223-286 (1964)

Wyman, J.: Speculations concerning possible allosteric effects in membranes and other extended biological systems. Nobel Symposia 11, 266-282 (1969)

The Oxygen Equilibrium of *Murex trunculus* Haemocyanin

J. V. BANNISTER, A. GALDES, AND W. H. BANNISTER

Introduction

The oxygen-binding properties of *Murex trunculus* haemocyanin were described briefly by Wood and Dalgleish (1973) in a previous study from this laboratory. In the absence of divalent ions in the pH range 7.2-9.2, noncooperative oxygen binding and a small reverse Bohr effect were observed. In the presence of Mg^{2+}, cooperative oxygen binding (with a Hill coefficient as large as 3 or more in some cases) and a more pronounced Bohr effect were seen. In a later study 50,000 molecular weight fragments ("Oxygen-binding domains") of the haemocyanin, obtained by subtilisin digestion, were shown to bind oxygen noncooperatively, without a Bohr effect and with significantly decreased oxygen affinity with respect to untreated haemocyanin (Bannister et al., 1975). We describe here a systematic study of the oxygen equilibrium of *Murex trunculus* haemocyanin under various conditions of pH, divalent ion concentration and ionic strength. Oxygen-binding data are analysed in relation to cooperativity and linkage phenomena.

Experimental Methods

Tris-HCl buffers (0.1 M) were used in the pH range 7.0-8.9, and 0.05 M borate buffers in the pH range 9.1-9.8. The ionic strength of these buffers was in the range 0.04-0.10 mol/l. Buffers of higher ionic strength were prepared by adding appropriate amounts of NaCl. Calcium chloride or $MgCl_2$ were added to the buffers as required, and Ca^{2+} and Mg^{2+} concentrations were checked by atomic absorption spectrophotometry.

Preparation of haemocyanin, sedimentation velocity and oxygen-binding experiments (at 20 ± 2°C) were performed as described previously (Bannister et al., 1975). The experimental error of percentage oxygen saturation values was about 3% and that of P_{O_2} values was about 0.3 mm Hg.

Protein samples were brought to the required pH for oxygenation experiments by dialysis against three changes of the appropriate buffer. For experiments at pH 10.0 in the presence of Mg^{2+}, a haemocyanin pellet was dissolved in a small volume of 0.05 M borate buffer, pH 10.0 containing 10 mM Mg^{2+}, to obtain a stock solution of haemocyanin in this buffer. Protein solutions containing the appropriate concentration of Mg^{2+} in 0.05 M borate buffer, pH 10.0, were prepared by diluting the stock solution just before use.

Oxygen-binding curves (Y vs. P_{O_2}) were fitted by least squares with a polynomial of order $n \leq 6$ in the region $10\% \leq Y \leq 90\%$. The P_{50} was estimated from the value of the fitted polynomial at Y = 50%. The Hill plot (log [Y/(100-Y)] vs. log P_{O_2}) was fitted by the logarithmic form of the least squares polynomial, and the Hill coefficient, n_H, was estimated from the slope of the plot at Y = 50%. The free energy per site for saturating the haemocyanin with oxygen, ΔG, was estimated from the ligand equilibrium curve (Y vs. log P_{O_2}) (Wyman, 1964). For the computation of the ΔG value, the ligand equilibrium curve was integrated numerically using Simpson's rule. Values of Y at equidistant log P_{O_2} values were obtained by Lagrangian interpolation between the experimental data points.

Results

Oxygenation in the Absence of Divalent Ions. The Ca^{2+} and Mg^{2+} concentrations of haemocyanin dialyzed against buffers not containing these ions were less than 0.01 mM. Under these conditions dissociation of wholes (100 S) into halves (60 S) and twentieths (9 S) started at pH 7.6 and was nearly complete at pH 8.2 (less than 10% wholes). The observed oxygen-binding properties of the haemocyanin are summarized in Table 1. A

Table 1. Oxygen-binding data for *Murex trunculus* haemocyanin at low ionic strength and virtual absence of divalent cations

pH	P_{50} mm Hg	n_H	ΔG kcal mol^{-1}
7.0	2.8 ± 0.3	1.4 ± 0.1	4.83 ± 0.08
7.6[a]	2.6 ± 0.1	2.3 ± 0.6	4.66 ± 0.08
8.2	4.0 ± 0.3	1.3 ± 0.2	4.90 ± 0.05
9.2	4.9 ± 0.5	1.2 ± 0.1	4.95 ± 0.05

Values are mean ± S.E.M. of four determinations except where indicated.

[a] Three determinations.

small but statistically significant reverse Bohr effect was found ($P < 0.01$). Most of the change in P_{50} with pH occurred in the pH range 7.6-8.2 like the dissociation of the haemocyanin. The value of ΔG increased slightly on increasing the pH. The value of n_H at pH 7.0 indicates the possibility of weak cooperative interactions in wholes under these conditions ($P \approx 0.05$). The n_H decreased with pH and was not significantly different from unity at pH 9.2 ($P > 0.10$).

Oxygenation in the Presence of Ca^{2+} Ions

Constant Ca^{2+} Concentration. In the presence of approximately 10 mM Ca^{2+} (range 9-12 mM) the haemocyanin did not dissociate up to pH 8.7. Traces of halves (but no twentieths) appeared around pH 9.3 and increased with further increase in pH. Oxygenation curves are given in Figure 1. Cooperative behaviour was observed over the whole pH range investigated. The value of n_H was highest around pH 7.0 and decreased with increase in pH (Table 2). The P_{50} and ΔG increased with pH to a maximum value at pH 8.2, and then decreased at higher pH values. Thus, the haemocyanin showed a reverse Bohr effect in the pH range 7.0-8.2, and a normal Bohr effect above pH 8.2. The P_{50} increased sharply with pH in the pH range 7.6-8.2, where dissociation of the haemocyanin would have occurred in the absence of Ca^{2+}.

Variable Ca^{2+} Concentration. At pH 7.0 the haemocyanin remained as wholes at all Ca^{2+} concentrations investigated (0-50 mM). At pH 8.2 it showed halves and twentieths at Ca^{2+} concentrations below 5 mM. The effect of Ca^{2+} on oxygen binding is summarized in Table 3. Calcium ions increased the P_{50}, ΔG and n_H at both pH values. Relatively large changes

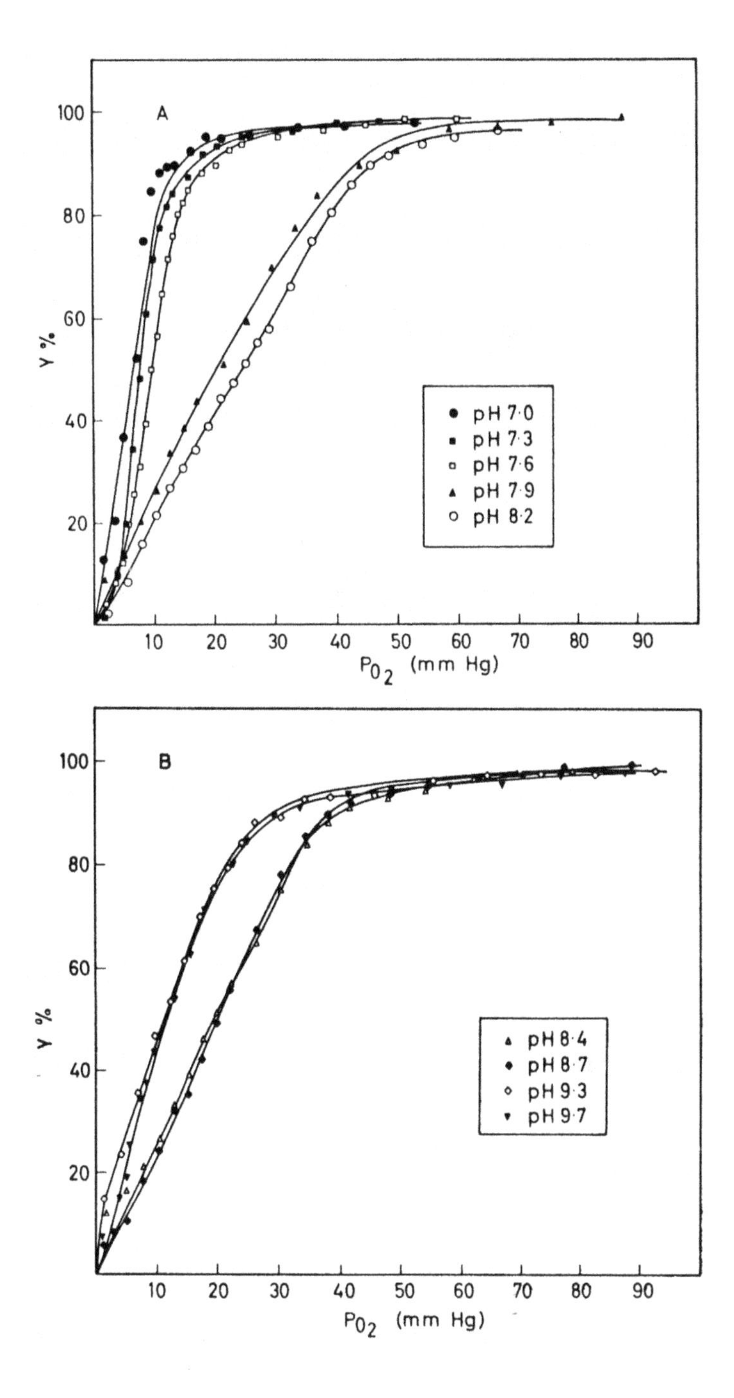

Fig. 1A and B. Oxygen-binding curves for *Murex trunculus* haemocyanin in the presence of approximately 10 mM Ca^{2+}. (A) pH range 7.0-8.2 showing reverse Bohr effect; (B) pH range 8.4-9.7 showing normal Bohr effect

were observed at relatively low Ca^{2+} concentrations. The n_H increased markedly at pH 7.0 but not at pH 8.2.

Table 2. Oxygen-binding data for *Murex trunculus* haemocyanin at low ionic strength in the presence of approximately 10 mM Ca^{2+}

pH	P_{50} mm Hg	n_H	ΔG kcal mol^{-1}
7.0	5.5 ± 0.5	3.1 ± 0.8	5.17 ± 0.07
7.3	7.2 ± 0.3	3.0 ± 0.3	5.34 ± 0.09
7.6	9.4 ± 0.2	3.2 ± 0.1	5.48 ± 0.04
7.9	20.4 ± 0.2	1.7 ± 0.0	5.71 ± 0.04
8.2	25.8 ± 0.7	2.1 ± 0.1	5.99 ± 0.03
8.4	21.3 ± 0.9	1.9 ± 0.1	5.75 ± 0.03
8.7	19.8 ± 0.6	2.1 ± 0.1	5.78 ± 0.03
9.3	11.1 ± 0.2	1.4 ± 0.1	5.27 ± 0.02
9.7	11.5 ± 0.1	1.6 ± 0.1	5.39 ± 0.03

Values are mean ± S.E.M. of three determinations.

Table 3. Oxygen-binding data for *Murex trunculus* haemocyanin at low ionic strength in the presence of various total Ca^{2+} concentrations

(Ca^{2+}) mM	P_{50} mm Hg	n_H	ΔG kcal mol^{-}
(a) pH 7.0			
0.01[a]	2.8 ± 0.3	1.4 ± 0.1	4.83 ± 0.08
0.32	5.2 ± 0.3	3.0 ± 0.2	5.03 ± 0.09
0.89	6.1 ± 0.3	3.1 ± 0.2	5.28 ± 0.04
2.80	7.4 ± 0.1	3.1 ± 0.2	5.29 ± 0.05
4.82	5.5 ± 0.4	2.8 ± 0.2	5.14 ± 0.06
8.96	5.5 ± 0.5	3.1 ± 0.8	5.17 ± 0.07
19.58	6.5 ± 0.3	3.7 ± 0.3	5.23 ± 0.03
46.33	6.5 ± 0.3	4.0 ± 0.3	5.29 ± 0.05
(b) pH 8.2			
0.01[a]	4.0 ± 0.3	1.3 ± 0.2	4.90 ± 0.05
0.77	14.2 ± 0.4	1.6 ± 0.0	5.52 ± 0.03
1.67	19.6 ± 1.6	1.9 ± 0.1	5.76 ± 0.06
2.50	22.7 ± 0.4	1.6 ± 0.1	5.90 ± 0.04
5.13	24.1 ± 1.2	1.7 ± 0.2	5.77 ± 0.08
9.69	25.8 ± 0.7	2.1 ± 0.1	5.99 ± 0.03

Values are mean ± S.E.M. of three determinations except where indicated.
[a]Four determinations.

Effect of Ionic Strength on Oxygenation

Constant High Ionic Strength. The haemocyanin was studied at an ionic strength of approximately 0.5 mol/l. This shifted the pH range of dissociation in the absence of Ca^{2+} by about 0.3 pH unit towards the alkaline side and increased the alkaline stability of the haemocyanin. Thus, halves and twentieths appeared at pH 7.9, and at pH 8.2 80% wholes were observed in contrast to less than 10% wholes for solutions at low ionic strength. Oxygen-binding curves are presented in Figure 2. The P_{50}, ΔG and n_H values in the absence of Ca^{2+} (Table 4) were higher than those observed at low ionic strength (Table 1), and definite cooperative interactions were present in the pH range 7.0-7.9 in consistence with the increased stability of wholes. In the presence of Ca^{2+},

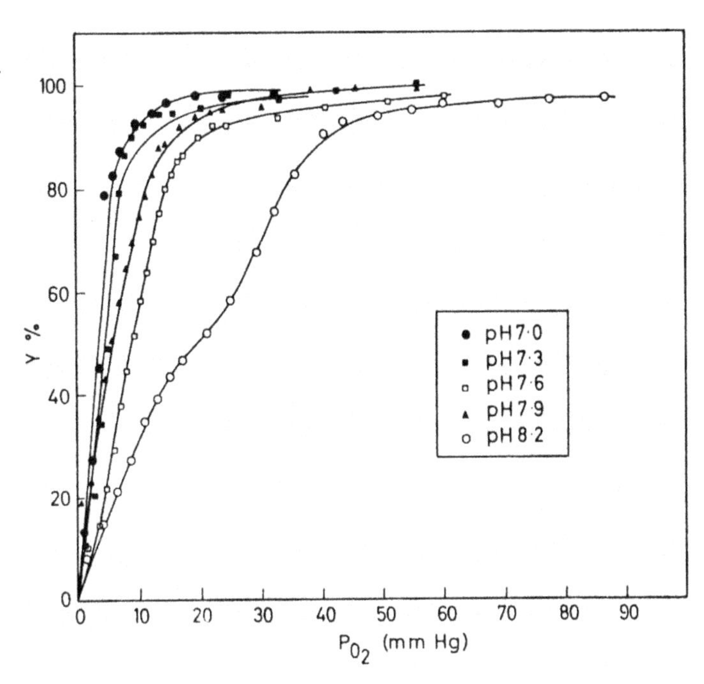

Fig. 2. Oxygen-binding curves for *Murex trunculus* haemocyanin at approximately 0.5 mol/l ionic strength in absence of Ca^{2+}

Table 4. Oxygen-binding data for *Murex trunculus* haemocyanin at approximately 0.5 mol/l ionic strength

pH	P_{50} mm Hg	n_H	ΔG kcal mol^{-1}
(a) In absence of Ca^{2+}			
7.0	4.1 ± 0.2	2.1 ± 0.3	4.88 ± 0.01
7.3	4.4 ± 0.6	3.1 ± 0.0	4.92 ± 0.10
7.6	7.1 ± 1.0	2.3 ± 0.2	5.04 ± 0.10
7.9	6.0 ± 0.4	1.6 ± 0.1	4.99 ± 0.09
8.2	18.9 ± 0.8	1.3 ± 0.2	5.62 ± 0.04
(b) In presence of approximately 10 mM Ca^{2+}			
7.0	4.8 ± 0.6	2.5 ± 0.5	4.82 ± 0.07
7.3	5.7 ± 0.2	3.3 ± 0.9	5.09 ± 0.06
7.6	7.2 ± 0.5	2.6 ± 0.2	5.21 ± 0.03
7.9	21.4 ± 0.7	2.3 ± 0.2	5.75 ± 0.07
8,2	26.2 ± 0.9	1.4 ± 0.1	5.86 ± 0.04

Values are mean ± S.E.M. of three determinations.

however, the oxygenation properties (Table 4) did not differ appreciably from those at low ionic strength (Table 2).

Variable Ionic Strength at Constant pH. In the presence of Ca^{2+} at pH 8.2, the haemocyanin existed as wholes at all ionic strengths investigated. In the absence of Ca^{2+} it showed various degrees of dissociation. The percentage of twentieths decreased steadily with increase in ionic strength, while the percentage of wholes increased. At very high ionic strength (0.90 mol/l) some dissociation of whole to halves was observed

Table 5. Oxygen-binding data for *Murex trunculus* at various ionic strengths at pH 8.2

Ionic strength mol/l	P_{50} mm Hg	n_H	ΔG kcal mol^{-1}
(a) In absence of Ca^{2+}			
0.03[a]	4.0 ± 0.3	1.3 ± 0.2	4.90 ± 0.05
0.26	6.1 ± 0.5	1.9 ± 0.1	5.22 ± 0.05
0.34	6.0 ± 0.2	2.4 ± 0.0	5.13 ± 0.06
0.39	8.8 ± 0.7	2.3 ± 0.1	5.28 ± 0.03
0.43	9.1 ± 0.3	2.2 ± 0.1	5.41 ± 0.02
0.47	18.9 ± 0.8	1.3 ± 0.2	5.62 ± 0.04
0.90	20.5 ± 0.2	1.5 ± 0.0	5.73 ± 0.05
(b) In presence of approximately 10 mM Ca^{2+}			
0.06[a]	25.8 ± 0.7	2.1 ± 0.1	5.99 ± 0.03
0.29	19.5 ± 1.4	1.6 ± 0.1	5.67 ± 0.11
0.42	19.4 ± 0.6	1.6 ± 0.1	5.69 ± 0.01
0.46	21.5 ± 0.8	2.0 ± 0.0	5.79 ± 0.04
0.50	26.2 ± 0.9	1.4 ± 0.1	5.86 ± 0.04
0.93	27.9 ± 1.6	1.3 ± 0.1	5.71 ± 0.06

Values are mean ± S.E.M. of three determinations except where stated.

[a]Four determinations.

but the halves did not exceed 20%. The observed oxygen-binding properties are summarized in Table 5. In the absence of Ca^{2+}, the P_{50} and ΔG increased systematically with ionic strength. The P_{50} showed a rather abrupt rise around 0.4 mol/l ionic strength. The n_H at first increased somewhat, but decreased again at ionic strengths higher than 0.4 mol/l. It is possible that some structural (conformational) change may have occurred around 0.4 mol/l ionic strength, since little change in molecular species (about 75% wholes, 5% halves, and 20% twentieths) was evident in this region. In the presence of Ca^{2+}, the P_{50} and ΔG initially decreased slightly with increase in ionic strength, reaching a minimum around 0.3-0.4 mol/l ionic strength, and then regained their original values with further increase in ionic strength. The turning point again suggests some structural change around 0.4 mol/l ionic strength. The behaviour of n_H was to decrease steadily with increase in ionic strength.

Dissociation at pH 10.0 and Oxygenation Properties of Wholes and Halves. The dissociation of the haemocyanin in 0.05 M borate buffer, pH 10.0 containing 10-55 mM Mg^{2+}, is shown in Figure 3. It is seen that wholes predominated at high, and halves predominated at low, Mg^{2+} concentrations. Oxygen-linked dissociation of wholes to halves (Wood and Dalgleish, 1973) was estimated from the composition of reoxygenated samples determined within one hour of the readdition of oxygen. Since Wood and Dalgleish (1973) have reported that reassociation is slow, and in this study it was ascertained that complete reassociation takes over 96 hours, the one-hour composition of reoxygenated samples was taken to represent closely that of deoxygenated samples. Figure 3 shows oxygen-linked dissociation at 31 to 11.6 mM Mg^{2+}, but not at 53 mM Mg^{2+}.

Since dissociation occurs rapidly on deoxygenation (Wood and Dalgleish, 1973), whilst reassociation on reoxygenation is slow, it may be expected

Fig. 3. Dissociation diagram for *Murex trunculus* haemocyanin at various Mg^{2+} concentrations at pH 10.0. O, ●: wholes; Δ, ▲: halves; *empty symbols:* oxygenated haemoxyanin; *filled symbols:* haemocyanin after deoxygenation

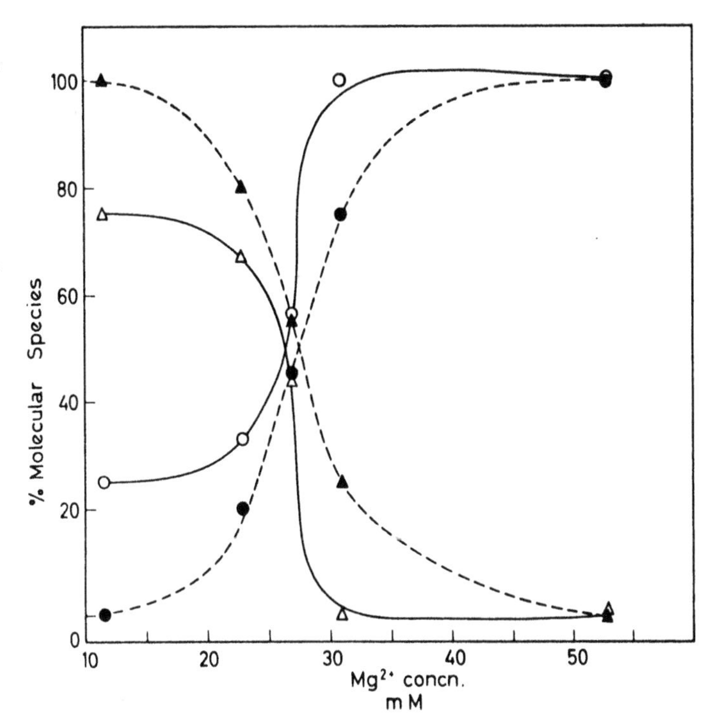

that the oxygen-binding parameters obtained from an oxygenation experiment, as in the present work, relate to the composition of the deoxygenated sample and not to that of the fresh oxygenated sample. Thus, samples containing 11.6 mM Mg^{2+}, which go completely to halves on deoxygenation (Fig. 3), and preparations of pure halves at the same Mg^{2+} concentration, obtained from fresh oxygenated haemocyanin by sedimentation in a partition cell in the analytical ultracentrifuge, showed equivalent oxygen-binding parameters with no significant difference in P_{50}, ΔG and n_H values ($P > 0.10$).

Table 6. Oxygen-binding data for *Murex trunculus* haemocyanin in 0.05 M borate buffer, pH 10.0, at different percentages of wholes, the rest of the mixtures being composed of halves

Wholes %	P_{50} mm Hg	n_H	ΔG kcal mol^{-1}	No. of determinations
100	20.1 ± 0.7	1.7 ± 0.1	5.77 ± 0.02	3
75	19.3 ± 0.7	1.6 ± 0.1	5.75 ± 0.04	3
45	15.9 ± 1.0	1.6 ± 0.1	5.64 ± 0.04	2
20	16.6 ± 0.5	1.5 ± 0.2	5.54 ± 0.15	2
0	14.0 ± 0.6	1.5 ± 0.1	5.50 ± 0.03	8

Values are mean ± S.E.M. (or mean deviation from mean).

Oxygen-binding data for various proportions of wholes and halves are given in Table 6. Both wholes and halves showed cooperative oxygen binding with no significant difference between n_H values ($P > 0.10$). The P_{50} and ΔG were significantly higher in wholes than in halves ($P < 0.001$). The oxygen-binding curves of the mixtures of wholes and halves could be reproduced by linear combination of the curves obtained for pure wholes and pure halves according to the equation:

$$Y_M = \Phi_W Y_W + \Phi_H Y_H, \quad (1)$$

where Y_M is the predicted oxygen saturation of the mixture at each value of P_{O_2}; Y_W and Y_H are the respective oxygen saturations of pure wholes and pure halves at the same P_{O_2}; and Φ_W and Φ_H are the respective fractions of wholes and halves in the mixture. This finding (Fig. 4) indicates that wholes and halves in this system do not interact, but bind oxygen independently of each other and of Mg^{2+} concentration (above 11.6 mM) which is a hidden assumption in Equation (1). Invariance of n_H with Mg^{2+} concentration is seen in Table 6.

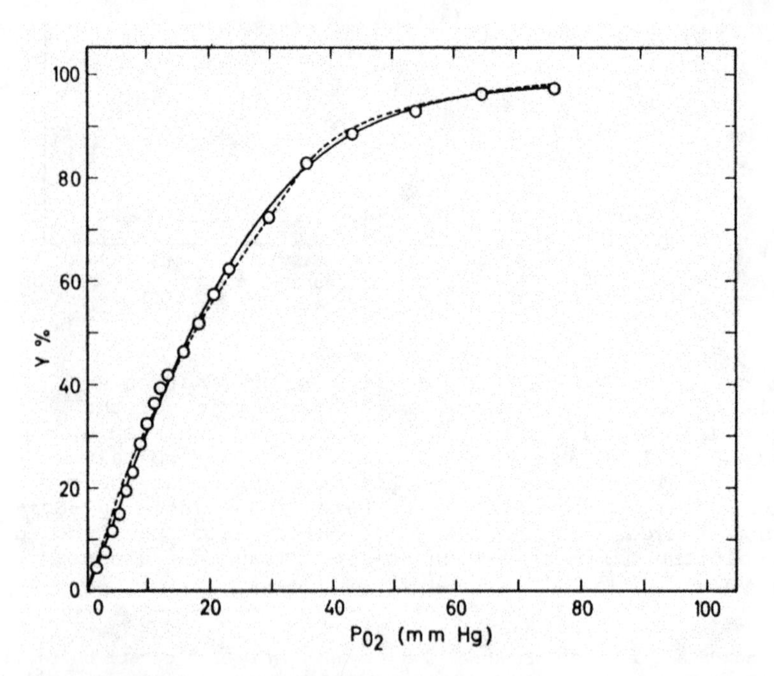

Fig. 4. Experimental (*solid line*) and predicted (*interrupted line*) oxygen-binding curve for *Murex trunculus* haemocyanin at pH 10. The solution contained 45% wholes and 55% halves

Discussion

Cooperativity. As with other haemocyanins cooperativity was enhanced, if not induced, by Ca^{2+}. The effect on wholes (Table 3; pH 7.0) indicates that the action of Ca^{2+} is not merely to prevent dissociation. A specific effect of divalent ions has often been suggested. However, the results at elevated ionic strength in the absence of Ca^{2+} (Table 4) indicate that the presence of divalent ions is not essential for cooperative behaviour. Cooperativity in the absence of divalent ions has been reported by Van Driel and Van Bruggen (1974) for *Helix pomatia* α-haemocyanin at pH 8.2 and an ionic strength of 1.1 mol/l. The cooperativity was associated with an oxygen-linked association of tenths to halves on deoxygenation. It was not possible to follow rapid oxygen-linked equilibria in the present study, and hence it is not known whether the observed cooperativity at high ionic strength was due to an oxygen-linked association-dissociation equilibrium.

The smallest haemocyanin subunit found capable of cooperative oxygen binding was the half molecule. At pH 10.0 in the presence of Mg^{2+} (Table 6) halves showed the same degree of cooperativity as wholes. This indicates that no homotropic interactions between the halves occur in the fully associated molecule at least under the conditions of observation.

A detailed interpretation of the cooperative mechanism in the presence of divalent ions or high ionic strength is not possible. Certainly Ca^{2+} (or Mg^{2+}) ions might reduce the net negative change of the protein, and in fact Ca^{2+} was more effective in mediating cooperativity at pH 7.0 than at pH 8.2 (Table 3). This mertis assessment in the different haemocyanins. Divalent ions might bind to specific electronegative sites distinct from those responsible for association. Thus, at pH 10.0, 11.6 mM Mg^{2+} was sufficient to mediate maximum cooperativity but not to prevent the dissociation to halves (Table 6). Of course, this does not exclude association, however mediated, from being another essential requirement for cooperativity. High ionic strength might be considered to mediate cooperativity by nonspecific charge reduction. At pH 8.2 in the absence of Ca^{2+}, n_H (and the oxygen affinity) decreased at ionic strengths above 0.43 mol/l (Table 5), despite the reduction in net negative charge. Charge reduction might be necessary, but not sufficient, for cooperativity to take place. For a given degree of cooperativity, the oxygen affinity in the presence of Ca^{2+} was lower than that at high ionic strength, especially at pH 8.2 (cf. Table 2 and 4). Possibly Ca^{2+} and high ionic strength mediate cooperativity by different mechanisms.

Interpretation on Monod-Wyman-Changeux Theory. The cooperativity exhibited by molluscan haemocyanins in the presence of divalent ions has been qualitatively explained by various authors in terms of a two-state model, usually according to a simple Monod-Wyman-Changeux (MWC) concerted model (Van Driel, 1973; Vannoppen-Ver Eecke and Lontie, 1973; Van Driel and Van Bruggen, 1974).

Assuming that oxygen binding at low ionic strength in the presence of Ca^{2+} follows a two-state model, the P_{50} values of the T- and R-states for *Murex trunculus* haemocyanin can be calculated to be approximately 42 and 2 mm Hg, respectively, from the intercepts of the Hill asymtotes at Y = 50%. It should be noted that the P_{50} value of the postulated R-state is very close to that of solutions at low ionic strength in the absence of divalent ions, indicating that these are in the R-configuration.

The n_H values in the presence of Ca^{2+} exhibit *some* systematic trend when plotted against log P_{50}, passing through a maximum and approaching unity at certain limiting values of log P_{50} as shown in Figure 5. These limits agree quite well with the P_{50} values calculated for the T- and R-states. This is good evidence that a two-state cooperative mechanism is indeed operating under these conditions (Miller and Van Holde, 1974). On the other hand, the n_H value of solutions at high ionic strength shows no obvious systematic behaviour, indicating that cooperativity under these conditions cannot be explained on a simple two-state model.

Therefore, the oxygen binding of *Murex trunculus* haemocyanin at low ionic strength can be qualitatively interpreted according to the MWC model as follows: In the absence of divalent ions, the haemocyanin exists mainly in the R-state and hence binds oxygen non-cooperatively at all

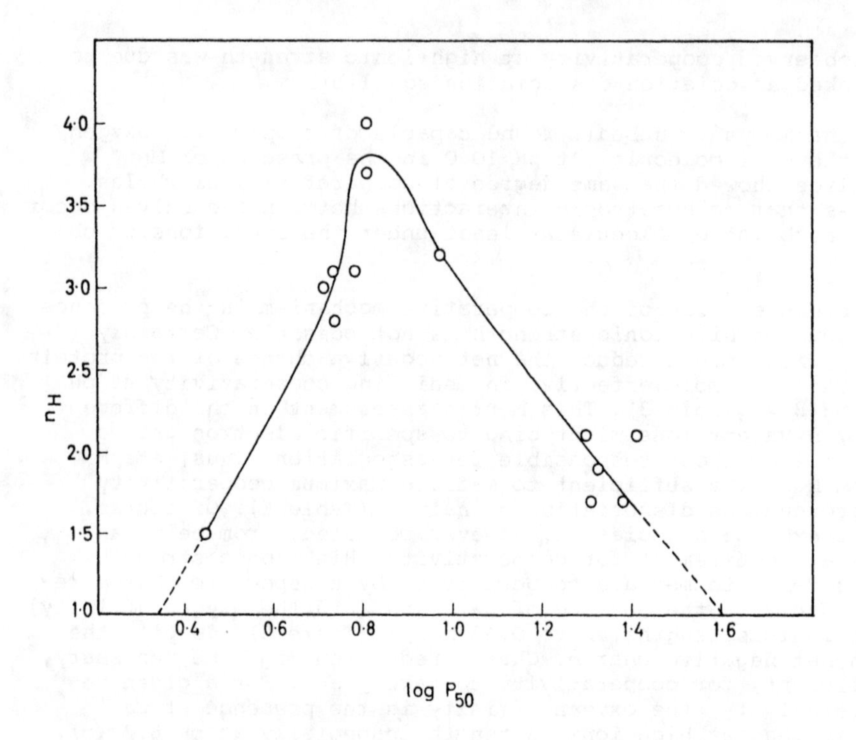

Fig. 5. Variation of n_H with log P_{50} for *Murex trunculus* haemocyanin whole molecules in the presence of Ca^{2+}

pH values. On addition of Ca^{2+} (or Mg^{2+}) to the haemocyanin dissociation is prevented and the divalent ions bind at specific sites of the deoxygenated, associated molecule, constraining it to the low oxygen affinity T-state. On binding of oxygen these divalent ions are gradually liberated until at a certain saturation value the constraint is released and the asociated molecule reverts to the R-state. The T- and R-states bind oxygen non-cooperatively with P_{50} values of approximately 42 and 2 mm Hg, respectively, the observed cooperativity being entirely due to the transition between the T- and R-states.

Further insight into the cooperative mechanism is possible by making a semiquantitative analysis in the framework of the MWC model, as described by Colosimo et al. (1975) for *Spirographis spallanzanii* chlorocruorin. According to these authors, it can be proved that when the number of interacting sites, r, on a macromolecule is sufficiently large, and when the T → R transition occurs in the region of Y = 50%, then the total apparent interaction free energy is given by

$$\Delta G_I = RT \ln \alpha, \qquad (2)$$

where α is the ratio of the binding constant for a site in the R-form to that for a site in the T-form.

The maximum value of n_H encountered in this study was 4.0 at Y = 50%, which occurred at pH 7.0 in the presence of 46 mM Ca^{2+}. A Hill plot for a sample under these conditions is given in Figure 6. The minimum value of ΔG_I can be estimated from this plot to be 2320 cals (Wyman, 1964; Colosimo et al., 1975). This value of ΔG_I corresponds, according to Equation (2), to a minimum value of α of 52.5. From computer-simu-

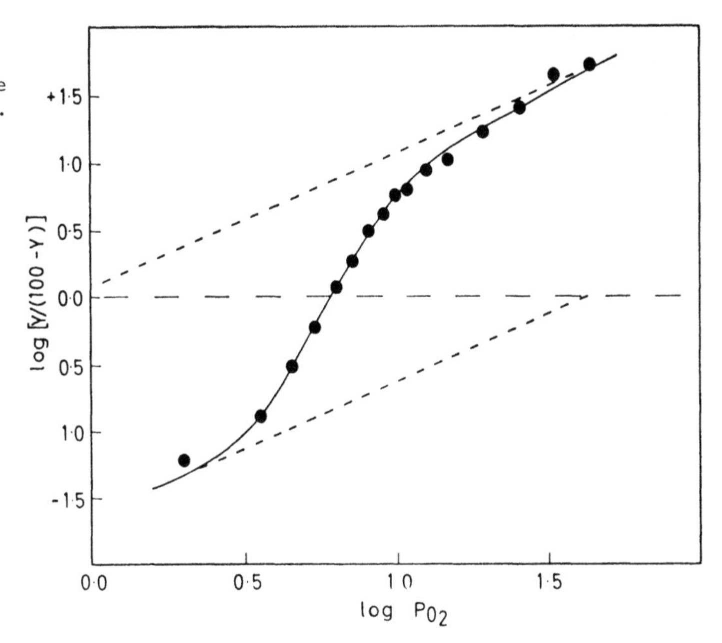

Fig. 6. Hill plot for *Murex trunculus* haemocyanin at pH 7.0 in the presence of 46 mM Ca^{2+}. *Interrupted lines:* asymptotes

lated data reported by Colosimo et al. (1975) it can also be seen that a value of ΔG_I of 2320 cal corresponds to a value of n_H/r of 0.57. Hence, the minimum value of n_H expected is 0.57/140 to 0.57/160 or 80 to 90, compared to a maximum value of 4.0 observed. Obviously these results are incompatible with the assumption that the whole molecule, with its 140 to 160 sites, behaves as a simple MWC model.

As the next simplest hypothesis it can be assumed that the sites in the whole molecule interact in independent constellations, each containing r' sites (Colosimo et al., 1975). The value of n_H observed would then be the same as that for a single constellation. The value of r' can be calculated from the relationship:

$$r' = (n_H + 2\lambda - 2)/(2\lambda - 1), \qquad (3)$$

where, $\lambda = (1 + \alpha)/(1 + \alpha^{1/2})^2$.

Using Equation (3) for $n_H = 4$ and $\alpha = 52.5$, r' is found to be 6.2. Furthermore, by using the above procedure on all the data for samples at low ionic strength in the presence of Ca^{2+}, it turns out that in all cases r' lies close to 6. This value is very close to the number of sites (7 to 8) in a twentieth molecule, indicating that this species is the basic cooperative unit in the haemocyanin.

Such a value of r', being based on a minimum value of α, only provides the minimum number of interacting sites, and hence does not rule out the possibility of interactions between the twentieths. Indeed, Van Driel (1973) has obtained some evidence that interactions between the tenths in *Helix pomatia* haemocyanin whole molecules are essential for cooperative behaviour. Nevertheless, the low r' value for *Murex trunculus* haemocyanin suggests that any such interactions between twentieths are of a secondary nature, the primary interactions occurring between sites on the same twentieths. The twenthieths must probably be constrained within the framework of a whole or half molecule before these cooperative interactions are possible.

Free Energy of Oxygenation. The total free energy change, ΔG, per mol of oxygen bound, was found to lie in the range 4.5-6.0 kcal. The values are close to those reported by Er-el et al. (1972) for *Levantina hierosolima* haemocyanin, and lower by about 2 kcal than those normally observed in haemoglobins (Antonini and Brunori, 1971). This ΔG represents the total free energy change per site for saturation of the haemocyanin with oxygen, and includes the free energy of interaction between sites in cooperative conditions. The higher the value of ΔG the more difficult is the oxygenation process, and vice versa. In fact the ΔG was found to follow closely the trend in oxygen affinity of the haemocyanins.

Linkage: H^+, Ca^{2+}. During oxygenation the change in the total number of protons (ΔH^+) or Ca^{2+} ions (ΔCa^{2+}) bound to haemocyanin can be calculated from the following equations:

$$d \log x_m / d\,pH = \Delta H^+/r, \tag{4}$$

$$d \log x_m / d \log [Ca^{2+}] = -\Delta Ca^{2+}/r, \tag{5}$$

where x_m is the median ligand activity and r denotes the total number of oxygen binding sites (Wyman, 1964).

Table 7. Total change in the number of bound protons (ΔH^+) during the oxygenation of *Murex trunculus* haemocyanin

Sample	pH	ΔH^+
(a) Low ionic strength		
Haemocyanin less Ca^{2+}	7.6-8.2	40 to 50
Haemocyanin plus 10 mM Ca^{2+}	7.6-8.2	90 to 100
Haemocyanin plus 10 mM Ca^{2+}	8.2-8.7	-36 to -41
(b) High ionic strength		
Haemocyanin less Ca^{2+}	7.6-8.2	100 to 115
Haemocyanin plus 10 mM Ca^{2+}	7.6-8.2	115 to 130

The value of ΔH^+, obtained from Equation (4), under various conditions is given in Table 7. It can be seen that in the pH range 7.6-8.2, H^+ ions are always bound (reverse Bohr effect) as oxygenation proceeds. Furthermore, it can be seen that the total number of H^+ ions bound can be very large indeed. On the other hand in the pH range 8.2-9.3 in the presence of Ca^{2+} *Murex trunculus* haemocyanin shows a normal Bohr effect, and H^+ ions are liberated during oxygenation. The presence of both a reverse and a normal Bohr effect has also been observed for a number of other haemocyanins including *Helix pomatia* haemocyanin (Vannoppen-Ver Eecke et al., 1973), and is also well known in haemoglobins (Antonini and Brunori, 1971). Therefore, the Bohr effect in haemocyanins can probably be explained, in analogy with that of haemoglobins, as being due to two oxygen-linked groups whose pK_i is modified on oxygenation. From the titration data for *Octopus vulgaris* haemocyanin reported by Salvato et al. (1974), the oxygen-linked group responsible for the reverse Bohr effect (pH 7.0-8.2) can be tentatively assigned as histidine imidazole, while that responsible for the normal Bohr effect (pH 8.2-9.3) can be tentatively assigned as imidazole or α-amino.

The linkage between Ca^{2+} binding and oxygenation can be explained on the same lines as that between H^+ binding and oxygenation. Equation (5) indicates that at low ionic strength 28-32 Ca^{2+} ions are liberated during oxygenation of *Murex trunculus* haemocyanin, both at pH 7.0 and pH 8.2. As mentioned above, the release of these Ca^{2+} ions may be instrumental in mediating the conformational change from the T-state to the R-state. Since the same number of Ca^{2+} ions are liberated at both pH 7.0 and pH 8.2, the number of oxygen-linked Ca^{2+} binding groups probably does not increase with pH in this pH range. This, in turn, indicates that the amino acid residues responsible for bonding oxygen-linked Ca^{2+} are already fully ionized at pH 7.0. When this is coupled with the fact that Ca^{2+} ions generally bind at oxygen anion sites (Williams, 1970), it seems probable that the oxygen-linked Ca^{2+}-binding groups are carboxylic groups.

Acknowledgments. W.H.B. thanks the Wellcome Trust for financial support.

References

Antonini, E., Brunori, M.: Hemoblobin and Myoglobin in Their Reactions with Ligands. Amsterdam: North Holland, 1971

Bannister, J.V., Galdes, A., Bannister, W.H.: Isolation and characterization of two-copper subunits from *Murex trunculus* haemocyanin. Comp. Biochem. Physiol. 51B, 1 4 (1975)

Colosimo, A., Brunori, M., Wyman, J.: Concerted changes in an allosteric macromolecule. In: Protein Ligand Interactions. Sund, H., Blauer, G. (eds.). Berlin: de Gruyter and Co., 1975, pp. 3-14

Er-el, Z., Shaklai, N., Daniel, E.: Oxygen-binding properties of haemocyanin from *Levantina hierosolima*. J. Mol. Biol. 64, 341-352 (1972)

Miller, K., Van Holde, K.E.: Oxygen binding by *Callianassa californiensis* hemocyanin. Biochemistry 13, 1668-1674 (1974)

Salvato, B., Ghiretti-Magaldi, A., Ghiretti, F.: Acid-base titration of hemocyanin from *Octopus vulgaris* Lam. Biochemistry 13, 4778-4783 (1974)

Van Driel, R.: Oxygen binding and subunit interactions in *Helix pomatia* hemocyanin. Biochemistry 12, 2696-2698 (1973)

Van Driel, R., Van Bruggen, E.F.J.: Oxygen-linked association-dissociation of *Helix pomatia* hemocyanin. Biochemistry 13, 4079-4083 (1974)

Vannoppen-Ver Eecke, T., D'Hulster, R., Lontie, R.: Influence of the saturation with carbon monoxide and of deoxygenation on the alkaline dissociation of *Helix pomatia* haemocyanin. Comp. Biochem. Physiol. 46B, 499-507 (1973)

Vannoppen-Ver Eecke, T., Lontie, R.: The effect of alkaline earth ions on the cooperativity of the oxygenation of *Helix pomatia* haemocyanin. Comp. Biochem. Physiol. 45B, 945-954 (1973)

Williams, R.J.P.: The biochemistry of sodium, potassium, magnesium and calcium, Quart. Rev. 24, 331-365 (1970)

Wood, E.J., Dalgleish, D.G.: *Murex trunculus* haemocyanin. 2. The oxygenation reaction and circular dichroism. Europ. J. Biochem. 35, 421-427 (1973)

Wyman, J.: Linked functions and reciprocal effects in hemoglobin: a second look. Advan. Protein Chem. 19, 223-286 (1964)

Properties of the Oxygen-Binding Domains Isolated from Subtilisin Digests of Six Molluscan Hemocyanins

J. BONAVENTURA, C. BONAVENTURA, AND B. SULLIVAN

Abstract

Molluscan hemocyanins appear to be composed of polypeptide subunits of unusually large size. It is known that these large subunits can be fragmented without denaturation by treatment with the proteolytic enzyme subtilisin. Some of the chromatographic zones are homogeneous, as evidenced by single bands in regular and SDS-gel electrophoresis. These artificial subunits or domains have molecular weights in the range of 50,000. Functional heterogeneity of these artificial subunits is indicated by differences in their oxygen affinities and in their pH and NaCl sensitivities. In contrast, subtilisin digestion of an arthropodan hemocyanin, that of *Limulus polyphemus*, cleaves peptide bonds, but fragments are not observed unless the molecule is denatured.

Introduction

The functional properties of hemocyanins found in arthropods and mollusks are qualitatively similar (Van Holde and Van Bruggen, 1971; Bonaventura et al., 1976). Both kinds of molecule exhibit comparable homotropic and heterotropic allosteric effects. There is nothing in these properties which suggests underlying structural dissimilarity. It is known, however, that both the molecular architecture and subunit structure of these two kinds of hemocyanin are quite different. One common theme in the hemocyanins is the ratio of copper to bound molecular oxygen - 2:1. Beyond this the structural similarities appear to stop, although comparison of their primary structures is not yet possible. Arthropodan hemocyanin subunits are relatively small (70-85,000 daltons). and their oligomeric structures are generally hexamers and dodecamers of 450,000 and 900,000 daltons. Molluscan hemocyanins have subunits whose size is comparable to hexamers of arthropod hemocyanin. For *Helix pomatia* hemocyanin the subunit is reported to have a molecular weight of 265,000-285,000 (Brouwer and Kuiper, 1973; Siezen and Van Bruggen, 1974; Waxman, 1975). The copper content of molluscan hemocyanins indicates that a subunit of this size possesses five to six pairs of copper atoms. Knowledge that the binding of molecular oxygen by hemocyanins is associated with pairs of copper atoms raises a number of questions concerning the molluscan hemocyanin subunit. Are the copper sites of molluscan hemocyanins regularly distributed in the 365,OO-dalton subunit or is there a specific site on the subunit with a cluster of oxygen-binding sites? Is there some regularity or repeating nature to the copper sites? Are the subunits identical? Data being gathered in a number of laboratories are answering these questions. The emerging picture of molluscan hemocyanin subunits is one of a flexible string of beads, or domains, strung end to end. Each bead possesses two copper atoms and can bind one molecule of oxygen.

It has been known for several years that digestion of native molluscan hemocyanins with subtilisin causes the release of functional "domains" whose molecular weight is of the order of 50,000 (Lontie et al., 1973; Pearson and Wood, 1974). Electron microscopic observation of molluscan hemocyanin subunits has shown a beaded arrangement in which the individual beads have sizes consistent with a 50,000-daltons bead, pearl, or domain (Siezen and Van Bruggen, 1974). Recent experiments (Gielens et al., 1975; Brouwer et al., 1976) have confirmed the string-of-pearls arrangement of the molluscan hemocyanin subunit. In this report, we address ourselves to the structure-function relationships in molluscan hemocyanins and examine the functional properties of the molluscan hemocyanin domains released upon subtilisin digestion. We have also subjected arthropodan hemocyanin subunits to subtilisin digestion and find that their digestion products differ substantially from those of molluscan hemocyanins.

Materials and Methods

Octopus sp., *Busycon carica, Littorina littorea,* and *Murex fulvescens* were collected in the area around Beaufort, North Carolina. *Octopus dofleini* were collected in Puget Sound, Washington. *Nautilus pompilius* were collected aboard the R/V ALPHA HELIX. Hemolymph from the latter two was kindly supplied by Dr. Arthur W. Martin, Department of Zoology, The University of Washington, Seattle, Washington. Hemolymph from the other species was obtained by direct cardiac puncture or sampling from hemocoels. Since more than 90% of the hemolymph protein of these species is hemocyanin,no further purification was undertaken. Digestion of the hemocyanins was done with a 0.1 weight percentage of subtilisin (Schwarz-Mann lot No. X 3865) in 0.05 M borate pH 8 (Lontie et al., 1973). After digestion for 4 h at room temperature, the digest mixture was dialyzed vs. 0.05 M Tris, 0.01 M EDTA pH 8.9, in the cold for 24 h. Chromatography of the digest mixtures was done on columns of DEAE-Sephadex A50. The ion exchanger was equilibrated with 0.05 M Tris, 0.01 M EDTA pH 8.9 and column development was effected with a linear gradient in NaCl from zero to 0.5 M salt. Fractions were monitored at both protein and copper-oxygen absorption bands (280 nm and 338-346 nm respectively, the latter wavelength depending on the particular hemocyanin). Regular and SDS gel electrophoresis was performed as previously described (Sullivan et al., 1976). Oxygen equilibria were performed using a tonometric procedure (Riggs and Wolbach, 1956). The rate of dissociation of oxygen from the oxyhemocyanin was observed by rapid mixing of air-equilibrated protein with dithionite-containing buffers in a Durrum stopped-flow spectrophotometer. The photomultiplier output was processed by an Aminco analog-digital converter and transient recorder (DASAR) coupled to a PDP-11 minicomputer and Tektronix graphics terminal. Difference spectra were obtained with an Aminco DW-2A spectrophotometer.

Results and Discussion

Molluscan Hemocyanin Digests. We have digested a number of molluscan hemocyanins with the proteolytic enzyme subtilisin. All of the digests so far examined show electrophoretic heterogeneity. This heterogeneity is apparent on regular and SDS gel electrophoresis. When the digests are applied to ion-exchange columns and developed with salt gradients as described in Materials and Methods, separation of many of the lectrophoretic zones is achieved. Figures 1-6 illustrate the chromatographic separation of subtilisin digests of *Octopus dofleini, Octopus* sp., *Nautilus pompilius, Littorina littorea* and *Murex fulvescens*, respectively. Some of the chromatographic zones are homogeneous as evidenced by the appearance of single bands on regular and SDS gels. Regular polyacrylamide gels for the zones of Figures 1-6 are shown in Figure 7. Given these results, we may question the specificity of the subtilisin digestion.

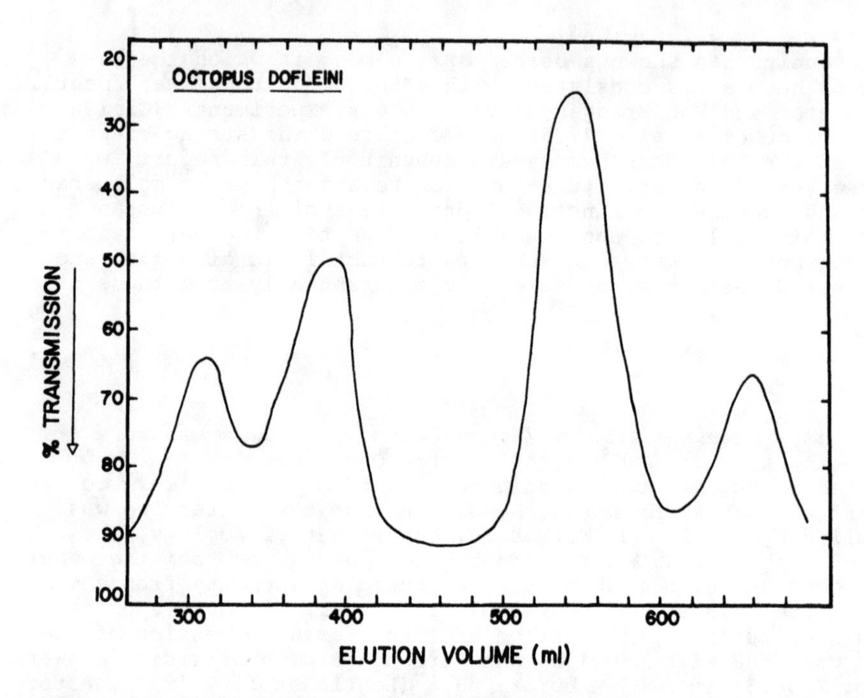

Fig. 1. Chromatogram obtained from DEAE-Sephadex chromatography of a subtilisin digest of *Octopus dofleini* hemocyanin. The fractions were monitored at a wavelength of 340 nm

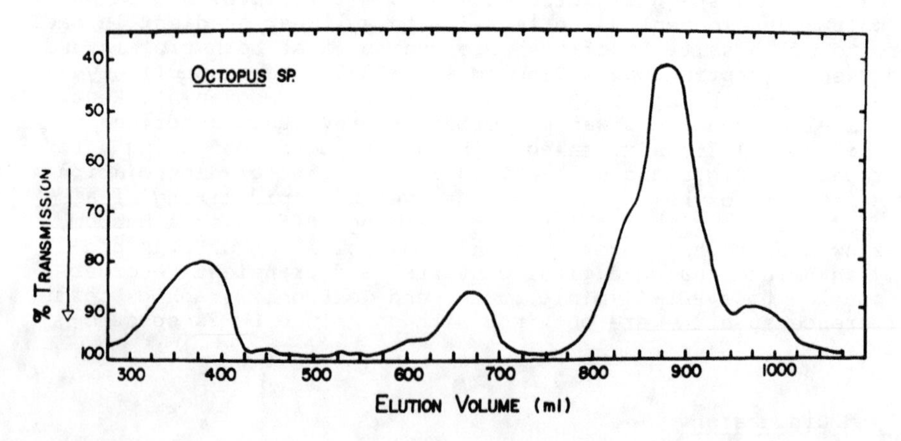

Fig. 2. Chromatogram obtained from DEAE-Sephadex chromatography of a subtilisin digest of *Octopus* sp. (North Carolina) hemocyanin. The fractions were monitored at a wavelength of 338 nm

In the case of subtilisin digests of *Octopus* sp. hemocyanin there is little variation between digests done at different times. Elution patterns from chromatographies of different digestions are similar in character, and electrophoretic mobilities of the separated zones are similar from one chromatography to the next. Thus, even though subtilisin may digest random coil polypeptides nonspecifically, the subtilisin-prepared hemocyanin fragments may result from rather specific cleavage of the polypeptide chain.

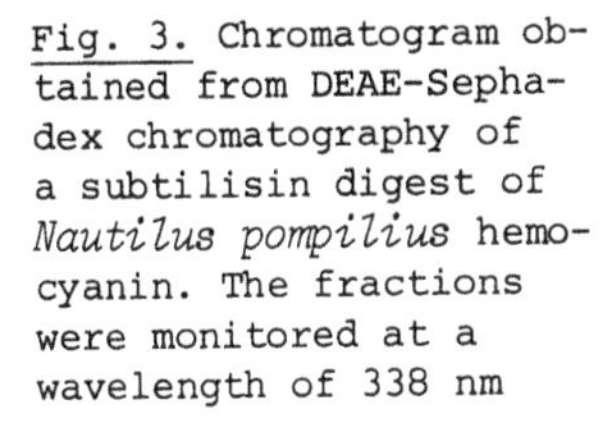

Fig. 3. Chromatogram obtained from DEAE-Sephadex chromatography of a subtilisin digest of *Nautilus pompilius* hemocyanin. The fractions were monitored at a wavelength of 338 nm

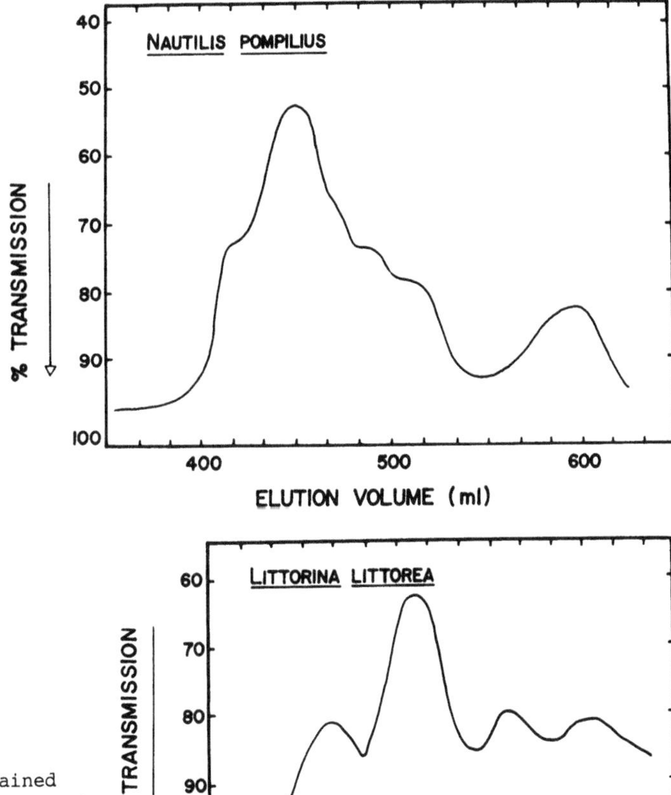

Fig. 4. Chromatogram obtained from DEAE-Sephadex chromatography of a subtilisin digest of *Littorina littorea* hemocyanin. The fractions were monitored at a wavelength of 346 nm

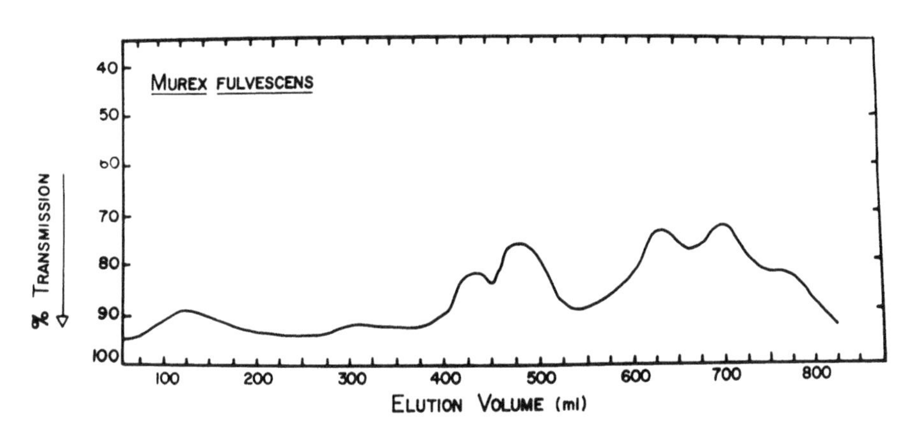

Fig. 5. Chromagrogram obtained from DEAE-Sephadex chromatography of a subtilisin digest of *Murex fulvescens* hemocyanin. The fractions were monitored at a wavelength of 338 nm

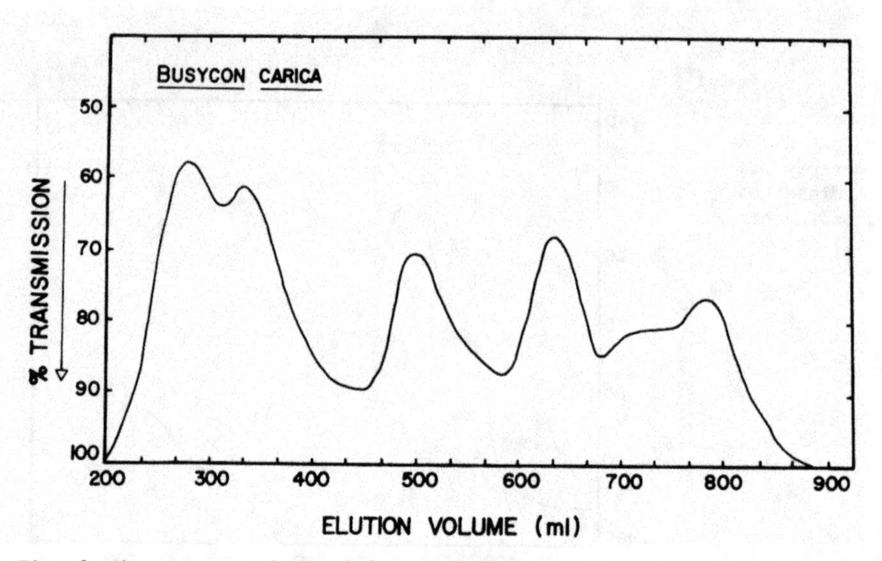

Fig. 6. Chromatogram obtained from DEAE-Sephadex chromatography of a subtilisin digest of *Busycon carica* hemocyanin. The fractions were monitored at a wavelength of 340 nm

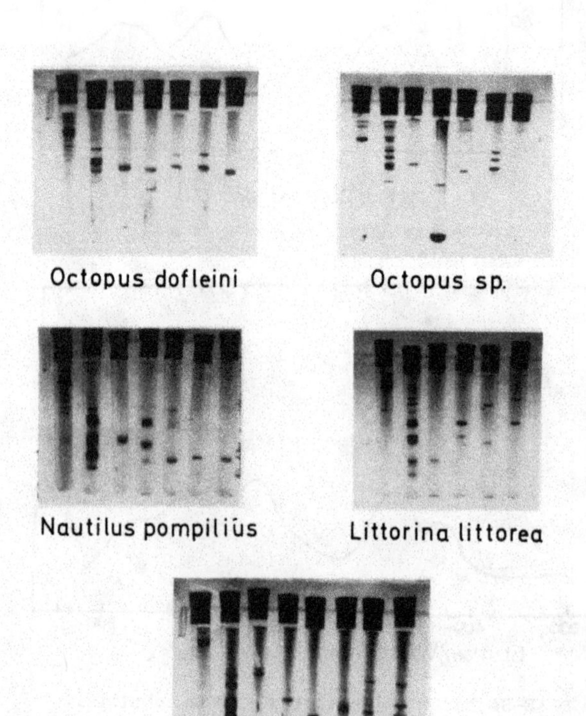

Fig. 7. Photographs of regular disc gels. In all cases the undigested hemocyanin is on the *extreme left*. *Immediately to the right* is the whole subtilisin digest mixture followed by the various chromatographic zones

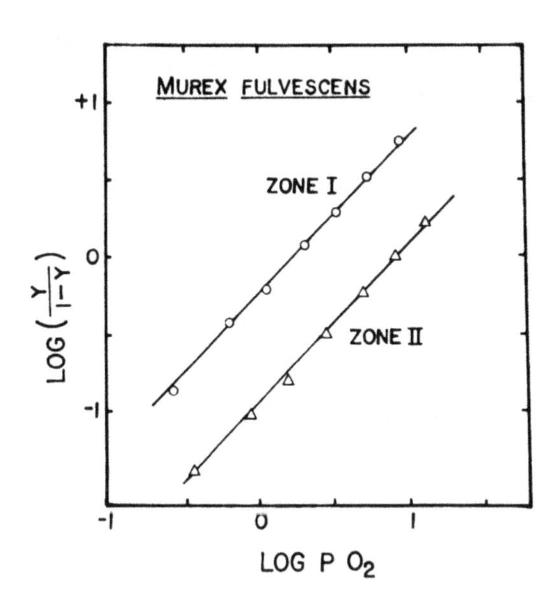

SDS Gels

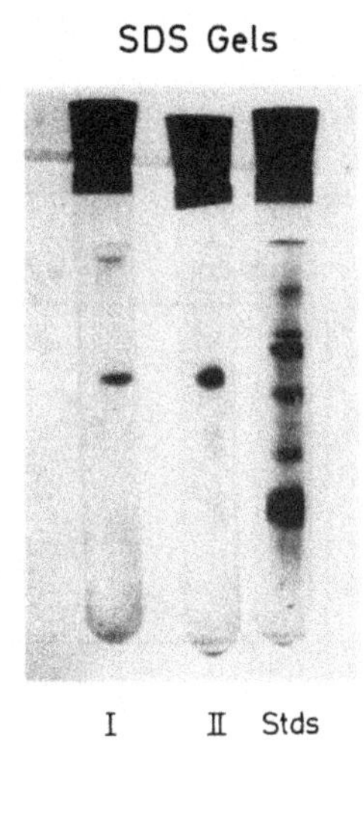

Fig. 9

Fig. 8

Fig. 8. Hill plots obtained from oxygen equilibrium experiments done with *Murex fulvescens* subtilisin zones I and II. The equilibria were done in 0.05 M Tris 0.01 M EDTA, pH 9 and 20°C. *Line drawn through data points:* n = 1.0

Fig. 9. SDS-gel electrophoresis of DEAE-Sephadex zones I and II derived from subtilisin-digested *Murex fulvescens* hemocyanin. Proteins of known molecular weight are shown in the gel on the right. *Murex* zones I and II migrate with mobilities between those of ovalbumin and serum albumin (43,000 and 68,000 daltons respectively)

The chromatographic fractionation of subtilisin-digested molluscan hemocyanins into more or less purified zones makes it possible to examine their functional properties. Subtilisin digestion and subsequent chromatographic separation of the digests into zones having varying electrophoretic mobilities does not destroy the oxygen-binding capability. Figure 8 shows Hill plots for oxygen binding to two chromatographic zones obtained from a subtilisin digest of *Murex* hemocyanin. These two zones (I and II) give fairly homogeneous distinct bands on regular and SDS disc gels (Figs. 7 and 9). Comparison of the electrophoretic mobilities of *Murex* I and II on SDS gels with standards as shown in Figure 9 indicates that zones I and II of the *Murex* hemocyanin digest have molecular weights near 50,000. It is apparent from the Hill plots of Figure 8 that the domains corresponding to zones I and II of the *Murex* hemocyanin chromatography have very different oxygen affinities. Moreover, the Hill plots for oxygen binding are fairly well represented by a Hill coefficient of 1.

The oxygen-binding properties of the other zones obtained from subtilisin digests of the hemocyanins (shown in Figs. 1-6) were also studied. The oxygen affinity of the hemocyanins found in the various chromatographic zones often showed variations. Values of the Hill coefficient were frequently less than 1. This thus suggests the presence of more than one species of oxygen-binding material within some of the chromatographic zones.

The pH dependence of oxygen binding by the chromatographically purified zones varied appreciably. Bohr effects were sometimes positive, sometimes negative, and sometimes negligible in domains obtained from a hemocyanin which normally possesses a strong Bohr effect. Figures 10 and 11 illustrate this point, and also another feature of the chro-

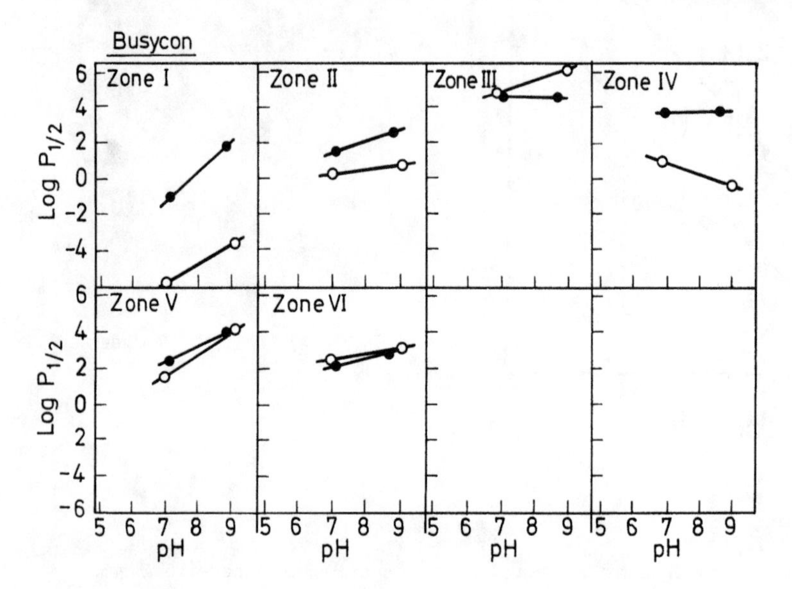

Fig. 10. The pH dependence of oxygen binding by *Busycon carica* hemocyanin subtilisin zones I through VI in the absence (O) and presence (●) of 3 M NaCl. Experiments performed in 0.05 M Bis-Tris and Tris buffers at 20°C

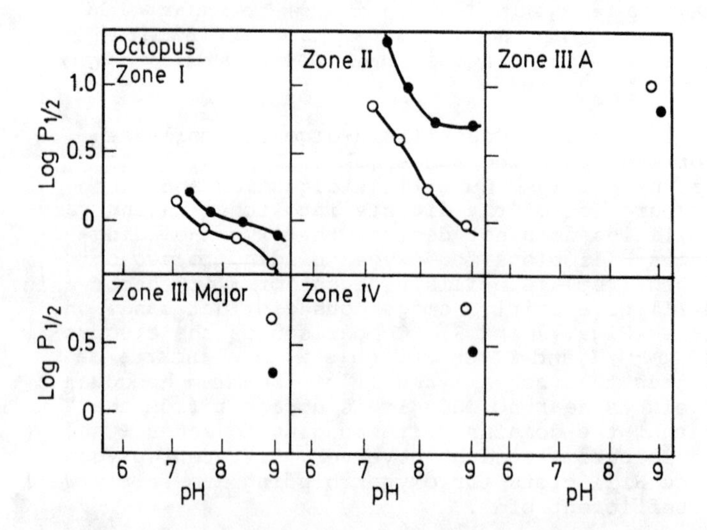

Fig. 11. The pH dependence of oxygen binding by *Octopus* sp. hemocyanin subtilisin zones I through IV in the absence (O) and presence (●) of 3 M NaCl. Experiments performed in 0.05 M Bis-Tris and Tris Buffers at 20°C

matographically separated zones. The different zones have variations in their sensitivities to sodium chloride. This has not been investigated with all of the hemocyanins thus far digested with subtilisin, but it is clearly the case for both *Octopus* and *Busycon* hemocyanin domains as illustrated in Figures 10 and 11. The pH dependence of the oxygen affinities of the chromatographically separated zones from subtilisin digests of the hemocyanins from four other molluscan species is given in Table 1. Inspection of this table shows that *Busycon* and *Octopus* hemocyanin digests are not unique in having zones or domains which differ in their oxygen affinities and pH sensitivities.

Table 1. Oxygen affinities of the chromatographic zones isolated from subtilisin digests of various molluscan species

Protein	Zone	P 1/2 (mm Hg) pH 8	pH 9
Octopus dofleini	I	4.7	2.7
	II	7.9	5.0
	III	12.9	6.2
	IV	44.7	15.3
Murex fulvescens	I	2.4	1.7
	II	9.6	7.9
	III	3.4	3.5
	IV	8.5	8.5
	V	-	5.6
	VI	7.2	9.1
Nautilus pompilius	I	6.8	5.9
	II	3.1	2.75
	III	4.5	4.2
	IV	4.7	4.3
	V	5.2	5.0
Littorina littorea	I	8.3	10
	II	1.8	3.2
	III	2.9	3.2
	IV	2.6	2.9

Experiments done in 0.05 M Tris, 0.01 M EDTA, pH 8 and 9 at 20°C.

The oxygen kinetics of many of the chromatographic zones were examined by measurement of oxygen dissociation by the rapid-mixing technique. In the domains thus far examined, there are large variations in the rates of oxygen dissociation. The results of the stopped-flow experiments is given in Table 2. The differences in the oxygen dissociation velocity constants do not parallel the differences in oxygen affinity. Thus there may also be significant differences in the second-order oxygen combination velocity constants of the domains.

Spectrophotometric measurements made in the course of oxygen-binding experiments showed variations in the absorption spectra of the oxygenated domains. This phenomenon was examined in more detail with three of the zones obtained from a subtilisin digest of *Octopus* sp. hemocyanin. The absolute spectra of zones I, II and III show differences in the wavelength maxima of their oxygenated forms. Figure 12 presents difference spectra which show the spectral heterogeneity of these domains.

Digests of Arthropodan Hemocyanin Subunits. Contrasting results were obtained for subtilisin digests of the subunits of an arthropodan hemocyanin. *Limulus* hemocyanin components, I-Vb, were separated by DEAE-Sephadex

Table 2. Apparent rates of oxygen dissociation from chromatographic zones of subtilisin digests of various molluscan hemocyanins in 0.05 M Tris, 0.01 M EDTA, pH 8 at 20°C

Protein	Zone	Rate (s^{-1})
Octopus sp.	I	4.3
	II	8.3
	IIIA	80.2
	III major	70.1
Octopus dofleini	I	40
	II	62
	III	60
	IV	98
Nautilus	I	40
	II	21
	III	32
	IV	34
	V	37
Murex	I	46
	III	27
	IV	35
	V	9
	VI	6

The apparent rates do not pertain to any portion of the reaction lost in the 2.3 ms dead-time of the stopped-flow apparatus.

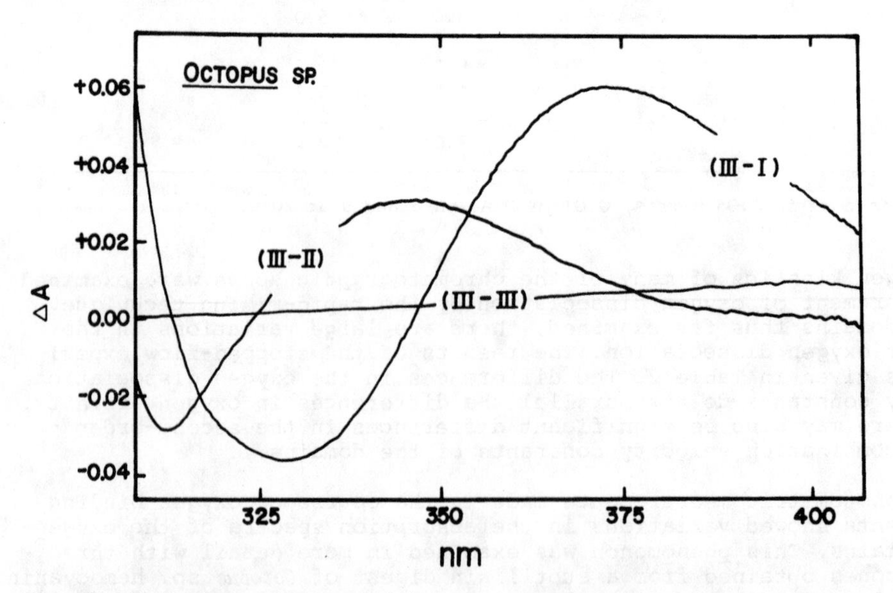

Fig. 12. Difference spectra of *Octopus* sp. hemocyanin zones I, II and III. Matched 1 cm quartz spectrophotometer cells were filled with oxyhemocyanin in 0.05 M Tris buffer at pH 8, 20°C. The optical densities were matched (± 0.005A) at 280 nm prior to obtaining the spectra

chromatography as previously described (Sullivan et al., 1974). Each component (80-100 mg) was dialyzed vs. 0.1 M borate buffer at pH 8. Digestion was started by the addition of 0.075 mg of subtilisin and was continued for 4h at 32°C. The digests were then dialyzed vs. 0.05 M Tris, 0.01 M EDTA, pH 8.9. This is essentially the same procedure as was used for the molluscan hemocyanins.

Regular polyacrylamide disc gel electrophoresis of the subtilisin digests of the various *Limulus* hemocyanin components showed little evidence of proteolytic cleavage. *Limulus* hemocyanins I and IV, which are single bands on regular disc gels, remain unchanged in their electrophoretic behavior after subtilisin digestion. These results are shown in Figure 13. In contrast, SDS gel electrophoresis of subtilisin digests of Hcy

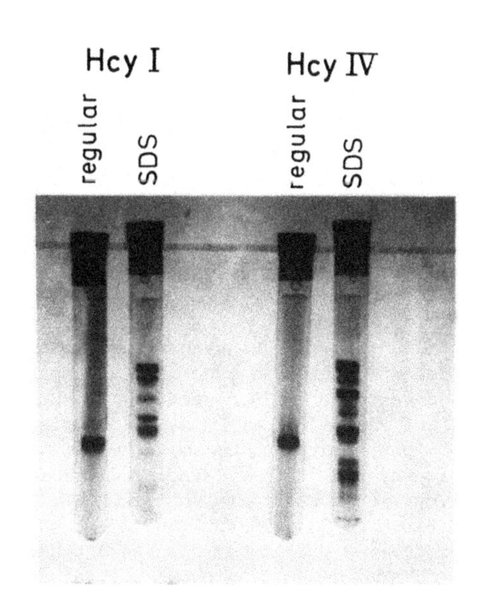

Fig. 13. Photograph of regular and SDS-gel electrophoresis patterns obtained from subtilisin digested *Limulus* hemocyanin I and IV respectively. Note that multiple bands do not appear until the protein is denatured with SDS even though the polypeptide has been cleaved by subtilisin

I and IV reveals extensive cleavage of these polypeptide chains. The SDS gels of the subtilisin-digested *Limulus* hemocyanin components I and IV are shown in Figure 13. Hcy I and IV normally migrate as single bands on SDS gels with molecular weights in the range 65-70,000 (Bonaventura et al., 1975; Sullivan et al., 1976). It is notable that the cleavage products due to subtilisin digestion separate only after denaturation by SDS. Similar behavior has been noted for digestion products of native *Limulus* hemocyanin components obtained after tryptic digestion (Bonaventura et al., 1975; Sullivan et al., 1976).

Concluding Remarks. From the foregoing, it is apparent that functionally distinct domains of molluscan hemocyanins can be obtained as a result of subtilisin digestion. In the case of *Octopus* hemocyanin, spectral differentiation between the chromatographic zones has also been shown to exist. It appears that the digestion with subtilisin does not in itself induce the heterogeneity. The significance of the heterogeneity between domains is still unresolved. It has not been clearly established that the subunits of molluscan hemocyanin are in fact all alike. Therefore the isolated domains may not represent fragments from a single molluscan hemocyanin subunit. It is possible that the domains which we have characterized are fragments of different hemocyanin subunits, although this seems less likely. Although more extensive experiments are clearly needed, our results lead us to the hypothesis that molluscan hemocyanin subunits are made up of domains whose oxygen-binding sites differ to some extent in oxygen affinity and sensitivity to pH and ionic conditions. The possible physiological significance of this kind of a molecular structure is not clear, nor is it obvious to what extent the differences which we report for the domains would be expressed within the hemocyanins subunit or the multisubunit aggregate found in vivo. The pssibility of preparing domains does, however, provide a means of functionally dissecting the large molluscan hemocyanin subunits. Cross-linking experiments presently in progress may make it possible to determine which domains are adjacent to each other in the native polypeptide. As a further comment, it seems that having electrophoretically homogeneous 50,000 molecular weight domains may make it possible to assess realistically the differences in primary structure between arthropodan and molluscan hemocyanins.

Acknowledgments. Supported in part by Grant HL 15460 from the National Institutes of Health, Grants BMS 73-01695 A01 and BMS 71-01432 A02 from the National Science Foundation. We thank Shirley Bourne, Giulia Ferruzzi, Gerald Godette, Louise Pennell and Marjorie Ryan for their extremely competent technical assistance. Joseph Bonaventura is an Established Investigator of the American Heart Association.

References

Bonaventura, J., Bonaventura, C., Sullivan, B.: Hemoglobins and hemocyanins. Aspects of structure and function. J. Exptl. Zool. 194, 155-174 (1975)

Bonaventura, J., Bonaventura, C., Sullivan, B.: Non-heme oxygen transport. In: Oxygen and Physiological Function. Jobsis, F. (ed.). Dallas, Texas: Professional Information Library (in press, 1976)

Brouwer, M., Kuiper, H.A.: Molecular weight analysis of *Helix pomatia* alpha hemocyanin in guanidine hydrochloride, urea and sodium dodecyl sulphate. Europ. J. Biochem. 35, 428-435 (1973)

Brouwer, M., Wolters, M., Van Bruggen, E.F.J.: Proteolytic fragmentation of *Helix pomatia* α-hemocyanin. Structural domains in thepolypeptide chain. Biochemistry (in press, 1976)

Gielens, C., Preaux, G., Lontie, R.: Limited trypsinolysis of beta hemocyanin of *Helix pomatia*. Europ. J. Biochem. 60, 271-280 (1975)

Holde, K.E. Van, Bruggen, E.F.J. Van: The hemocyanins. In: Timasheff, S.N., Fasman, G.D. (eds.). Biological Macromolecules Series. New York: Marcel Dekker, 1971, Vol. V, pp. 1-53

Lontie, R., DeLey, M., Robberecht, H., Witters, R.: Isolation of small functional subunits of *Helix pomatia* hemocyanin after subtilisin treatment. Nature (New Biol.) 242, 180-182 (1973)

Pearson, J.S., Wood, E.J.: Attempts to obtain small functional subunits of the haemocyanins from *Buccinum undatum* and *Neptunea antiqua*. Biochem. J. 2, 333-336 (1974)

Riggs, A., Wolbach, R.A.: Sulphydryl groups and the structure of hemoglobin. J. Gen. Physiol. 39, 585-605 (1956)

Siezen, R.J., Bruggen, E.F.J. Van: Structure and properties of hemocyanins XII. Electron microscopy of dissociation products of *Helix pomatia* alpha hemocyanin. Quaternary structure. J. Mol. Biol. 90, 77-89 (1974)

Sullivan, B., Bonaventura, C., Bonaventura, J.: Functional differences in the multiple hemocyanins of the horseshoe crab, *Limulus polyphemus* L. Proc. Natl. Acad. Sci. 71, 2558-2562 (1974)

Sullivan, B., Bonaventura, J., Bonaventura, C., Godette, G.: Hemocyanin of the horseshoe crab. I. Structural differentiation of the isolated components. Submitted to J. Biol. Chem. (1976)

Waxman, L.: The structure of arthropod and mollusc hemocyanin. J. Biol. Chem. 250, 3796-3806 (1975)

Oxygen-Binding of Associated and Dissociated *Octopus vulgaris* Hemocyanin

B. SALVATO AND L. TALLANDINI

Introduction

The oxygen-binding properties of several hemocyanins (Hcy) have been studied for many years (Ghiretti, 1962; Redmond, 1971).

It is well known that the oxygenation curves may be cooperative or not, depending on the experimental conditions and on the aggregation state of the protein.

In recent years, new interesting studies on the oxygen equilibria have been carried out in order to demonstrate whether a relationship exists between the cooperative effect and some particular structure of the protein molecule (Konings et al., 1969; De Phillips et al., 1969, 1970; Van Driel, 1973; Er-El et al., 1972; Klarman et al., 1975; Van Driel and Van Bruggen, 1975).

From these studies two hypotheses have been put forward. The first suggests that an interaction between native 1/10 subunits is essential for cooperative oxygen binding (Van Driel, 1973, experiments carried out on *Helix pomatia* Hcy).

According to the second hypothesis, the cooperative effect is due to two particular Ca^{2+} ions out of the 20 Ca^{2+} ions linked per copper-oxygen site (Klarman and Shaklai, 1972). A diphosphoglycerate-like effect has been assumed for these two Ca^{2+} ions (Shaklai et al., 1975, experiments carried out on *Levantina hierosolima* Hcy).

As reported in the workshop of 1974, the Hcy of *Octopus vulgaris* entirely dissociates into functional subunits under the action of urea. In this communication we present data on the molecular weight of the functional subunit and on its oxygen-binding properties. The oxygen equilibria of *Octopus* Hcy have been studied in the presence and in the absence of Ca^{2+} ions.

Experimental Section

Materials. The hemolymph was collected from living animals and centrifuged at 3000 r.p.m.; the serum was then diluted with H_2O 1:1 v/v and Hcy was precipitated twice with ammonium sulphate between 40 and 52% saturation.

The protein solution containing about 40 mg of Hcy per ml was dialyzed against 0.02 M Tris-HCl, Ca^{2+} 0.01 M buffer, pH 7.5 then frozen in vials after addition of sucrose (18% w/v). When stored at -30°C the protein maintains its properties unchanged for months, as shown by the absorption of the copper band.

Before use Hcy was dialyzed against the medium required for experiment. All the reagents were analytical grade and were used without further purification.

Sedimentation Velocity. Sedimentation velocity experiments were carried out with a Beckman model E analytical ultracentrifuge using schlieren optics.

The determinations were performed at 40.000 r.p.m., 20°C, with the AN-D rotor; the Hcy was in Tris-HCl 0.1 M plus urea 3 M and methylammonium HCl 0.1 M, pH 7.5. The protein concentration was determined spectrophotometrically ($E^{1\%}_{288}$ in 0.5 M KOH = 1.67; $E^{1\%}_{278}$ in Tris-HCl 0.1 M pH 8.0 = 1.43). Experimental S values were extrapolated to zero protein concentration and used for calculating the $S^{o}_{20,w}$.

The molecular weight was determined using the intrinsic viscosity value (η intr.) in the Flory and Mandelkern equation:

$$M^{2/3}\ \beta\ (1 - \Phi'\rho) = N.S^{o}\ \eta^{1/3}\ \eta^{o}$$

where η is the intrinsic viscosity, S^{o} the sedimentation coefficient at zero protein concentration, N Avogadro's number, η^{o} the viscosity of the solvent, M the molecular weight, Φ' the apparent partial specific volume and ρ the solution density. The constant β was assumed to be $2.16.10^{6}$. The partial specific volume was calculated from amino acid analysis (McMeekin and Marshall, 1952).

Gel Chromatography. Gel filtration was carried out on a column of Biogel A 1.5 M (200-400 mesh) 1.8 × 110 cm with the same buffer used for ultracentrifugation.

Cytochrome c, chymotrypsinogen A, ovoalbumin, bovine serum albumin, horse-γ-globulin, Blue Dextran 2000 were employed for reference. The sample was loaded on the column together with the standard proteins, dissolved in one ml of elution buffer, at a concentration of 6-7 mg/ml. Elution was carried out at the rate of fifty drops.

The data are presented as V_e/V_o (V_e = elution volume, V_o = blue dextran elution volume) vs. log MW of the proteins.

Intrinsic Viscosity. The viscosity of Hcy was measured in the same buffer used for ultracentrifugation, as well as in 6 M Gu HCl Tris-HCl 0.2 M buffer, pH 7.5. A viscosimeter provided with a helical capillary 90 cm in length, was used (McKie and Brandts, 1957) and the determinations were carried out at 20 ± 0.01°C; the time flow for water was equal to about 360 s.

The intercept at zero protein concentration of the plot η spec/C vs. C gives the η int. value.

Oxygenation Curves. Oxygenation curves have been carried out with a tonometer similar to that described by Rossi-Fanelli and Antonini (1958) and known quantities of air were added according to the method suggested by the same authors. The tonometer was thermostated at 20 ± 0.1°C. 5 ml of Hcy solution containing 4 mg of protein per ml were introduced into the tonometer. A 2 l vacuum flask, containing about 1/2 l of the buffer used for the Hcy solution, was placed between the pump and the titrator; the buffer was vigorously stirred to avoid excessive evaporation in the tonometer.

Deaeration of the Hcy solution was obtained by gradual evacuation. Known amounts of air were added in the tonometer.

The saturation degree of the protein was measured from the absorption of the 348 nm copper-oxygen band read in a thermostated Perkin-Elmer 402 recording spectrophotometer.

Spectra between 310 and 430 nm were taken and corrected for light scattering.

All determinations were carried out at 20 ± 0.1°C and the concentration of Hcy was controlled at the end of each experiment.

Oxygenation curves have been determined in the conditions reported in Table 1. The oxygen solubility in the media used was determined by means of the Winkler method (Standard Methods, APHA, N.Y., 1971).

Table 1.

Buffer	S_{20} (Svedb.)	O_2 Solub. mg/l	$p^{1/2}$ mm Hg	n
Tris-HCl pH 8.0	13.65	8.8	10.8	1
Tris-HCl Ca^{2+} [a] pH 8.0	13.65; 50	8.8	2.8	4
Tris-HCl 3 M urea pH 8.0	8.0	6.77	8.1	2.5
Tris-HCl 3 M urea pH 9.0	-	-	5.5	2.0
Tris-HCl 3 M urea Ca^{2+} [a] pH 8.0	8.0	6.77	6.3	4.5

[a] Ca^{2+} = 0.02 M; Tris-HCl always 0.1 M.
All values have been taken at 20°C.

Results

Sedimentation Velocity. In the presence of 3 M urea, Hcy sediments as a single symmetrical peak.

In Figure 1 values of S_{20} vs. protein concentration are reported.

Extrapolation to zero protein concentration gives a value of 7.92 S. This value, reported to the standard conditions give a 10.5 $S^{o}_{20,w}$.

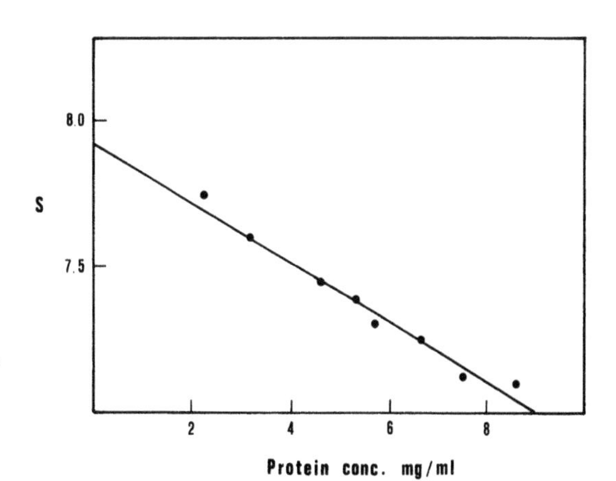

Fig. 1. Sedimentation velocity: The S values vs. protein concentration are reported. *O. vulgaris* Hcy in Tris-HCl 0.1 M urea 3 M, methylammonium-HCl 0.1 M pH 7.5 buffer

The sedimentation values for Hcy in the conditions used for the oxygen-binding experiments are reported in Table 1.

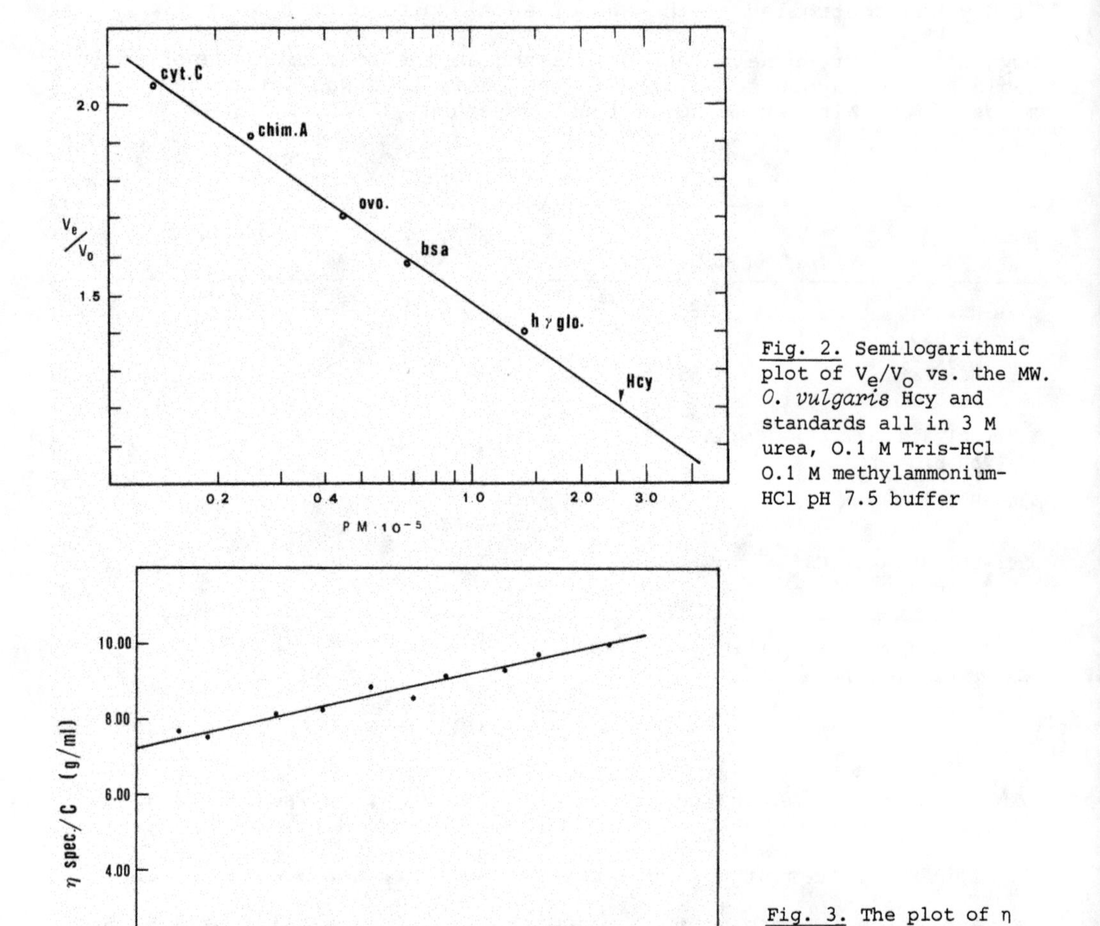

Fig. 2. Semilogarithmic plot of V_e/V_o vs. the MW. *O. vulgaris* Hcy and standards all in 3 M urea, 0.1 M Tris-HCl 0.1 M methylammonium-HCl pH 7.5 buffer

Fig. 3. The plot of η specific/C values vs. protein concentration. *O. vulgaris* Hcy in 3 M urea 0.1 M Tris-HCl 0.1 M methylammonium-HCl pH 7.5

Gel Chromatography. In Figure 2 the plot of V_e/V_o vs. logarithm of protein MW is shown. The calculated molecular weight for the Hcy is about 250,000 daltons.

Intrinsic Viscosity. In Figure 3 the plot of η specific/C vs. protein concentration is reported. Extrapolation to zero protein concentration gives a value of 7.20 ml/g which agrees with the values obtained by Costantino et al. (1971) on the same Hcy.

Intrinsic viscosity in 6 M Gu HCl resulted equal to 119-118 ml/g. This value is in agreement with the data on *Octopus vulgaris* Hcy of Costantino et al. (1971) for these conditions and with the data of Brouwer and Kuiper (1973) on the lighter component of *Helix pomatia* Hcy.

This value was treated by the equation $(\eta) = 0.716\ n^{0.66}$ where (η) is the observed intrinsic viscosity, n is the number of amino acid residues of the polypeptide chain (Tanford et al., 1967). Assuming a value of 118 for the mean residue MW calculated from the amino acid analysis, a value of 266,000 daltons was found in these conditions.

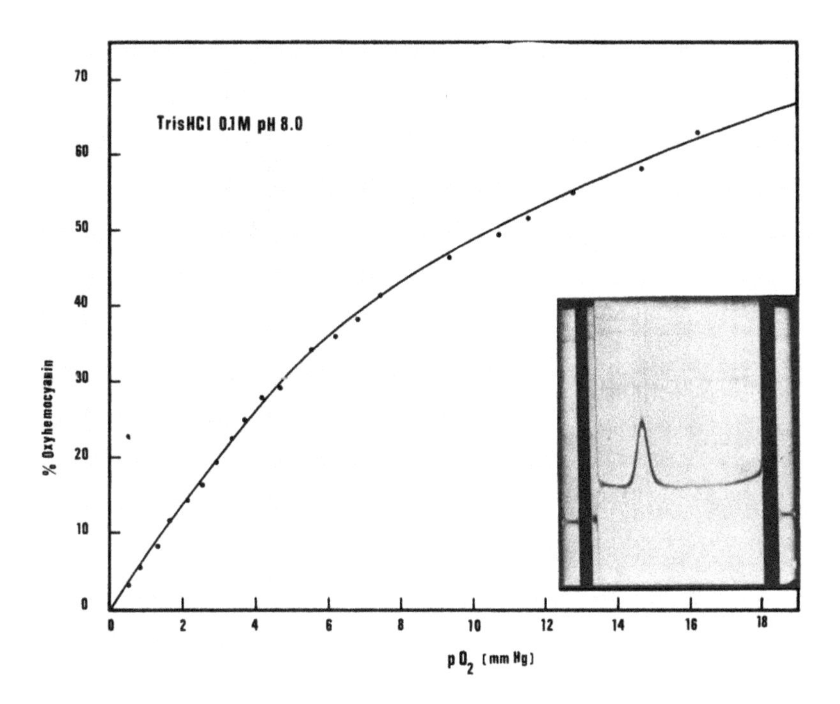

Fig. 4. Oxygenation curve of *O. vulgaris* Hcy in Tris-HCl 0.1 M pH 8.0 buffer. *Ordinate*: O_2-saturation percentage, abscissa pO_2 (mm Hg). *On right side:* Schlieren sedimentation pattern of the oxygenated Hcy in the same conditions. Time 32 min, 48,000 r.p.m., AN-D rotor, S = 13.65

Oxygen Binding. Figures 4 and 9a show the oxygen-binding curve and the Hill plot of *O. vulgaris* Hcy in 0.1 M Tris-HCl pH 8.0 buffer.

The curve is hyperbolic and the Hill plot is linear, with a slope n = 1.

The $p^{1/2}$ value is equal to 10.8 mm Hg.

In Figure 4 the sedimentation pattern at the ultracentrifuge is also reported: there is only one peak, with S value 13.65.

Figures 5 and 9b, 6 and 9c show the oxygen-binding curves and Hill plots of *O. vulgaris* Hcy in 0.1 M Tris-HCl 3 M urea pH 8.0 and pH 9.0 buffers. The $p^{1/2}$ are equal to 8.2 and 5.3 mm Hg (at 8.0 and 9.0 pH values respectively). In both cases the oxygenation curves are sigmoidal and the Hill plots have slope n = 1 for saturation values close to zero, and higher than 50% and 60% (at pH 8.0 and 9.0 respectively). The slope, n, is higher than 1 at intermediate saturation values.

In Figure 6 is also reported the sedimentation pattern at the analytical ultracentrifuge. Data are reported upon. The S value is equal to 8.

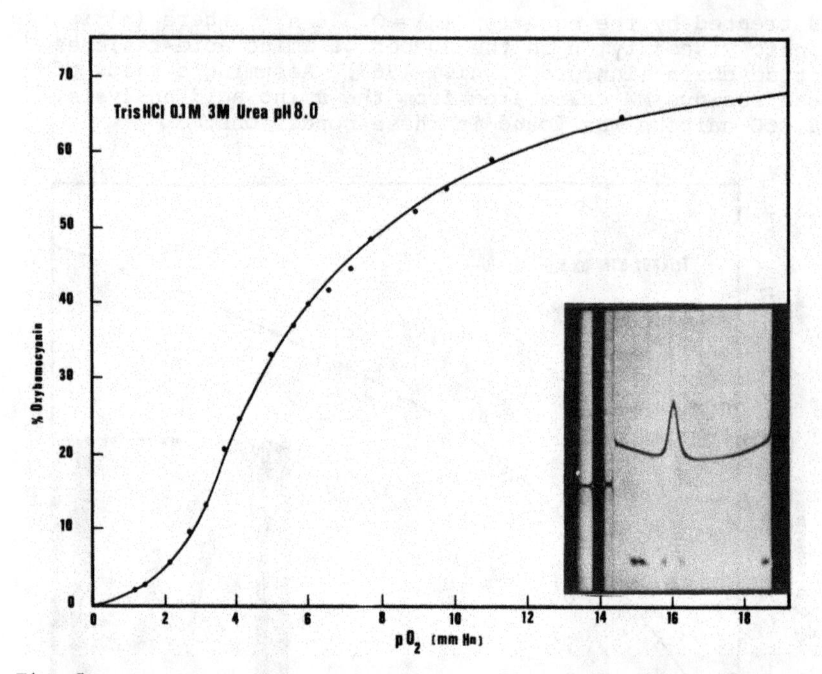

Fig. 5. Oxygenation curve of *O. vulgaris* Hcy in 3 M urea, 0.1 M Tris-HCl pH 8.0 buffer. *Abscissa* and *ordinate*: as in Figure 4. *On right side*: Schlieren sedimentation pattern of the oxygenated Hcy in the described condition. Time 56 min, 48,000 r.p.m., AN-D rotor, S = 8.0

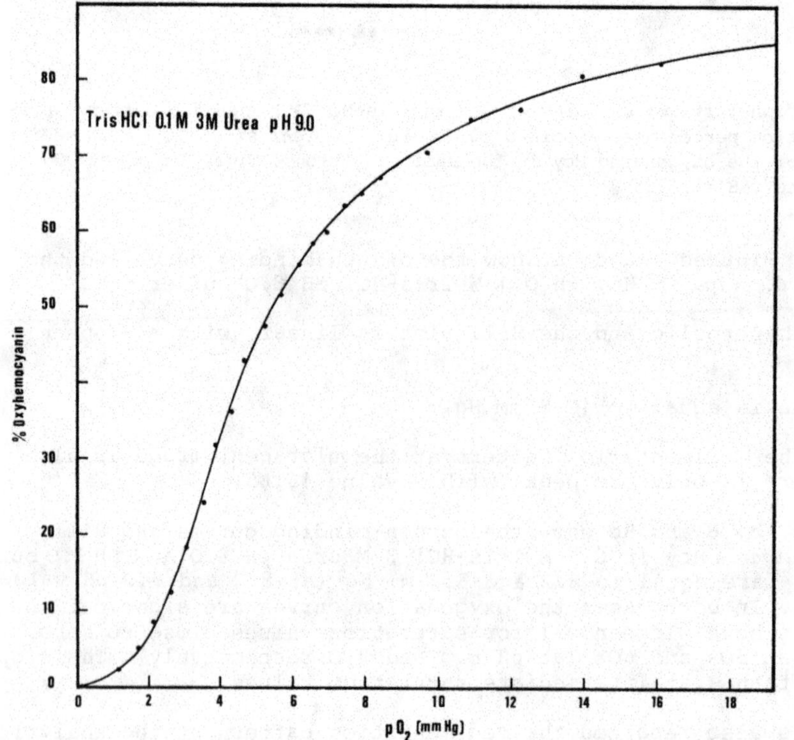

Fig. 6. Oxygenation curve of *O. vulgaris* Hcy in 3 M urea 0.1 M Tris-HCl pH 9.0 buffer

Fig. 7. Oxygenation curve of *O. vulgaris* Hcy in 0.1 M Tris-HCl 0.02 M Ca^{2+} 3 M urea pH 8.0 buffer. *Abscissa* and *ordinate*: as in Figure 4. *On right side*: Schlieren sedimentation pattern of the oxygenated Hcy in the described conditions. Time 40 min, 52,000 r.p.m., AN-D rotor

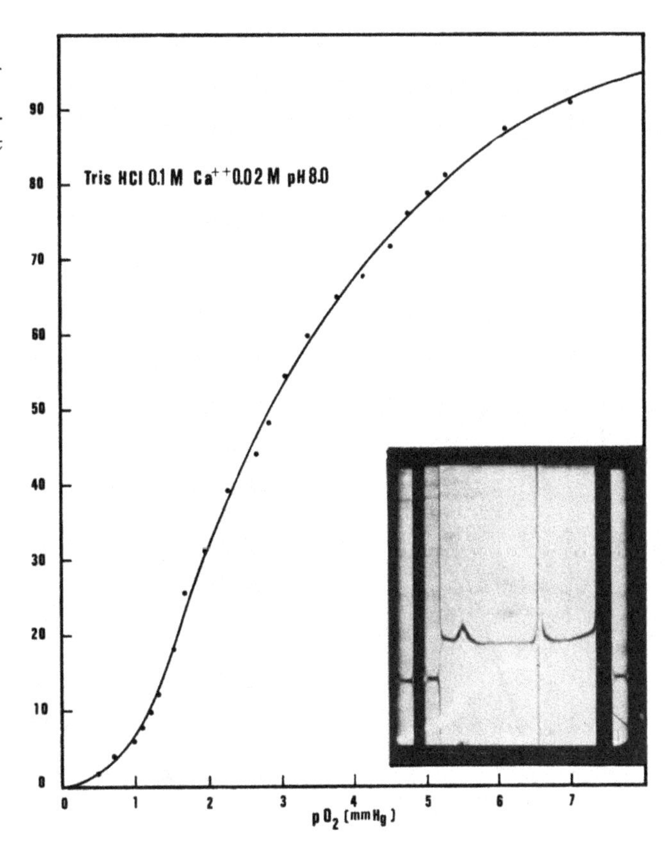

Figures 7 and 9d show the oxygen binding and Hill plot for *O. vulgaris* Hcy in 0.1 M Tris-HCl 0.02 M Ca^{2+} pH 8.0. The presence of Ca^{2+} strongly increases the oxygen affinity for the protein: the Hill plot shows a complex situation. The $p^{1/2}$ is equal to 2.8 mm Hg.

In Figure 7 the sedimentation pattern at the analytical ultracentrifuge is also shown. The S values for the two peaks resulted 13.64 and 50 (the smaller component is present in very little quantity).

Figures 8 and 9e show the oxygenation curve and the Hill plot for Hcy in 0.1 M Tris-HCl 3 M urea 0.02 M Ca^{2+} pH 8.0 buffer. Ca^{2+} ion appears to increase the cooperative effect observed in the presence of 3 M urea alone.

Figure 8 also shows the sedimentation pattern at the ultracentrifuge. It is apparently a single symmetrical peak. The calculated S value, 8.0, is not modified with respect to that from the Hcy in 3 M urea alone. The slope of the Hill plot is close to two for saturation values close to zero and higher than 30%.

All the data are summarized in Table 1.

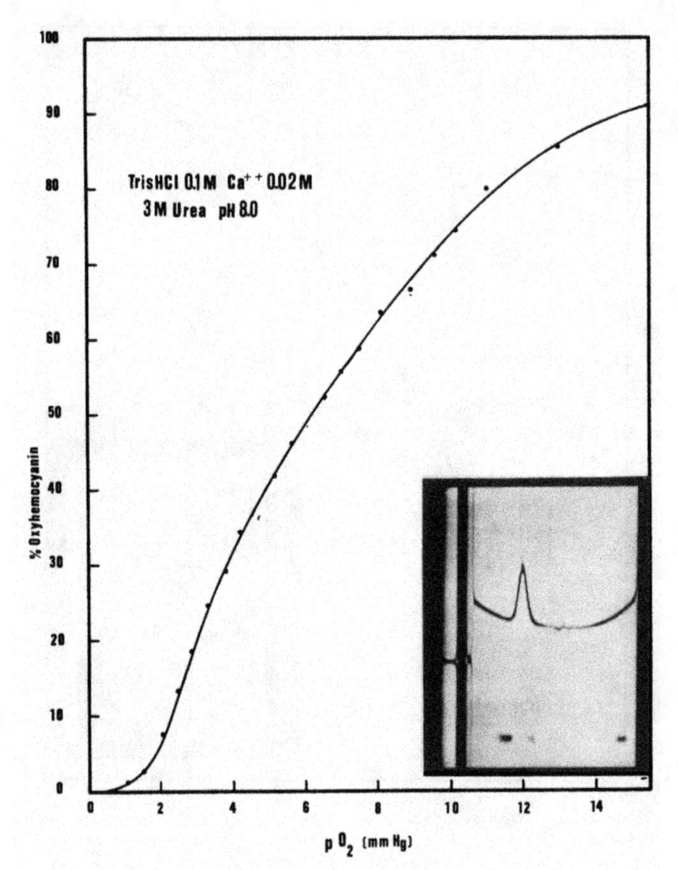

Fig. 8. Oxygenation curve of *O. vulgaris* Hcy in 0.1 M Tris-HCl 0.02 M Ca^{2+} pH 8.0 buffer. *On right side*: Schlieren sedimentation pattern of the oxygenated Hcy in the described conditions. Time 8 min, 52,000 r.p.m., AN-D rotor, S = 13.65, S = 50. From the photo it is clear that the lighter compoment is present in very small percentage

Discussion

Hemocyanin of *Octopus vulgaris* in 3 M urea completely dissociates into minimal, functional subunits. The dissociation is fully reversible. Actually, after removal of urea by dialysis against Ca^{2+} or Mg^{2+} containing buffer, Hcy completely reassociates, as we observed by electron micrographs and sedimentation velocity patterns.

All the molecular weights calculated from our data are close to 250,000 daltons. As demonstrated by the coincidence of the values obtained in 3 M urea and under stronger denaturing conditions (6 M GuHCl), this is the minimal subunit which can be obtained without breaking covalent bonds. In our opinion, this subunit corresponds to the 11 S sedimenting species observed for molluscan Hcy ($S^{o}_{20,w}$ = 10.5). It is important to remember that molluscan Hcy give subunits much smaller than 250,000 daltons (50,000-25,000) when treated with a large excess of acetic anhydride (Salvato et al., 1972), or exposed to pH values higher than 12 (Cox et al., 1972) and to 70% formic acid (Dijk et al., 1970). As previously hypothesized (Ghiretti-Magaldi et al., 1975) a possible interpretation of these data is that the sugar moiety plays an important role in linking the polypeptide chains of 25,000 daltons into aggregates of 250,000 daltons.

The oxygenation curves carried out with Tris-HCl buffer are cooperative in the presence of Ca^{2+} 0.02 M and non-cooperative in the absence of

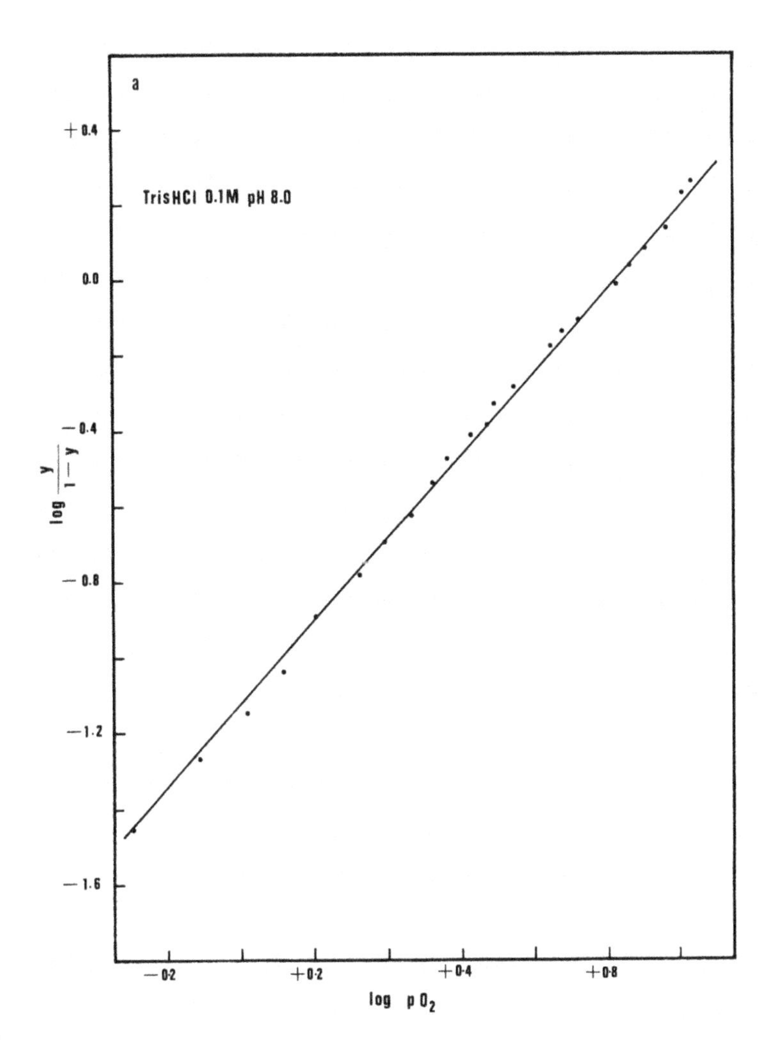

a

Fig. 9a-e. Hill plots of the oxygenation curves reported in Figures 4-8. Ordinate: $\log \frac{y}{1-y}$; abscissa: $\log pO_2$

this ion. As known, under these two sets of experimental conditions the aggregation state of the protein is different.

In the presence of 3 M urea, when Hcy is completely dissociated into 250,000 dalton subunits, the oxygenation curve is cooperative. This fact indicates that cooperativity in the absence of Ca^{2+} ions can be obtained. The oxygenation curves obtained in 3 M urea should be comparable with those in buffer alone, since fluorescence (Jori et al., Chap. 20, this vol.), differential spectrophotometry (unpublished results), and CD studies (Tamburro et al., 1976) show that only subtle differences, if any, exist between the conformations of the protein in the two media.

The addition of 0.02 M Ca^{2+} to the hemocyanin in 3 M urea increases the oxygen affinity and sligthly the cooperativity. It seems unlikely

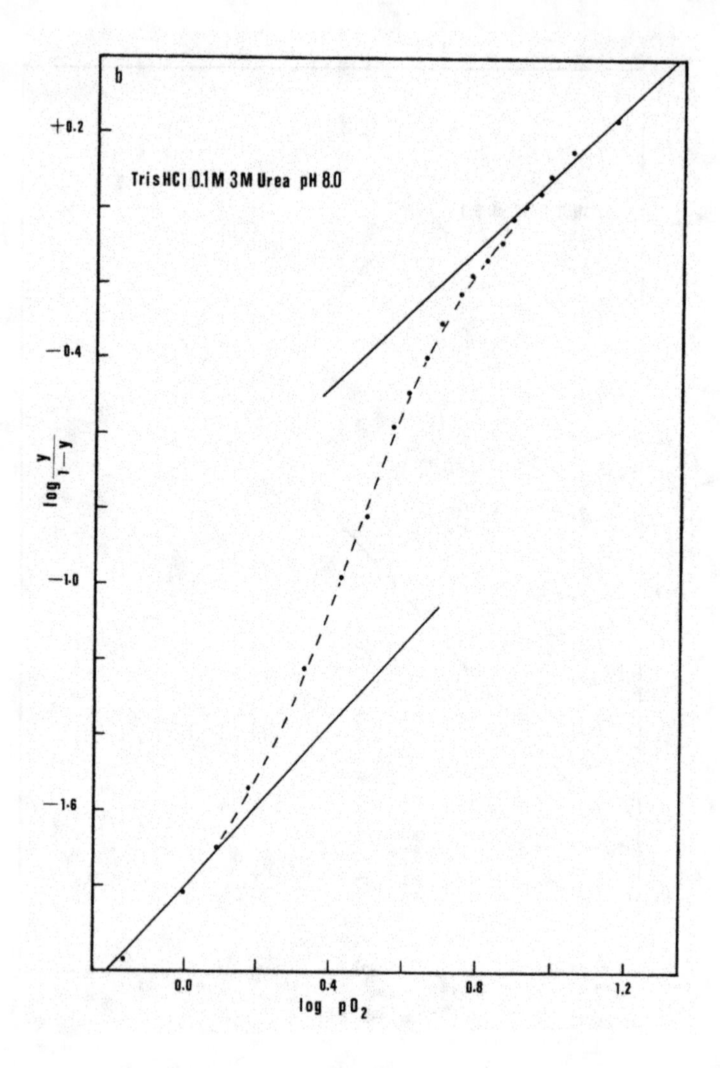

Fig. 9b

that the effect of urea on the oxygenation behaviour of the Hcy can be ascribed to a very limited number of specific sites which are responsible for cooperativity [in the case of Ca^{2+} two specific sites per 50,000 daltons have been proposed (Shaklai et al., 1975)].

More probably urea exerts a general effect on the protein by means of a limited conformational change, which increases the importance of site-site interactions. We hypothesize that the Ca^{2+} effect is essentially analogous to the urea effect. Both these reagents modify the site-site interactions. This effect, however, is not linked to some particular Ca^{2+} ions out of the 20 which are bound per copper-oxygen site (Klarman and Shaklai, 1972). On the other hand, Shaklai and coworkers were unable to differentiate it on the basis of stability constants.

These are the first data produced in our laboratory on oxygenation curves. Further studies are in progress.

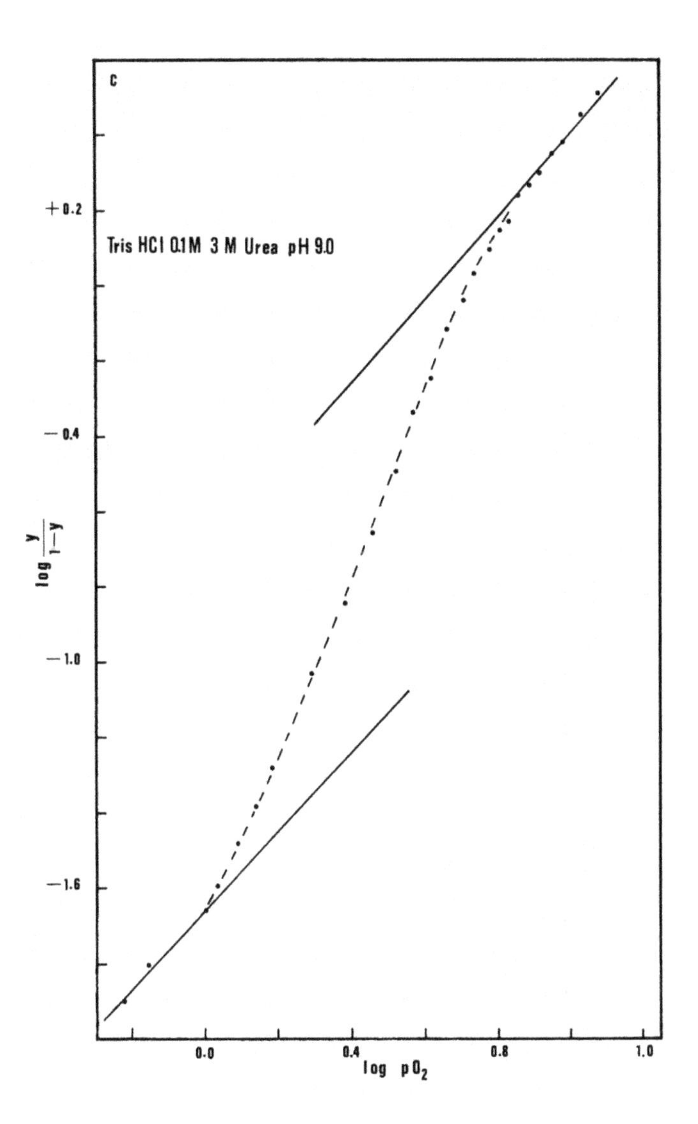
c
Tris HCl 0.1M 3 M Urea pH 9.0
+0.2
-0.4
-1.0
-1.6
log $\frac{y}{1-y}$
0.0
0.4
0.8
1.0
log pO_2

Fig. 9c

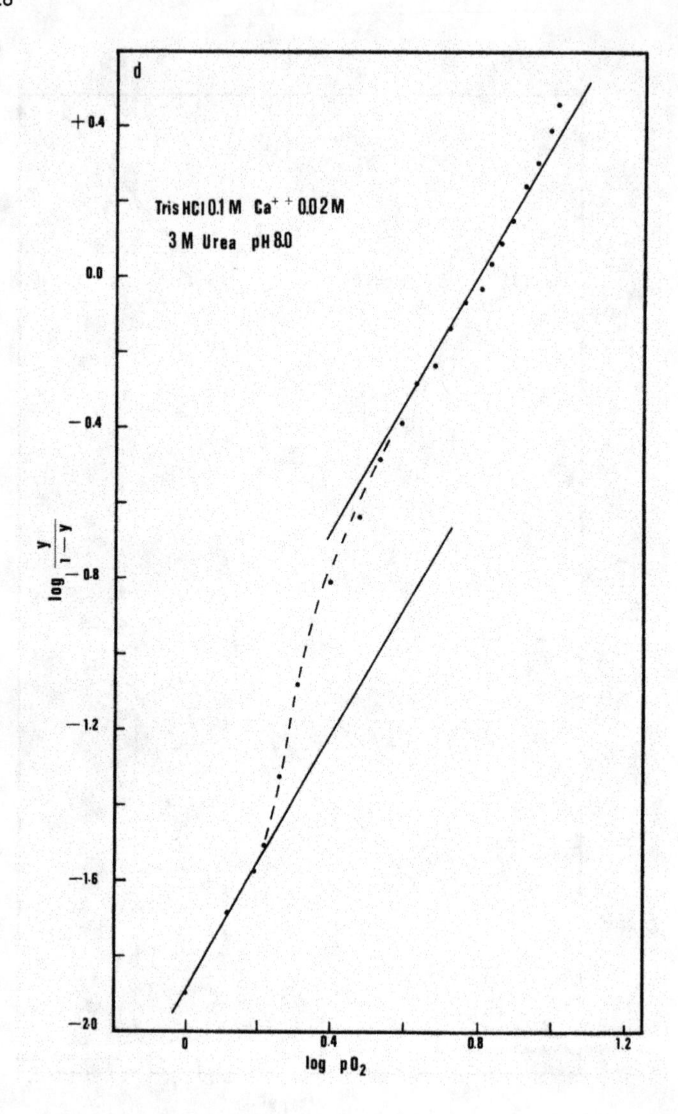
d
+0.4
0.0
−0.4
−0.8
−1.2
−1.6
−2.0
Tris HCl 0.1 M Ca⁺⁺ 0.02 M
3 M Urea pH 8.0
log y/1−y
0
0.4
0.8
1.2
log pO2

Fig. 9d

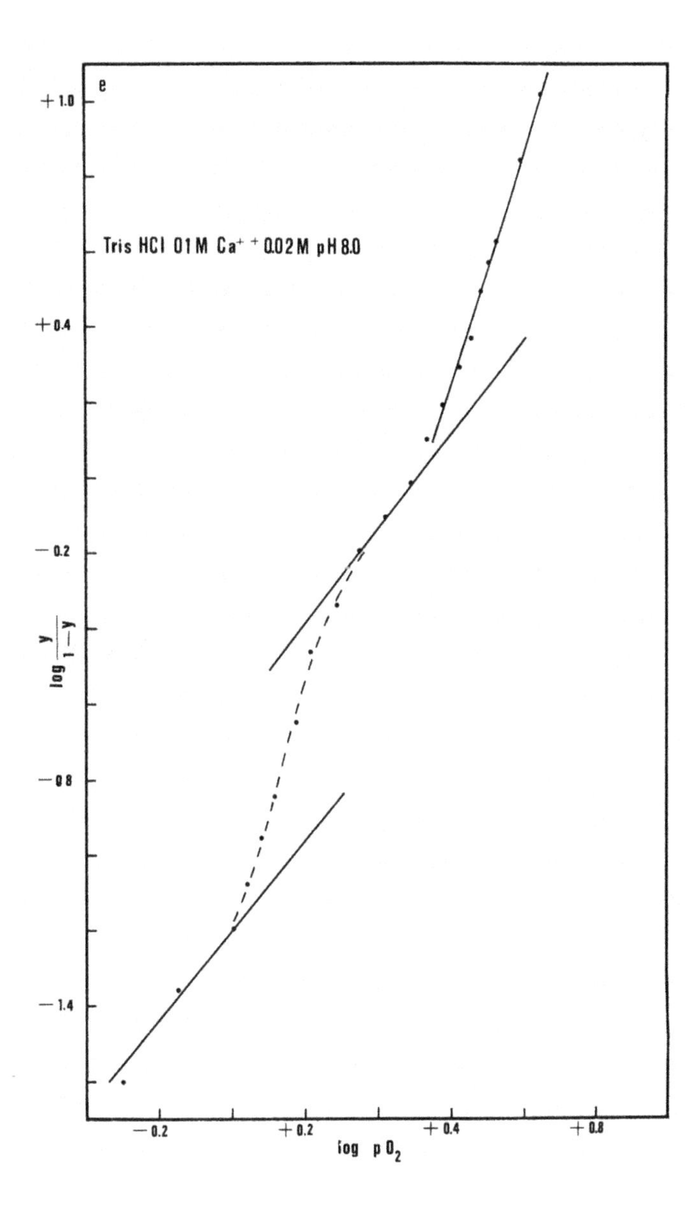

Fig. 9e

References

Brouwer, M., Kuiper, H.A.: Molecular weight analysis of *Helix pomatia* α-hemocyanin in guanidine-hydrochloride, urea and sodium dodecylsulfate. Europ. J. Biochem. 35, 478-485 (1973)

Costantino, L., Vitagliano, V., Wurzburger, S.: Proc. 1st Meet. Ital. Soc. Rheology. 2, 219 (1971)

Cox, J., Witters, R., Lontie, R.: The quaternary structure of *Helix pomatia* hemocyanins as determined by alkali treatment and succinylation. Intern. J. Biochem. 3, 283-293 (1972)

De Phillips, H.A., Nickerson, K.W., Johnson, M., Van Holde, K.E.: Physical studies of hemocyanins. IV. Oxygen-linked disassociation of *Loligo pealei* hemocyanin. Biochemistry 7, 3665-3672 (1969)

De Phillips, H.A., Nickerson, K.W., Van Holde, K.E.: Oxygen binding and subunit equilibria of *Busycon* hemocyanin. J. Mol. Biol. 50, 471-479 (1970)

Dijk, J., Brouwer, M., Coert, A., Gruber, M.: Structure and function of hemocyanins. VII. The smallest subunit of α and β hemocyanin of *Helix pomatia*: size, composition, N-, and C-terminal amino acids. Biochim. Biophys. Acta 221, 467-479 (1970)

Er-El, Z., Shaklai, N., Daniel, E.: Oxygen-binding properties of haemocyanin from *Levantina hierosolima*. J. Mol. Biol. 64, 341-352 (1972)

Ghiretti-Magaldi, A., Tamino, G., Salvato, B.: The monophyletic origin of hemocyanins on the basis of the amino acid composition. Structural implications. Boll. Zool. 42, 167-179 (1975)

Ghiretti, F.: Hemerythrin and hemocyanin. In: The Oxygenases. Hayaishi, O. (ed.). New York: Academic Press, 1962, pp. 540-542

Klarman, A, Shaklai, N.: The binding of Ca^{++} ions to Hcy from *Levantina hierosolima* at physiological pH. Biochim. Biophys. Acta 257, 150-157 (1972)

Klarman, A., Shaklai, N., Daniel, E.: Oxygen binding by hemocyanin from *Levantina hierosolima*. I. Exclusion of subunit interaction as a basis for cooperativity. Biochemistry 14, 102-104 (1975)

Konings, W.N., Van Driel, R., Van Bruggen, E.F.J., Gruber, M.: Structure and properties of hemocyanins. V. Binding of oxygen and copper in *Helix pomatia* hemocyanin. Biochim. Biophys. Acta 194, 55-66 (1969)

McKie, J., Brandts, J.F.: High precision capillary viscosimetry. In: Methods in Enzymology. Colowick, S.P., Kaplan, N.O. (eds.). 1957, 4, pp. 257-287

McMeekin, R.L., Marshall, K.: Specific volumes of proteins and the relationship to their amino-acid contents. Science 116, 142 (1952)

Redmond, J.R.: Blood respiratory pigments. Arthropods. In: Chemical Zoology. Florkin, M. (ed.). New York: Academic Press, 1971, 6, pp. 133-135

Rossi-Fanelli, A., Antonini, E.: Studies on the oxygen and carbon monoxide equilibria of human myoglobin. Arch. Biochem. Biophys. 77, 478-492 (1958)

Salvato, B., Sartore, S., Rizzotti, M., Ghiretti-Magaldi, A.: Molecular weight determination of polypeptide chains of molluscan and arthropod hemocyanins. FEBS Lett. 22, 5-7 (1972)

Shaklai, N., Klarman, A., Daniel, E.: Oxygen binding by hemocyanin from *Levantina hierosolima* II. Interpretation of cooperativity in terms of ligand-ligand linkage. Biochemistry 14, 105-108 (1975)

Standard Methods, 1971. Oxygen, dissolved. New York: APHA, 1971, pp. 477-481

Tamburro, A.M., Salvato, B., Zatta, P.: A circular dichroism study of some hemocyanins. Comp. Biochem. Physiol. 55(3B), 347-356 (1976)

Tanford, C., Kawahara, K., Lapanje, S.: J. Am. Chem. Soc. 89, 729-736 (1967)

Van Driel, R.: Oxygen binding and subunit interactions in *Helix pomatia* hemocyanin. Biochemistry 12, 2696-2698 (1973)

Van Driel, R., Van Bruggen, E.F.J.: Functional properties of chemically modified hemocyanin. Fixation of hemocyanin in the low and the high oxygen affinity state by reaction with a bifunctional imido-ester. Biochemistry 14, 750-755 (1975)

Interactions of Sulfur-Containing Ligands with Arthropod and Molluscan Hemocyanins

I. Y. Y. Lee, N. C. Li, and K. T. Douglas

Introduction

It has been known for some time (Rombauts et al., 1960a, b, 1968) that addition of thiourea to a number of hemocyanin causes a decrease in the absorption band ≈ 346 nm. The loss of absorbance occurs in two stages; a more-or-less rapid primary reaction is followed by a much slower secondary reaction (Rombauts, 1968). This is shown schematically in Figure 1 in which ΔA_0 is the change in absorbance caused by the primary reaction after correction for the secondary reaction. This simple subtraction procedure is obviously only possible in those cases wherein the rate constants for the two reactions are very different. The primary reaction corresponds to expulsion of oxygen from the active site, whereas the secondary reaction is accompanied by oxygen uptake by the system (Rombauts, 1960b, 1968) and this was ascribed to steric hindrance. Thiocyanate ion was also active, but at much higher concentrations and showed no secondary reaction (Rombauts, 1968). The secondary reactions varied also with the nature of the hemocyanin, being almost negligible for the protein from *Octopus vulgaris* (Rombauts, 1968).

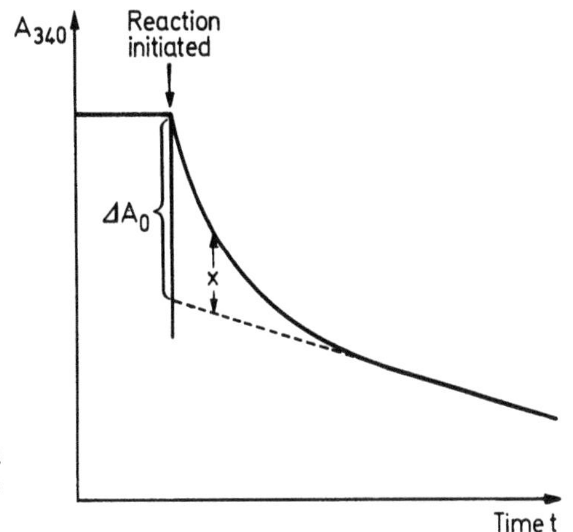

Fig. 1. Schematic representation of the biphasic decrease in absorbance at 340 nm when a sulfur-nitrogen ligand is added to a buffered solution of hemocyanin

It was proposed that the primary reaction was explicable as an equilibrial competition between oxygen and thiourea or thiocyanate for the copper active sites in hemocyanin and was strongly subject to steric effects (Rombauts, 1968). The secondary reaction was regarded as an oxidative process, possibly involving attack by thiourea on disulfide linkages in the protein, and indeed there is biochemical and chemical precedent in the literature for such a view (Toennies, 1937; Walker and Walker, 1960; Maloof and Soodak, 1961).

We now report a series of experiments which focuses primarily on the primary reactions in this area of nitrogen-sulfur ligands and hemocyanins from both arthropodan (*Limulus polyphemus*) and molluscan (*Busycon canaliculatum*) sources.

Experimental

Materials. Busycon and *Limulus* hemolymphs, obtained from the Marine Biological Laboratory, Woods Hole, Massachusetts, U.S.A., were treated as follows. Protein pellets, obtained by centrifugation of the hemolymphs (120,000 g at $4^{o}C$, Spinco Model L Ultracentrifuge) were dissolved in a small amount of water and dialyzed exhaustively against water for two days in the cold. Purified protein was stored at $4^{o}C$ and was characterized by means of absorbances at 280 nm (protein) and 345 nm (copper).

All ligands were commercially obtained and purified by recrystallization to sharp melting point before use. Borate buffers were prepared to have constant ionic strength of 0.1. Water used in these studies was glass distilled and deionized.

Kinetic Methods. Rate measurements were made by UV-Visible spectrophotometry using a GCA-McPherson 707-K Double Beam Kinetic Spectrophotometer, fitted with an automatic sample positioner for contemporaneous study of several reactions. The reaction chamber was thermostated to 25.00 $\pm$ $0.02^{o}C$ by means of water circulating from a Haake E52 Thermoregulator pump.

The following experimental procedure was generally adopted. A concentrated stock solution (0.6 M) of ligand in the appropriate buffer was prepared. An aliquot of this solution was added to a thermally equilibrated solution of hemocyanin in buffer in the cuvette to give a constant total volume (3.0 ml) and an appropriate ligand concentration. After mixing, the change in absorbance at 340 nm was recorded continuously as a function of time.

NMR Methods. NMR spectra were obtained with a Varian A-60 spectrometer, operated at 60 MHz and an ambient temperature of 31 $\pm$ 1°. Care was taken to keep the radiofrequency well below saturation. The observed line width, $\Delta\nu$, was measured from the full width at half-amplitude of the $-CH_2$ or $-CH_3$ proton peak, in the manner described by Ke et al. (1973).

Results

Kinetics. A collection of the ligands tested, and their activities, is presented in Table 1, along with some of Rombauts' data (Rombauts, 1968) for comparison. Activity is designated + or - depending on whether any absorbance change at 340 nm could be detected or not.

In Figure 2, the time courses of one of these ligands, thioacetamide, are shown for reactions with *Busycon* hemocyanin [Fig. 2 (a)] and *Limulus* hemocyanin [Fig. 2 (b)]. The secondary reaction is especially obvious for *Limulus* hemocyanin, but is also finite for *Busycon* hemocyanin. It is also apparent that the arthropodan hemocyanin reacts orders of magnitude more slowly in the primary reaction than molluscan hemocyanin, while the reverse appears to be the case for the secondary reaction.

The effects of several of the ligands tested at pH 9.08 can be seen more explicitly in Figure 3, which superimposes the time courses of

Table 1. Interactions of hemocyanins with sulfur-nitrogen ligands

Ligand	*Helix pomatia*[a]	*Limulus polyphemus*	*Busycon canaliculatum*
KSCN	+	+	+
$NH_2C(=S)NH_2$	+	+	+
$CH_3C(=S)NH_2$	b	+	+
$NH_2NHC(=S)NH_2$	-	-	possibly
$CH_3NHC(=S)NHCH_3$	-	b	-
$PhNHC(=S)NH_2$	b	b	+
Glycylglycine	b	-	-

[a]Rombauts (1968).
[b]Data not available.

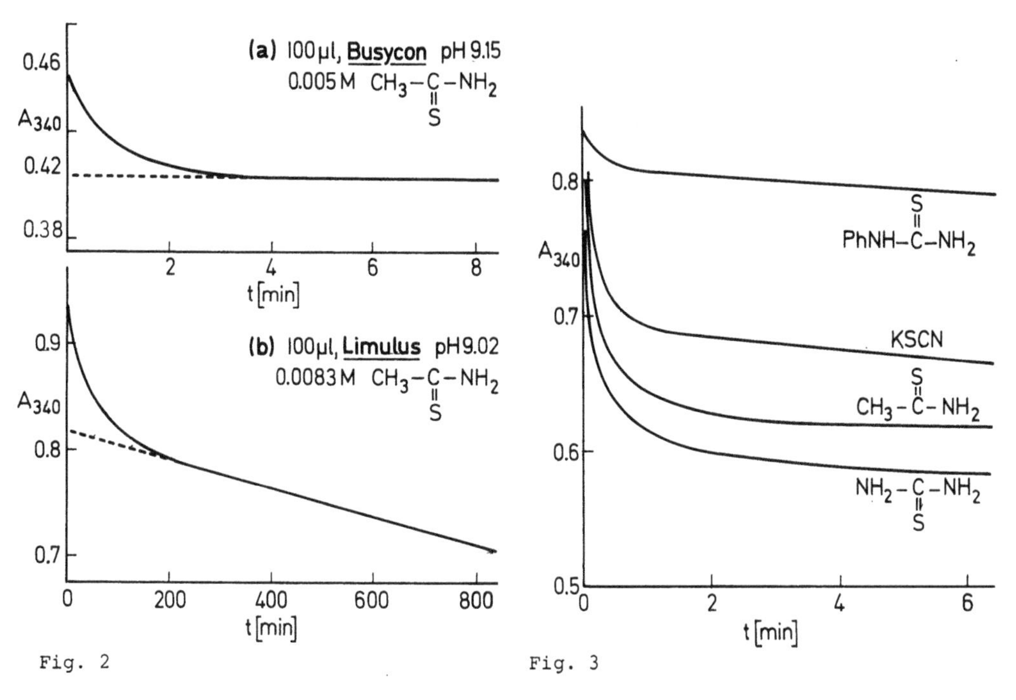

Fig. 2

Fig. 3

Fig. 2. Time courses of the reaction of thioacetamide with *Busycon* and *Limulus* hemocyanin

Fig. 3. Effect of the nature of various sulfur-nitrogen ligands on the reaction with *Busycon* hemocyanin at 25°C, pH 9.08 and a ligand concentration of 0.01 M

the reactions at constant total ligand and hemocyanin concentrations for *Busycon*. The maximal absorbance change (ΔA_o) occurs with thiourea. Substitution of one hydrogen by a phenyl group leads to a very marked diminution in the extent of the primary reaction (see Table 2). The presence of a substituent on each *nitrogen*, as in MeNH.CS.NHMe, com-

Table 2. Changes in ΔA_o at 340 nm for *Busycon canaliculatum* hemocyanin with sulfur-nitrogen ligands at 25°C in pH 9.08 borate buffer, ionic strength 0.1. Ligand concentrations were 0.01 M and hemocyanin concentration was 8.30×10^{-5} M (copper sites)

Ligand	(ΔA_o)	% Change
$NH_2C(=S)NH_2$	0.236 ± 0.004	28.4
$CH_3C(=S)NH_2$	0.216 ± 0.001	26.0
SCN^-	0.171 ± 0.010	20.6
$PhNHC(=S)NH_2$	0.032 ± 0.005	3.9
$CH_3NHC(=S)NHCH_3$	no change	0
Glycylglycine	no change	0

pletely eliminates detectable reactivity. Thiocyanate ion, at a concentration of 0.01 M, is only half as effective as thiourea, in agreement with the work of Rombauts on other hemocyanins (Rombauts, 1968). However, thioacetamide is almost as effective an agent in the primary reaction as thiourea. This observation led to the choice of thioacetamide as a probe and consequently to further study of its interaction with *Busycon* hemocyanin active-site. *Busycon* was chosen for its relatively rapid primary reaction and slow secondary reaction.

From Figure 4 we can see that the extent of decrease in A_{340} is a linear function of *Busycon* hemocyanin concentration in pH 9.15 borate buffer at constant thioacetamide concentration (0.01 M). The concentration of hemocyanin varied from $0.415\text{-}1.66 \times 10^{-4}$ M, expressed in terms of copper sites.

The dependence of ΔA_o on thioacetamide concentration at a constant level of hemocyanin is shown in Figure 5. There is a leveling off at high ligand concentrations and the data can be fitted to Equation (1)

$$A_o = \frac{(\Delta A_o)_{max} \cdot [CH_3CSNH_2]_o}{(K_{eq} + [CH_3CSNH_2]_o)} \tag{1}$$

Rearrangement of Equation (1) yields Equation (2)

$$\frac{1}{(\Delta A)_o} = \frac{1}{(\Delta A_o)_{max}} + \frac{K_{eq}}{(\Delta A_o)_{max}} \cdot \frac{1}{[CH_3CSNH_2]_o} \tag{2}$$

and K_{eq} and $(\Delta A_o)_{max}$ were determined as 0.00426 M and 0.337 respectively by an uneighted linear regression analysis (correlation coefficient =

0.9999) of $1/(\Delta A)_o$ *versus* $1/[CH_3CSNH_2]_o$. The significance of K_{eq} will be discussed later.

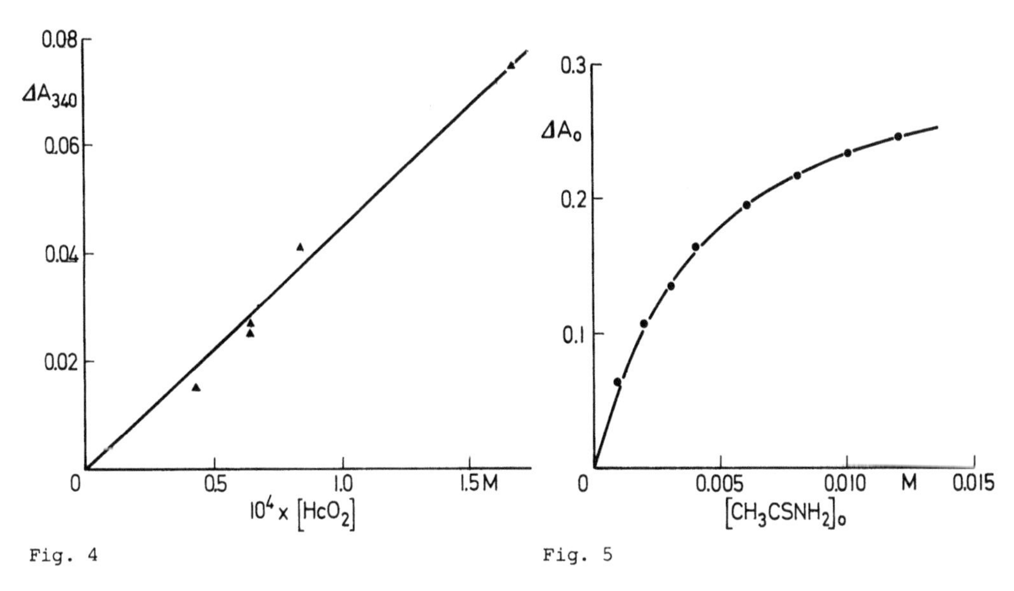

Fig. 4 Fig. 5

Fig. 4. Dependence of ΔA_o on *Busycon* hemocyanin concentration at pH 9.15; thioacetamide concentration was 0.01 M

Fig. 5. Saturation dependence of ΔA_o on thioacetamide concentration for the primary reaction with *Busycon* hemocyanin at pH 9.08 (borate buffer) at 25°C. The concentration of hemocyanin, in terms of copper sites, was 0.83×10^{-4} M. Points are experimental: line is theoretical for (ΔA_o) limiting of 0.337 and K_{eq} of 0.00426 M

Table 3a, b. (a) Line-width of CH_2- proton peaks of 0.5 M glycylglycine, GG, in H_2O

System	Δν (Hz) α-CH_2	β-CH_2	/ with 1 M Na_2SO_3
GG, pH 6.9	1.1	1.1	1.1
GG, 1.45% *Limulus* HcO_2, pH 6.9	1.1	1.8	1.2
GG, 2.42% *Limulus* HcO_2	1.1	2.2	1.3
GG, 1.63% *Limulus* ApoHc, pH 7.7	1.1	1.2	1.1
GG, 2.03% *Limulus* ApoHc	1.1	1.2	1.1

(b) Line-widths of $-CH_3$ proton peak of thioacetamide, TA, in pH 8.95 borate buffer (H_2O)

System	Δν (Hz)
0.12 M TA, pH 8.95	0.5
0.12 M TA, 0.876% *Busycon* HcO_2, pH 8.95	0.6 (3.5 min), 0.6 (6.5 min), 0.6 (40 min)

NMR Results. Table 3a lists values of line-widths of the α- and β-CH_2 proton peaks of glycylglycine in the absence and presence of *Limulus* oxyhemocyanin (HcO_2), deoxyhemocyanin (after addition of 1 M Na_2SO_3), and apohemocyanin (ApoHc). It is seen that the line-width of the α-CH_2 (adjacent to the carboxyl group) has the same value for all solutions, whereas that of the β-CH_2 (adjacent to the terminal amino group) increases with concentration of HcO_2. In the presence of deoxyhemocyanin and apohemocyanin the line-width of both -CH_2 signals of GG remain essentially the same as in the absence of the protein. These observations indicate that glycylglycine is bound to the Cu(II) in *Limulus* oxyhemocyanin. NMR results for thioacetamide are collected in Table 3b. No evidence of line broadening of the CH_3-proton signal was evident for *Busycon* hemocyanin even though line-widths were measured as a function of time for both species.

Discussion

Rombauts (1968) proposed that the primary reaction of thiourea with hemocyanins was an equilibrial competition between oxygen and the nitrogen-sulfur ligand for the copper active sites. In agreement with this he reported that higher oxygen pressures lead to decreased thiourea effects and that the effect is reversible. He added that the equilibrium is subject to steric effects as the following order of effectivity of thiourea analogs was observed.

thiourea > methly, allyl, > ethyl > *sym*-diethylthiourea

However, the above series is not highly appropriate per se to illuminate electronic effects in the ligand and we have found the following,

$$(NH_2)_2C{=}S \;\simeq\; (CH_3)(NH_2)C{=}S \;\gg\; (C_6H_5NH)(NH_2)C{=}S \;\gg\; (CH_3NH)_2C{=}S \;\simeq\; 0$$

The approximately equal powers demonstrated by thioacetamide and thiourea, electronically quite different but sterically comparable, argues in favor of Rombauts' steric hypothesis.

The inactivity of N,N[1] - dimethylthiourea, in addition to supporting the steric view, implies that for successful binding to the copper site a free $-\overset{S}{C}-NH_2$-group must be available. Blocking the sulfur, as S-methyl isothiourea, leads to inactivity indicating that the sulfur must be free (Rombauts, 1968).

Support for complexation of thioacetamide to the protein comes from Figure 5. One can calculate the change in absorbance which will occur when oxyhemocyanin (P) and thioacetamide (L) are mixed and reach equilibrium.

$$P + L \underset{K}{\rightleftarrows} (PL)$$

Defining, $K = \frac{(P)\ (L)}{(PL)}$

$$\text{Then, } \Delta A_o = [P_o] \cdot (\varepsilon_P - \varepsilon_{PL}) \cdot \frac{[L_o]}{(K + [L_o])} \qquad (3)$$

wherein $[P_o]$ = initial hemocyanin concentration

$[L_o]$ = initial thioacetamide concentration
ε_p, ε_{PL} are the extinction coefficients of oxyhemocyanin and thioacetamide - protein complex respectively at 340 nm.

Equation (3) is of the same form as Equation (1).

The K_{eq} of Equation (1) need not correspond directly to K in Equation (3) as more than a single equilibrium may be involved. However, K_{eq} is readily shown to be the concentration of ligand at which the protein is half-saturated and is an apparent dissociation constant. Consequently, at thioacetamide concentration of 0.01 M or greater a level of > 75% saturation is achieved.

Thioacetamide belongs to a class of ligands which binds to oxyhemocyanin with concomitant expulsion of oxygen. However, there is another class of ligands. Nuclear magnetic resonance line-broadening studies (Ke et al., 1973) have shown that ligands such as glycine and glycylglycine bind to the copper site *without* oxygen expulsion. We have substantiated the observation of Ke et al. (1973) that there is no change in the absorbance at 340 nm of oxyhemocyanin in the presence of 0.01 M glycylglycine with *Busycon* and extended it to include stripped *Limulus* hemocyanin.

The source of paramagnetic line-broadening for glycine derivatives is indicated in Formula 1 although no specific bonding interactions are implied. The glycine ligand binds to two coordination sites in the presence of oxygen: a comparable diagram for sulfur-nitrogen ligands is Formula 2.

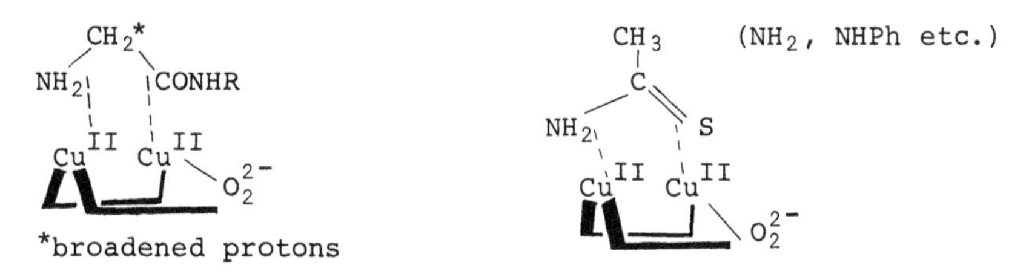

Formula 1 Formula 2

Not only do the glycine and thiourea-type ligands differ in their effects on oxygen-binding, but Table 3 shows that they differ also in their NMR properties: there is no observed line-broadening with the methyl protons of thioacetamide for *Busycon* hemocyanin, whereas both hemocyanins give rise to β-CH_2 line-broadening with glycylglycine.

The lack of broadening for nitrogen-sulfur ligands could be caused by

(1) reduction of paramagnetic Cu(II) to diamagnetic Cu(I) by thioacetamide or,

(2) sulfur being a weakly binding group so that the chelate is not strong enough for the broadening to be detectable or,

(3) binding of thioacetamide at a locus *other* than the copper-site i.e. an allosteric effect.

Summary

Addition of certain sulfur-nitrogen ligands to oxyhemocyanins from *Limulus polyphemus* of *Busycon canaliculatum* leads to diminution of the copper bands around 340 nm. The order of effectivity is:

$$(NH_2)_2C{=}S \approx CH_3(NH_2)C{=}S > SCN^- > PhNH(NH_2)C{=}S \gg (MeNH)_2C{=}S \approx 0$$

The reaction was biphasic: a rapid loss in absorbance at 340 nm (primary reaction) was followed by a much slower secondary reaction. The primary reaction for *Busycon*, with a given ligand, is orders of magnitude faster than that for *Limulus*. The change in absorbance caused by the primary reaction is linearly dependent on hemocyanin concentration but hyperbolically dependent on thioacetamide concentration. The apparent dissociation constant describing the thioacetamide-*Busycon* system is 4.26 mM.

No such effect on the absorbance at 340 nm was observed with glycylglycine, a ligand shown to bind to the copper-site of hemocyanin by means of the β-CH_2 proton NMR signal, which was paramagnetically broadened on addition of oxyhemocyanin (*Limulus*, pH 6.9, H_2O), but not on addition of apohemocyanin. No broadening was observed with thioacetamide (-CH_3 proton signal, pH 8.95, H_2O) with *Busycon* hemocyanin.

Acknowledgments. We are grateful to the National Science Foundation (Grant PCM 76-11744) for support of this project and also to Dr. Sam Guo and Mrs. Chi-Chang for assistance in some of the NMR work. We also thank Duquesne University for the GCA McPherson spectrophotometer.

References

Ke, C.H., Schubert, J., Lin, C.I., Li, N.C.: Nuclear magnetic resonance study of the binding of glycine derivatives to hemocyanin. J. Am. Chem. Soc. 95, 3375-3379 (1973)

Maloof, F., Soodak, M.: Cleavage of disulfide bonds in thyroid tissue by thiourea. J. Biol. Chem. 236, 1689-1692 (1961)

Rombauts, W., Lontie, R.: The expulsion of oxygen from hemocyanin by thiourea. Arch. Intern. Physiol. Biochem. 68, 230-231 (1960a)

Rombauts, W., Lontie, R.: The reaction of thiocyanate ion with hemocyanin. Arch. Intern. Physiol. Biochem. 68, 695-696 (1960b)

Rombauts, W.: The interaction of thiourea and thiocyanate ion with the oxygen-carrying groups of haemocyanin. In: Physiology and Biochemistry of Haemocyanins. Ghiretti, F. (ed.). London: Academic Press, 1968, pp. 75-80

Toennies, G.: Relations of thiourea, cysteine and the corresponding disulfides. J. Biol. Chem. 120, 297-313 (1937)

Walker, J.B., Walker, M.S.: Inhibitions of sulfhydryl enzymes by formamidine disulfide. Arch. Biochem. Biophys. 86, 80-84 (1960)

The Reaction of *Helix pomatia* Methaemocyanin with Azide and Flouride

R. WITTERS, M. DE LEY, AND R. LONTIE

Introduction

Haemocyanin, the copper protein of the haemolymph of *Helix pomatia*, reversibly binds oxygen in a ratio of 1 O_2 per two copper atoms (Guillemet and Gosselin, 1932).

Oxyhaemocyanin is characterized by copper-oxygen bands at 346 nm (ε = 8800) and 580 nm (ε = 540). These bands gradually decrease on storage (Heirwegh and Lontie, 1960) and disappear almost completely through displacement of oxygen, likely as peroxide, in slightly acid medium by azide or fluoride, resulting in the formation of methaemocyanin (Witters and Lontie, 1975). By the action of fluoride the deep-blue colour of oxyhaemocyanin vanishes, while by the action of azide an ochre complex is formed, which is reversibly decomposed on dialysis.

We have further investigated these reactions by circular-dichroic (CD) and by electron paramagnetic resonance (EPR) spectroscopy.

Materials and Methods

H. pomatia haemocyanin was prepared according to Heirwegh et al. (1961); the α- and β-components were not separated. The haemocyanin concentration was measured spectrophotometrically, A(0.1%, 1 cm, 278 nm) = 1.416 at pH 9.2.

A haemocyanin solution of 40 mg/ml was dialysed extensively against 0.1 M sodium acetate buffer, pH 5.0; it was separated into two equal fractions, which were rendered 25 mM in sodium azide and 0.1 M in potassium fluoride respectively. These solutions were kept at 37°C for 48 h in order to accelerate the methaemocyanin formation. The solutions were then made 10 mM in EDTA and dialysed separately against 0.1 M sodium acetate buffer, pH 5.0.

For absorption and circular-dichoric spectra dilutions to 1 and 6 mg/ml respectively were made. Circular-dichoric spectra were recorded with a Cary 61 spectropolarimeter (Cary, Monrovia, CA) at 25°C.

EPR measurements were carried out with an E-109 spectrometer (Varian, Palo Alto, CA) at -170°C, microwave frequency 9.12 GHz, field modulation amplitude 1 mT, microwave power 10 mW at g = 2 and 195 mW at g = 4. The haemocyanin solutions were concentrated in a preparative ultracentrifuge (Spinco Model L, rotor 30, 3 h at 27,500 r.p.m. at 4°C). The sedimented haemocyanin was dissolved in a minimum of solvent; haemocyanin concentrations of 150 mg/ml (6 mM Cu) were reached.

As the Visking dialysis membranes contained traces of reducing agents capable of regenerating methaemocyanin partially, they were heated for 1 h in 50% ethanol, treated with 10 mM $NaHCO_3$ and 1 mM EDTA and finally washed with distilled water according to McPhie (1971).

Results and Discussion

Treatment of *H. pomatia* haemocyanin with 25 mM sodium azide or 0.1 M potassium fluoride at 37°C for 48 h results in lowering of the negative CD band at 346 nm to 13 and 5% respectively of the original value (Witters and Lontie, 1975). A complete regeneration of the copper-oxygen bands was obtained after reduction with hydrogen peroxide or hydrogen sulphide (De Ley et al., 1975).

The EPR spectrum of both methaemocyanin preparations is shown in Figure 1 and compared to the spectrum of a Cu^{2+}-citrate solution at pH 7,

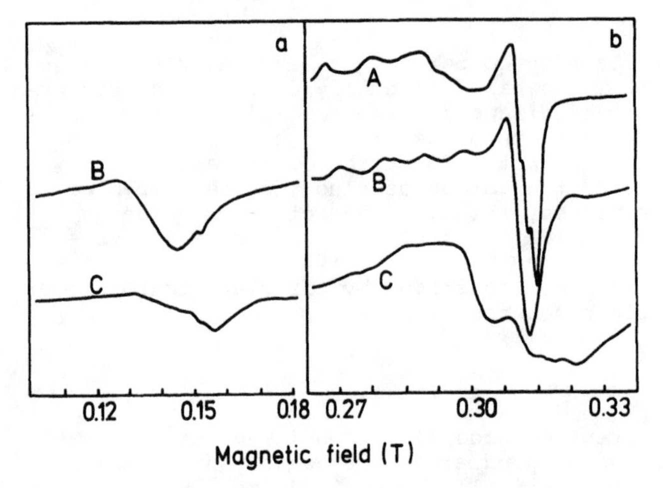

Fig. 1a and b. EPR spectra at g = 4 (a) and at g = 2 (b) of a solution of 1 mM Cu^{2+} in 0.5 M sodium citrate, pH 7.0 (converted to 6 mM Cu^{2+}) (*A*), of *H. pomatia* methaemocyanin (6 mM Cu) in 0.1 M sodium acetate buffer, pH 5.0, formed by the action of azide (*B*) or fluoride (*C*) as described in the text, after removal of the reagents by dialysis

converted to the same copper concentration and receiver gain. The latter EPR spectrum at g = 2 is broad and strongly reduced in intensity with respect to that of single Cu^{2+} as a result of dipole-dipole coupling in the Cu^{2+} dimers (Dunhill et al., 1966). The peak-to-peak intensity of the EPR signal at g = 2 of the azide generated methaemocyanin amounts to 70% of the Cu^{2+}-citrate signal, and to only 1% of that of single Cu^{2+} in 2 M $NaClO_4$, 10 mM HCl (Malmstrom et al., 1970). Although its signal amplitude is similar, the EPR spectrum of methaemocyanin, prepared by the action of fluoride, is much broader.

The addition of azide to methaemocyanin at pH 5.0 results in a rapid colour change to ochre. In the presence of small amounts of sodium azide (up to 5 mM) an absorption band centred at 380 nm has been detected. Increasing the azide concentration raises the absorption between 300 and 600 nm, showing, however, no pronounced maximum. The measurements at pH 5.0 were hampered by a large contribution of the light-scattering of undissociated haemocyanin molecules. The reaction of methaemocyanin with 50 mM azide, as measured at 450 nm in a stopped-flow apparatus (Durrum Model D-110, Palo Alto, CA) is complete in 10 s. The analysis according to pseudo-first-order kinetics showed upward curvature on a semi-logarithmic plot, which points to the complexity of the reaction.

Fig. 2. Molar circular dichroism $\Delta\varepsilon$ (l/mol/cm), expressed per Cu atom, of *H. pomatia* methaemocyanin (0.47 mM Cu) in 0.1 M sodium acetate buffer, pH 5.0, formed by the action of azide as described in the text, after removal of azide by dialysis (*blank*). *Other curves*: same solution in the presence of increasing amounts of sodium azide. All spectra have been corrected for the residual 13% oxyhaemocyanin in the blank

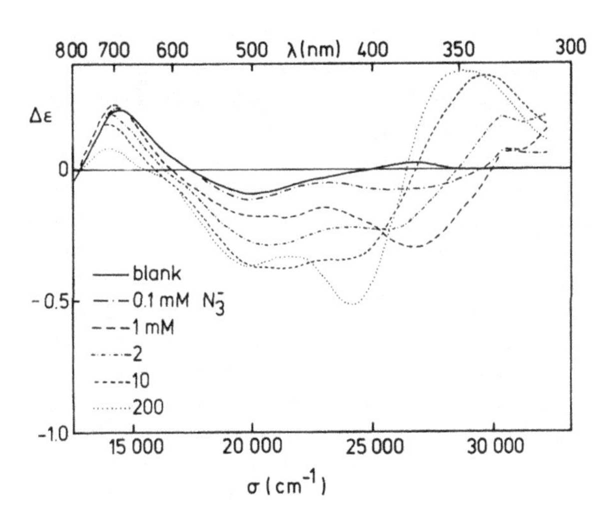

The reaction of methaemocyanin with sodium azide was also investigated at pH 5.0 by CD measurements between 300 and 800 nm. Figure 2 shows the difference CD spectra after subtraction of the contribution of the remaining 13% oxyhaemocyanin in the preparation. This residual spectrum of oxyhaemocyanin was calculated from the experimental value at 350 nm in the absence of azide, and the parameters of the CD spectrum of *H. pomatia* haemocyanin (Lontie and Vanquickenborne, 1974). Addition of small amounts of azide (up to 1 mM) gave rise to a small negative band at 380 nm, which on increasing the azide concentration to 10 mM was obscured by a positive band at 350 nm and a negative sprectrum between 400 and 600 nm. Further increasing the azide concentration to 200 mM lowered the positive band at 700 nm, sharpened the negative one at 420 nm and shifted the positive band at 350 nm to higher wavelengths. The small negative band at 380 nm points to the presence of traces of copper with a high affinity for azide, differing from the other copper sites. An apparent association constant (K = 2.23) was calculated from the spectral change at 510 and 430 nm (intersection points with the baseline when the residual spectrum of oxyhaemocyanin was not subtracted), according to the method applied by Byers et al. (1973) for the binding of azide to caeruloplasmin.

The intensity of the EPR signal of methaemocyanin at pH 5.0 was slightly affected by the presence of increasing amounts of azide. The value of the hyperfine splitting constant $A_{\parallel}$ decreased from 13.4 to 10.2 mK in the presence of azide, a change which was complete at 10 mM sodium azide (Fig. 3).

The distinctly different EPR spectra of fluoride and azide generated methaemocyanins are shown in Figure 1. Addition of increasing amounts of fluoride to the former did not yield additional absorption bands, the CD spectra only showed an increase and a shift towards higher wavelengths of the positive band at 700 nm. A more striking effect is the decrease of the EPR signal at g = 4 and g = 2, as shown by the difference spectra at g = 2 in Figure 4, which were obtained after subtraction from the observed spectra of the spectrum of methaemocyanin in the presence of 1 M potassium fluoride. By integrating these spectra an apparent association constant of about 1 mM^{-1} has been estimated.

All the preceding experiments were carried out in acetate buffer at pH 5.0. A colour change upon raising the pH indicated the destruction of the methaemocyanin azide complex. CD spectra of methaemocyanin were

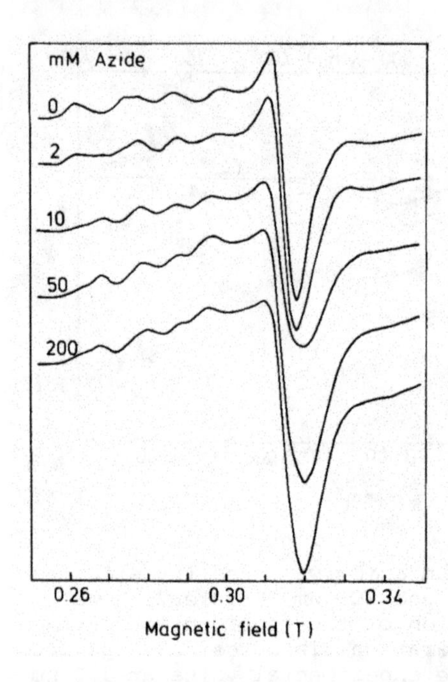

Fig. 3

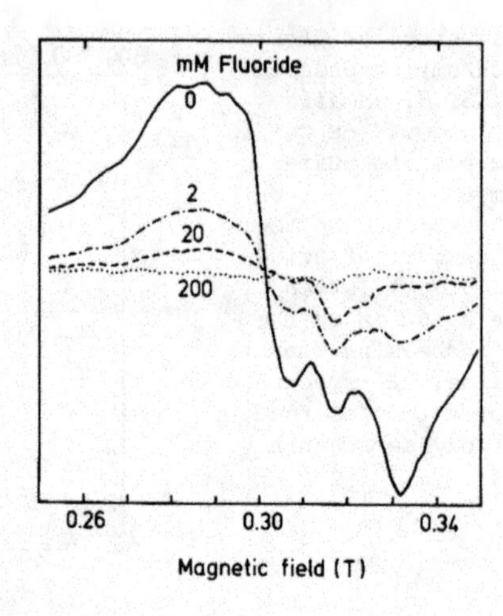

Fig. 4

Fig. 3. EPR spectra at g = 2 of *H. pomatia* methaemocyanin (6 mM Cu) in 0.1 M sodium acetate buffer, pH 5.0, formed by the action of azide as described in the text, after removal of azide by dialysis (*upper curve*). *Other curves:* the same solution in the presence of increasing amounts of sodium azide

Fig. 4. Difference EPR spectra at g = 2 of *H. pomatia* methaemocyanin (6 mM Cu) in 0.1 M sodium acetate buffer pH 5.0, formed by the action of fluoride as described in the text, after removal of fluoride by dialysis (*upper curve*). *Other curves:* the same solution in the presence of increasing amounts of potassium fluoride. From all the curves the residual spectrum of the methaemocyanin preparation in the presence of 1 M fluoride has been subtracted

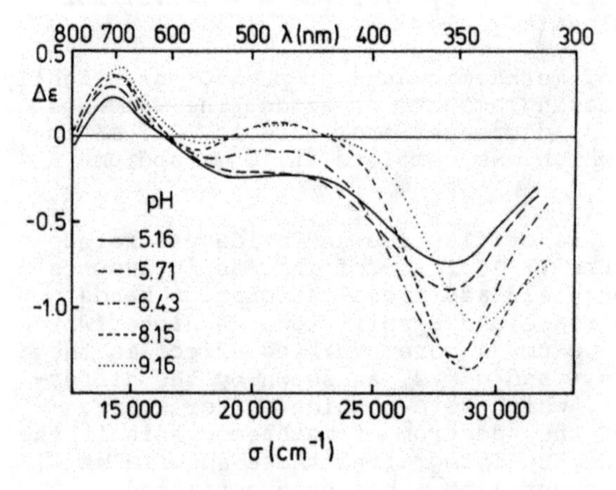

Fig. 5. The influence of pH on the molar circular dichroism Δε (l/mol/cm), expressed per Cu atom, of *H. pomatia* methaemocyanin (0.47 mM Cu) prepared by the action of azide as described in the text. The original solution was diluted four-fold with the appropriate buffers yielding a final azide concentration of 6.25 mM

Fig. 6. The influence of pH on the EPR spectrum of *H. pomatia* methaemocyanin (6 mM Cu), formed by the action of azide as described in the text, after removal of azide by dialysis. The pH of the solutions was adjusted and sodium azide added to a final concentration of 1.25 mM

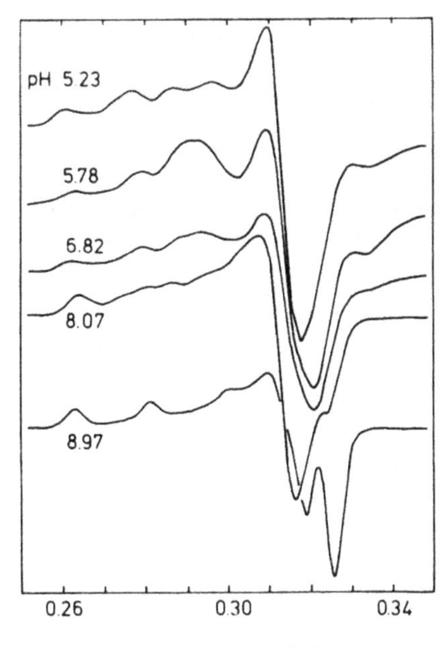

recorded in the presence of 6.25 mM sodium azide between pH 5.16 and 9.16 (Fig. 5). At high pH the spectrum between 300 and 600 nm resembles that of the remaining oxyhaemocyanin with negative bands at 350 and 580 nm, and a positive one at 480 nm. This displacement of azide at higher pH values may possibly be used to remove its last traces from methaemocyanin. Similar observations were made by EPR measurements (Fig. 6). Azide generated methaemocyanin solutions were rendered 1.25 mM in sodium azide and their pH adjusted in order to cover the region from pH 5.23 to 8.97. Besides an increase of $A_{\parallel}$ from 12.9 to 19.4 mK a definite change in shape was observed in the $g_{\perp}$ region.

Conclusions

The EPR spectra at g = 4 of methaemocyanin obtained by a treatment with azide or fluoride indicate the presence of copper pairs in accordance with the observed regeneration with hydrogen peroxide and with hydrogen sulphide.

The difference between these two preparations has to be investigated further and can probably not be interpreted completely by the traces of azide left in the methaemocyanin preparation obtained by treatment with azide, as shown by the EPR signal at g = 2.

The addition of azide to methaemocyanin, obtained with azide treatment, yields an absorption and a CD band at 380 nm, similar to the one described for type 2 copper, and seems, thus, to be due to traces of single copper atoms. At higher azide concentrations methaemocyanin develops a positive CD band at 340 nm and negative CD bands in the visible region. These bands are probably due to copper pairs. The binding of azide decreases with increasing pH.

The EPR spectrum of methaemocyanin, obtained by the action of fluoride, vanishes at g = 4 and at g = 2 by the addition of fluoride. The fluoride derivative of methaemocyanin, as far as the EPR spectrum is concerned, thus corresponds to oxyhaemocyanin, as fresh preparations show no EPR signal. As a working hypothesis a copper group, constituted of Cu(I), a bridging ligand, low spin Cu(III) and fluoride or peroxide, could be considered.

Acknowledgments. We wish to thank the National Fonds voor Wetenschappelijk Onderzoek for research grants and for the fellowship awarded to one of us (M.D.L.). This work was supported by the Fonds voor Collectief Fundamenteel Onderzoek (Contract No. 2.0016.76) and by the Fonds Derde Cyclus, Katholieke Universiteit te Leuven. The collaboration of Miss C. Houthuys is gratefully acknowledged.

References

Byers, W., Curzon, G., Garbett, K., Speyer, B.E., Young, S.N., Williams, R.J.P.: Anion-binding and the state of copper in caeruloplasmin. Biochem. Biophys. Acta 310, 38-50 (1973)

De Ley, M., Candreva, F., Witters, R., Lontie, R.: The fast reduction of *Helix pomatia* methaemocyanin with hydrogen sulphide. FEBS Lett. 57, 234-236 (1975)

Dunhill, R.H., Pilbrow, J.R., Smith, T.D.: Electron spin resonance of copper(II) citrate chelates. J. Chem. Phys. 45, 1474-1481 (1966)

Guillemet, R., Gosselin, G.: Sur les rapports entre le cuivre et la capacite' respiratoire dans les sangs hemocyaniques. Compt. Rend. Soc. Biol. 111, 733-735 (1932)

Heirwegh, K., Lontie, R.: Decrease of the copper bands of *Helix pomatia* haemocyanin. Nature (London) 185, 854-855 (1960)

Heirwegh, K., Borginon, H., Lontie, R.: Separation and absorption spectra of α- and β-haemocyanin of *Helix pomatia*.Biochem. Biophys. Acta 48, 517-526 (1961)

Lontie, R., Vanquickenborne, L.: The role of copper in hemocyanins. In: Metal Ions in Biological Systems, Vol. 3. High Molecular Complexes. Sigel, H. (ed.). New York: Dekker, 1974, pp. 183-200

Malmstrom, B.G., Reinhammar, B., Vanngard, T.: The state of copper in stellacyanin and laccase from the lacquer tree *Rhus vernicifera*. Biochim. Biophys. Acta 205, 48-57 (1970)

McPhie, P.: Dialysis. In: Methods in Enzymology, Vol. 22. Jakoby, W.B. (ed.). New York and London: Academic Press, 1971, pp. 23-32

Witters, R., Lontie, R.: The formation of *Helix pomatia* methaemoxyanin accelerated by azide and fluoride. FEBS Lett. 60, 400-403 (1975)

Kinetics of Reaction Between Hemocyanin and CN^- and of Reconstitution of Hemocyanin with $K_3Cu(CN)_4$

B. SALVATO AND P. ZATTA

Introduction

It is well established that the addition of CN^- to a native hemocyanin solution causes the release of oxygen and the blenching of the solution.

This reaction has been interpreted in two quite different ways. Felsenfeld (1954) supposed that the reaction could be ascribed to a displacement of the metal from the complex by a stronger ligand.

Pearson (1936) interpreted the blenching of the solution as being produced by a formation of a complex between hemocyanin and CN^-. De Ley and Lontie (1972) demonstrated that in the C.D. spectrum of *Helix pomatia* hemocyanin, new bands appear in the visible region following the addition of a small amount of CN^-. Actually we believe that this reaction is more complicated than has been thought so far.

In this communication we report the kinetics studies of the reaction between hemocyanin and CN^- and the reconstitution of hemocyanin by the reaction of apo-protein with $K_3Cu(CN)_4$.

All experiments were carried out using *Octopus vulgaris* and *Carcinus maenas* hemocyanins.

Experimental Section

Preparation of Hemocyanin. The hemolymph was collected from living animals and diluted 1/1, then the hemocyanin was prepared by double precipitation with ammonium sulphate. The pellet was dissolved in Tris-HCl buffer 0.01 M, pH 7.5 and dialyzed in the same buffer. Finally hemocyanin was frozen after the addition of 20% sucrose and stored at -20°C. The solution of hemocyanin was prepared by dialyzing the protein with the convenient buffer. Protein concentration was determined spectrophotometrically ($E^{1\%}_{278}$ = 12.4 in 0.1 M Tris-HCl pH 8.0 buffer; $E^{1\%}_{288}$ = 14.1 in KOH 0.3 M).

Reaction of Hemocyanin with CN^-. The reaction was carried out at 3°C in a thermostatic cell, using 10 ml of hemocyanin at 10 mg/ml, in Tris-HCl buffer 0.1 M, pH 7.5.

At the start of the experiment, 0.1 ml of freshly prepared KCN 1 M was added and the reaction mixture was stirred continuously. At different times, 0.5 ml of solution was removed and put in a Sephadex G-25 Superfine column (1 × 10 cm) equilibrated with Tris-HCl buffer 0.1 M pH 7.5. In the protein collected, the ratio between protein and copper was determined. Copper was determined with a model 300 Perkin-Elmer atomic absorption spectrophotometer equipped with a multiple lamp. The effectiveness of this separation was previously ascertained, using hemocyanin (native and copper-free) to which a known amount of copper was added as Cu^+ and Cu^{2+}.

The reaction between hemocyanin and CN^- was also carried out at 3^oC in a Sephadex G-25 Superfine column (2 × 80 cm) equilibrated with Tris-HCl 0.1 M, pH 7.5 containing 0.025 M KCN.

Fluorescence Measurements. The reaction with CN^- was also followed by measuring the time dependence of the fluorescence emission intensity. It is known that the quantum yield of copper-free hemocyanin is about 1.5 times higher than that of deoxy-hemocyanin.

Measurements were carried out with the spectrophotofluorimeter Perkin-Elmer model MPF-2A with recorder. The spectra were used without any correction.

Data are presented as fluorescence intensity, relative to that of zero time.

Reconstitution of Hemocyanin Using a Sephadex G-25 Superfine Column. The reconstitution of *O. vulgaris* hemocyanin was performed in a Sephadex G-25 Superfine column (2 × 40 cm) equilibrated with $NaHCO_3$ buffer 0.02 M, CH_3COONa 0.02 M, 2.25 µg Cu^+/ml as $K_3Cu(CN)_4$. This copper concentration corresponds to 1/10 of copper concentration in a solution of 10 mg/ml of *O. vulgaris* native hemocyanin.

Reconstitution of Hemocyanin, Dialyzing the Protein in a Bio-Rad Laboratories Mini-beaker. Reconstitution of hemocyanin was finally performed in a dialyzing mini-beaker purchased from Bio-Rad laboratories. We used the bio-filter 50 mini-beaker which retains molecular weight higher than 30,000 daltons.

Five ml of apo-hemocyanin 10 mg/ml in Tris-HCl μ = 0.1, pH 7.0 was introduced into the beaker.

A buffer solution containing Cu^+ as $K_3Cu(CN)_4$ at a flow rate of 4 ml/min, was circulated (at different times). Before removing the sample, buffer solution without copper was circulated for half an hour in the capillaries.

The rate of copper equilibrium had previously been measured with a blank containing only a buffer solution. In our conditions no differences berween the flow rate at 2 and 4.0 ml/min were observed. The half-time for the equilibration was 10 minutes and was independent of the copper concentration in the circulatory buffer.

Results

Reaction between Native Hemocyanin and CN^-. A few minutes after CN^- addition the Hcy solution is blenched. In Figure 1A, the time dependence of copper still bound, normalized to 100 mg of protein, is shown.

The curve presents three typical zones. In the first one there is a rapid loss of about 50% of copper which follows an apparent first order trend. In the second zone the copper still bound decreases slowly, while in the last part a new increase of rate of reaction appears.

The kinetics of this curve are clearly more complicated than those corresponding to the simple model proposed by Felsenfeld (1954) and Pearson (1936).

Time Dependence of Fluorescence Intensity during the Reaction between Hemocyanin and CN^-. In Figure 1C the fluorescence intensity normalized to that at

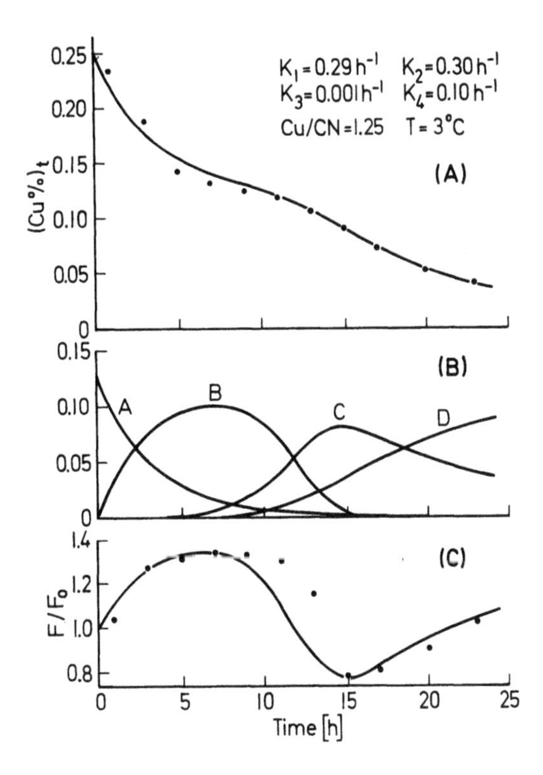

Fig. 1A-C. Kinetics of the reaction between *O. vulgaris* Hcy and CN^- at T = 3°C; Cu/CN^- = 1/25. (A) Percentage of copper still bound to protein, plotted against time. *Symbols*: experimental points. *Line*: simulated process according to proposed model. (B) Time dependence of individual species A, B, C, D, calculated according to proposed model. (C) Time dependence of relative fluorescence. *Symbols*: experimental points. *Line*: calculated from (B), assuming the following fluorescence ratios: $\frac{B}{A} = 1.45$; $\frac{C}{A} = 0.5$; $\frac{D}{A} = 1.45$

zero time is plotted against time. The fluorescence intensity shows modifications corresponding to the three zones of Figure 1A. This indicates the presence of intermediates in the reaction, concording with the fact that the kinetics are complicated.

Reaction of Apo-Protein and CN^- in a Sephadex G-25 Superfine Column. In Figure 2 the elution profile of a Sephadex G-25 Superfine column shows three peaks.

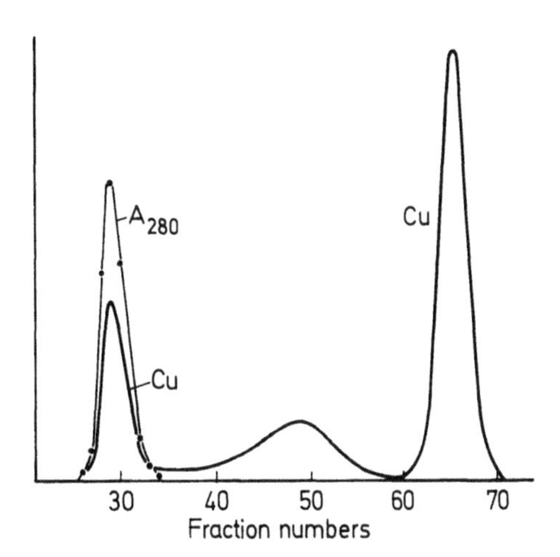

Fig. 2. Elution profile of *O. vulgaris* Hcy from Sephadex G-25 column equilibrated with CN^--containing buffer

The first one emerges with the protein and represents the copper still bound, the other two represent the copper released.

The narrow peak contains about 50% of the total copper and is released in the first instances of elution.

These data agree well with those reported in the results above, and confirm the presence of two types of kinetically different copper at the hemocyanin active site.

Reconstitution of Hemocyanin Using a Sephadex G-25 Superfine Column. In Figure 3B the elution profile of copper in the reconstitution experiments in a Sephadex G-25 Superfine column equilibrated with a copper-containing buffer is shown.

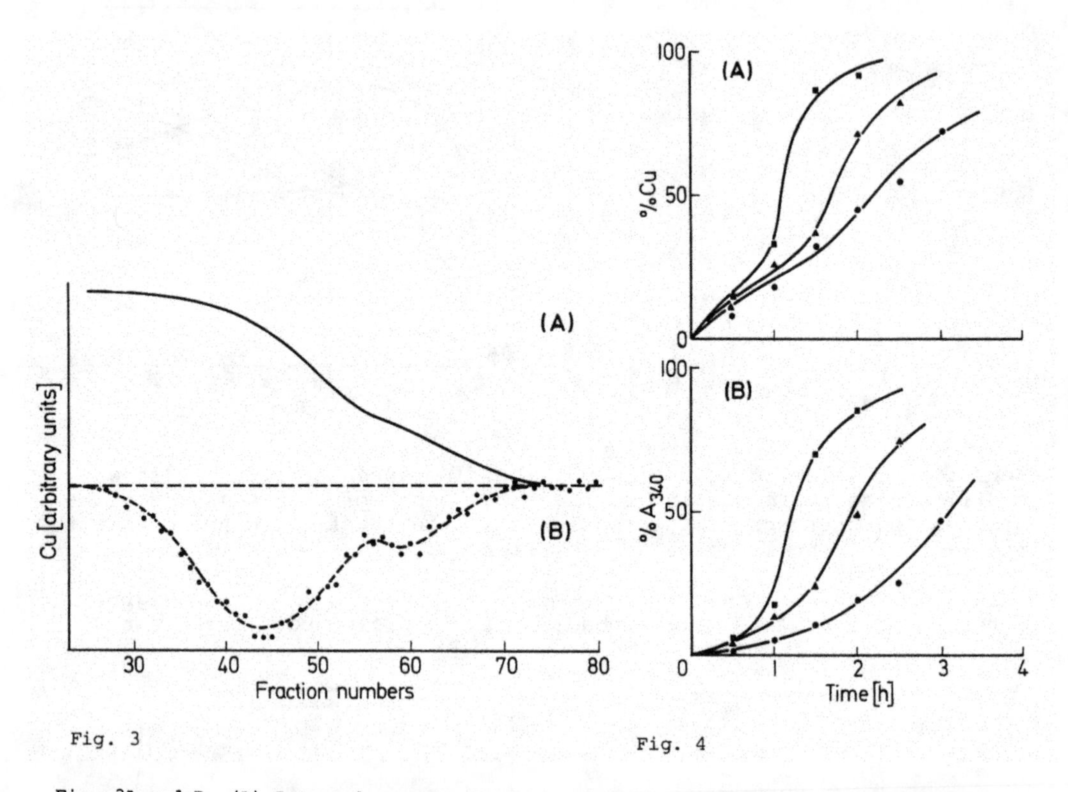

Fig. 3A and B. (A) Integral curve calculated from B. (B) Elution profile of reconstitution of *O. vulgaris* Hcy in a column of Sephadex G-25 equilibrated with copper-containing buffer

Fig. 4A and B. Reconstitution of *C. maenas* Hcy in Bio-Rad mini-beaker. (A) Time dependence of percentage of introduced copper. (B) Time dependence of recovered copper band. Symbols represent different ratios between copper concentration in flowing buffer and in native Hcy at same concentration: ●: 1/10; ▲: 1/5; ■: 1/1

At the initial time the uptake of copper is slow, then a great increase occurs.

Figure 3A represents the integral curve of the experimental copper elution profile shown in Figure 3B.

Even the reaction of reconstitution appears to be complex and probably corresponds to the reverse of the reaction between hemocyanin and CN^-.

Reconstitution of Hemocyanin, Dialyzing the Apo-Protein in a Bio-Rad Laboratories Mini-beaker. Figure 4 demonstrates the kinetics of reconstitution of *C. maenas* hemocyanin. The data are consistent with those obtained with a Sephadex G-25 Superfine column. The reaction appears complex with a slow start and a rapid increase subsequently. Further the reconstitution appears to be cooperative. The cooperative part is dependent on the external copper concentration.

Analogous results were obtained with *O. vulgaris* hemocyanin. The major difference arises from the copper which is still bound in the starting hemocyanin (about 20%).

Figure 5 shows the reconstitution results where the Cu/Cu theor. is 0.5/10.

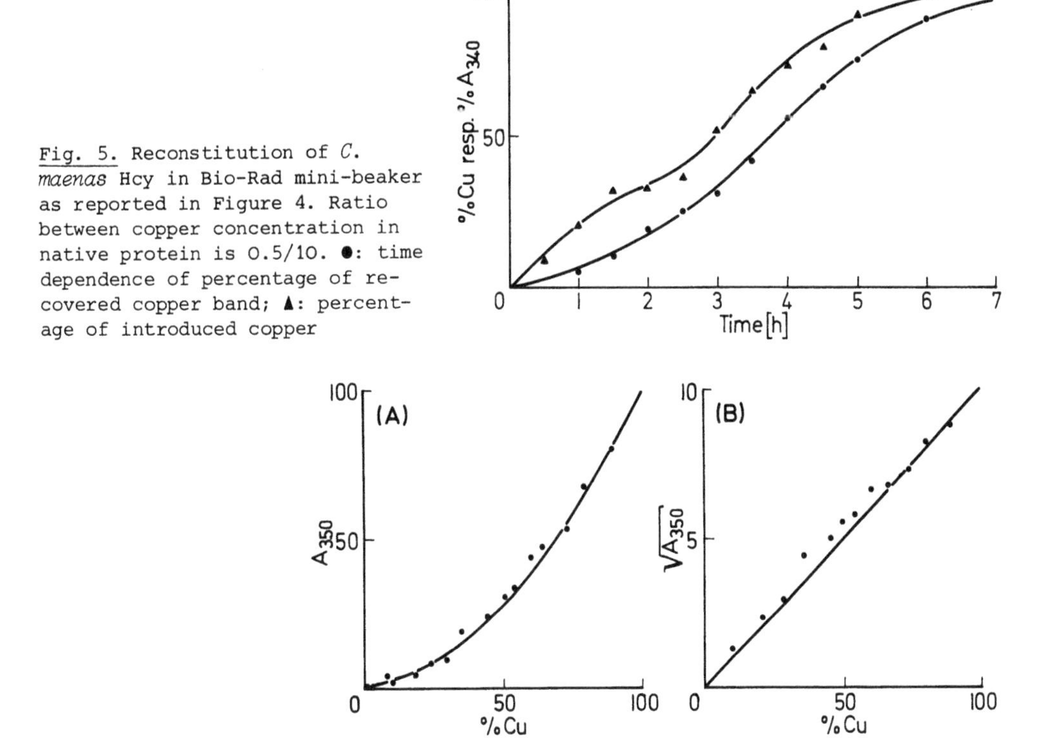

Fig. 5. Reconstitution of *C. maenas* Hcy in Bio-Rad mini-beaker as reported in Figure 4. Ratio between copper concentration in native protein is 0.5/10. ●: time dependence of percentage of recovered copper band; ▲: percentage of introduced copper

Fig. 6A and B. (A) Percentage of reconstituted copper band against percentage of copper bound. (B) Same data as square root of percentage of reconstituted copper band, against percentage of copper bound. Data calculated from those given in Figure 5

Figure 6A represents the percentage of reconstituted copper band at 340 nm against the percentage of copper bound.

These data were calculated from those shown in Figure 5, and this curve looks like that reported for the partially reconstituted *Helix pomatia* hemocyanin (Konings et al., 1969).

In Figure 6B the same data are shown as the square root of the percentage of reconstituted copper band at 350 nm against the percentage of copper bound. A straight line is obtained, as for the partially reconstituted *H. pomatia* hemocyanin.

Discussion

In the last hemocyanin symposium we delt with the kinetics of the reaction between hemocyanin and CN^- and we proposed a sequential model, which implies kinetic differences between the two copper atoms at the hemocyanin active site (Table).

In this model we considered:

1. A as a concentration of the active site containing two copper atoms;
2. B as a concentration of the active site containing one copper atom;
3. C as a conformer of B;
4. D as a concentration of copper-free sites.

Since CN^- is in large excess with respect to the copper, we considered all reactions of the first order.

We considered the rate of reaction from B to C as depending on the concentration of the copper-free sites.

Table

$$A \xrightarrow{k_1} B \xrightarrow{k_2} C \xrightarrow{k_4} D, \qquad B \xrightarrow{k_3} D$$

$$\begin{cases} \dfrac{dA}{dt} = -k_1A \\ \dfrac{dB}{dt} = k_1A - k_2BD - k_3B \\ \dfrac{dC}{dt} = k_2BD - k_4C \\ \dfrac{dD}{dt} = k_3B + k_4C \end{cases}$$

$$A_o = A + B + C + D \qquad \text{at } t_o \quad A = A_o$$

$$(Cu)_t = 2A + B + C \qquad B = 0, \; C = 0, \; D = 0$$

The assumption is justified taking into account the fact that hemocyanin is a very large molecule and thus the conformational modifications that can arise from the displacement of copper can affect the rate of reaction of the surrounding sites which are still copper-containing.

The system of differential equations which mathematically describes the proposed model was treated in a computer with a "continuous system modeling program".

In Figure 1A the curve represents the computer-calculated curve, assuming the values reported in the figure to be the constants.

In Figure 1B the calculated time dependence of A, B, C, D concentration is shown.

The curve in Figure 1C was calculated using the data of 1B, assuming the fluorescence ratios given in the legend.

It is interesting to note that the ratio of the energy yield of copper-free hemocyanin and deoxy-hemocyanin has the same value as assumed before. This model easily interprets the results obtained from the reaction between native hemocyanin and CN^- on a Sephadex G-25 Superfine column.

Experiments carried out by Cox and Elliott (1974) regarding the exchangeable copper^{64} with hemocyanin demonstrated that the exchange occurs only with Cu^+ complex in the presence of a small amount of CN^-. A very interesting aspect of these experiments is that only 50% of copper is exchangeable in all their experimental conditions. The recovery of the copper band is inversely dependent on the percentage of copper exchanged.

With our model these results can be forecast. In the isotope exchange experiments the total amount of copper bound remains constant and equal to the initial amount of copper. Thus, taking into account that the B-to-C conformational modification occurs after the removal of the majority of the first copper atom, if a modification from B to C does not occur, then the second copper atom cannot be released.

The dependence of the copper band on exchanged copper is a further indication that CN^- forms a complex with hemocyanin.

The constants evaluated by Felsenfeld (1954) with his model can represent, on the basis of the reported data, only an approximate evaluation of the instability constants of the complex between hemocyanin and CN^-. We evaluated (Salvato, unpublished data) the order of magnitude of the instability constant of the copper-hemocyanin complex by acid-base titration data, ranging between 1×10^{-26} and 1×10^{-27} mol/l; these constants are much more reasonable than those of Felsenfeld, considering that Cu^+ ligands with instability constant around 10^{-26} are unable to displace the copper from hemocyanin.

The reconstitution data are also consistent with the proposed model. Our conditions are in fact homogeneous with those used for the hemocyanin and CN^- reaction.

The data presented demonstrate the presence of two kinetically different types of copper, and thus a conformational modification is needed for a complete reconstitution.

Results reported in Figures 5 and 6 demonstrate that the system simulates the apparent equivalence between the two copper atoms at the active site of hemocyanin.

References

Cox, J.A., Elliott, F.G.: Isotopic copper exchange in *Pila* haemocyanin with three radioactive cupreous complexes. Biochim. Biophys. Acta 371, 392-401 (1974)

De Ley, M., Lontie, R.: The reversible reaction of cyanide with *Helix pomatia* hemocyanin. Biochim. Biophys. Acta 278, 404-407 (1972)

Felsenfeld, G.: The binding of copper by hemocyanin. J. Cell. Comp. Physiol. 43, 23-38 (1954)

Konings, W.N., Siezen, E.J., Gruber, M.: Structure and properties of hemocyanins. VI. Association-dissociation behavior of *Helix pomatia* hemocyanin. Biochim. Biophys. Acta 194, 376-385 (1969)
Pearson, O.H.: The reaction of cyanide with the hemocyanin of *Limulus polyphemus*. J. Biol. Chem. 115, 171-177 (1936)

Properties of Hemocyanin from *Limulus polyphemus* (Horseshoe Crab) Under Dissociating Conditions

M. Brunori and G. Amiconi

Introduction

Reversible oxygen binding by hemocyanins, the respiratory proteins of molluscs and arthropods, has been extensively investigated employing proteins from different sources under various conditions. Among other things it has been shown that oxygen binding is characterized by the presence of both homotropic and heterotropic interaction effects as with hemoglobins, a fact which stresses the general nature of the molecular mechanisms involved in determining the functional properties of these proteins (Wyman, 1967; Van Holde and Van Bruggen, 1971).

In the case of hemoglobins, studies on myoglobin or the isolated hemoglobin chains have provided valuable information for the interpretation of the properties of the assembled molecule (Antonini and Brunori, 1971). For hemocyanins, investigations of a similar nature on the minimum functional unit have recently received considerable attention, and the separation and characterization of various components from arthropod hemocyanins is being actively pursued (see other articles in this volume).

The minimum functional unit of molluscan hemocyanins (two copper atoms per 50,000 daltons) has never been obtained in the native state by conventional methods, but in the last few years evidence has been produced for the existence and the function of such species after limited proteolysis (Lontie et al., 1973). In the case of hemocyanins from arthropods, on the other hand, it is known that at alkaline pH (e.g. above 9) the molecule spontaneously dissociates into a subunit with a molecular weight of about 70,000 daltons, containing two copper atoms and capable of reversible and non-cooperative oxygen binding (Pickett et al., 1966; Sullivan et al., 1974).

This note reports the results of some observations on equilibrium and kinetics of oxygen binding by the hemocyanin from *Limulus polyphemus* (horseshoe crab) at alkaline pH, where such a molecule is known to dissociate into subunits and is represented by a mixture of different components (Sullivan et al., 1974).

Materials and Methods

Hemocyanin from *Limulus polyphemus* was prepared by centrifugation of the hemolymph and collection of the supernatant. The solution was then dialyzed against glycine or borate buffer, pH 9.2, containing 10^{-3} M EDTA. As shown by ultracentrifugation, the protein at pH $\geq$ 9 consists of essentially one component with a sedimentation coefficient of $S_{20,w} \simeq 5.0$, consistent with a molecular weight of 70,000 daltons. This finding is in agreement with what was reported previously (Sullivan et al., 1974).

Protein concentration was determined spectrophotometrically using the value of $E_{1\,cm}^{1\%} = 13.9$ at $\lambda = 280$ nm (Nickerson and Van Holde, 1971).

Oxygen equilibria were determined by the method of Rossi Fanelli and Antonini (1958). Oxygen kinetics were investigated using the temperature-jump technique (Eigen and De Maeyer, 1963; Brunori, 1971).

Results and Discussion

The oxygen binding of the minimum functional unit of hemocyanin from *Limulus* is non-cooperative; under all conditions explored the value of n in the Hill equation was found to be very close to 1.0. At 20°C and pH 9.2, 2% borate buffer, the $p^{1/2}$ is 3.5 mmHg. Thus under our experimental conditions the various components behave alike, although under other conditions they were shown to be functionally different (Sullivan et al., 1974).

The dependence of $p^{1/2}$ on pH and calcium concentration was explored at 20°C. The oxygen affinity is pH independent from 9 to 10, the value of n remaining always close to 1. Upon addition of calcium to the solution, the value of $p^{1/2}$ and n change as depicted in Figure 1. When the total concentration of calcium chloride is above ∿ 0.02 M, there is a noticeable effect on both parameters, which possibly indicates reassociation of the protein.

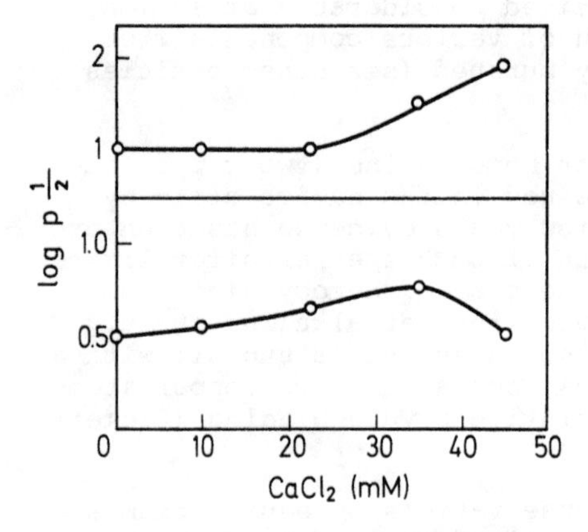

Fig. 1. Effect of total calcium chloride concentration on oxygen affinity and value of n of hemocyanin from *Limulus polyphemus*. Conditions: glycine buffer 0.2 M ionic strength pH 9.4, 20°C and a protein concentration of 13.5 mg/ml

The temperature dependence of the oxygen binding has been explored over the range 10 to 40°C at pH 9.2 both in borate and glycine buffers. Figure 2 depicts the dependence of the oxygen affinity on temperature reported in terms of a Van't Hoff plot. The overall ΔH is -11.5 Kcal/mol, which includes the heat at solution of oxygen.

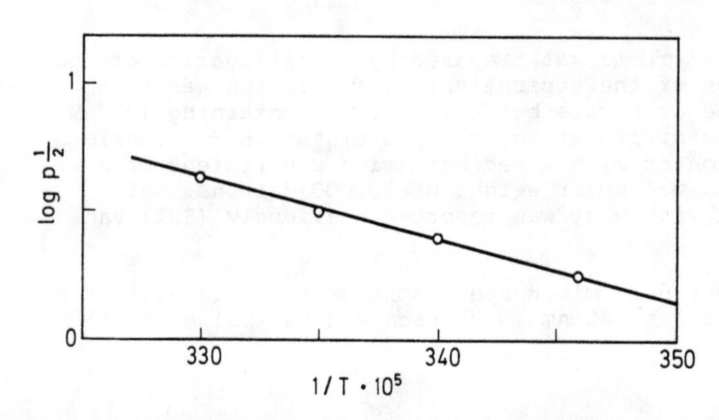

Fig. 2. Temperature dependence of oxygen affinity observed with hemocyanin from *Limulus polyphemus*, in glycine buffer 0.2 M ionic strength, pH 9.4 and a protein concentration of 13.5 mg/ml

Table 1. Thermodynamic parameters for oxygen binding by simple oxygen carriers at $20^{o}C$

Protein	pH	$K \times 10^{-5}$ (M^{-1})	$-\Delta G$ (Kcal/mol)	$-\Delta H^{a}$ (Kcal/mol)	$-\Delta S^{a}$ (e.u.)	References
Hemocyanin	9.4	2.76	7.3	8.5	4.1	this work
Hemerythrin	7.0	1.55	7.0	10.5	12.1	Bates et al. (1968)
Myoglobin	7.0	10.6	7.8	11.9	14.0	b

[a]For oxygen in solution.
[b]Amiconi, G., Antonini, E., Brunori, M., Magnusson, E. (unpublished results).

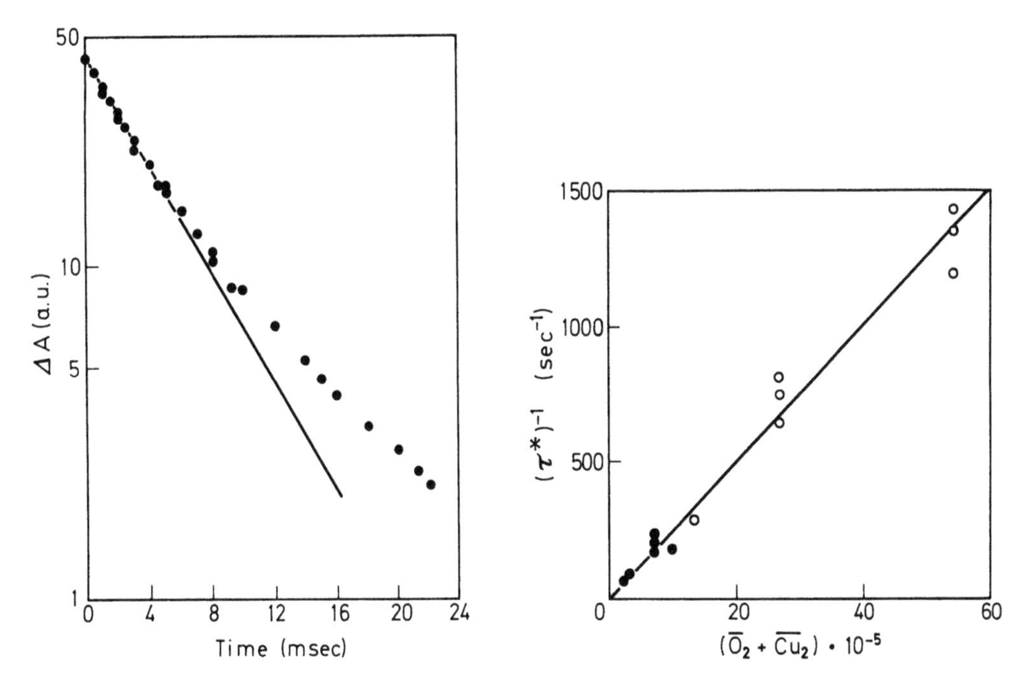

Fig. 3. Fig. 4

Fig. 3. Relaxation spectrum of reaction of hemocyanin from *Limulus polyphemus* with oxygen. Conditions: glycine buffer 0.2 M ionic strength, pH 9.4. Temperature $25^{o}C$ (after jump). Observation wavelength = 366 nm. Protein concentration = 3×10^{-5} binding equivalents/liter. Fractional saturation with oxygen $\sim$ 0.9

Fig. 4. Dependence of mean reciprocal relaxation time, $(\tau^*)^{-1}$, on concentration of free oxygen and free sites. Conditions as in Figure 3, except that protein concentration was varied from 3 to 10×10^{-5} binding equivalents/liter. Closed symbols indicate experiments in which oxygen fractional saturation was equal to or less than 90%

Table 1 compares the thermodynamic parameters for the binding of oxygen to hemocyanin from *Limulus polyphemus* (pH 9.4), to hemerythrin from *Sipinculus nudus* (pH 7) and to myoglobin from sperm whale (pH 7). It is of interest to note that the differences in affinity between the various proteins stem from differences in both the enthalpic and entropic terms. However, two points are remarkable from comparison of the data: (1) that O_2 binding to hemocyanin is associated with a small and negative entropy change, a value three to four times smaller than that observed for either hemerythrin or myoglobin; (2) that in all three cases the approximately similar free energies of binding are the results of entropy-enthalpy compensation effects (Lumry, 1971).

Temperature jump experiments, performed at pH 9.4 in glycine buffer, show a clearly discernible chemical relaxation effect in the millisecond time range. The direction of the optical density change observed upon increase in temperature is consistent with the negative enthalpy change and the spectral properties of oxy- and deoxy-hemocyanin. The relaxation cannot be described by a single exponential process, and the data were analyzed using an average relaxation time (τ^*), as defined by Swartz (1965) (Fig. 3). An analysis of the data in terms of a discrete number of chemical relevation effects was not considered worthwhile in view of the finding that the isolated components display different kinetic behavior (Sullivan et al., 1974; Bonaventura et al., Chap. 34, this vol.).

The dependence of the reciprocal τ^* on the concentration of the reagents is shown in Figure 4. The linear dependence implies that the observed relaxation effect reflects a bimolecular process, i.e. the binding of oxygen to the copper site. The apparent second order velocity constant obtained from the data in Figure 4 is $\vec{k} = 2.5 \times 10^6$/M/s.

This value is very close to the average of the combination velocity constants for binding of O_2 measured for the various components (Bonaventura et al., Chap. 34, this vol.). However in view of the kinetic heterogeneity of the system, apparent from this study and from the observations on the isolated components, any more complete analysis of the detailed mechanism of the O_2 reaction is hindered.

References

Antonini, E., Brunori, M.: Hemoglobin and Myoglobin in Their Reactions with Ligands. Amsterdam: North Holland, 1971

Bates, K., Brunori, M., Amiconi, G., Antonini, E., Wyman, J.: Studies on hemerythrin. I. Thermodynamics and kinetic aspects of oxygen binding. Biochemistry 1, 3016-3020 (1968)

Brunori, M.: Kinetics of the reaction of *Octopus vulgaris* hemocyanin with oxygen. J. Mol. Biol. 55, 39-48 (1971)

Eigen, M., De Maeyer, L.: Relaxation methods. In: Technique of Organic Chemistry. Friess, S.L., Lewis, E.S., and Weissberger, A. (eds.). New York: Interscience, 1963, Vol. VIII, Part 2, pp. 895-1054

Holde, K.E. Van, Van Bruggen, E.F.J.: The hemocyanins. In: Subunits in Biological Systems. Timasheff, N.S., Fasman, G.D. (eds.). New York: Marcel Dekker, 1971, Vol. V., pp. 1-54

Lontie, R., De Ley, M., Robberecht, H., Witters, R.: Isolation of small functional subunits of *Helix pomatia* haemocyanin after subtilisin treatment. Nature (New Biol.) 242, 180-182 (1973)

Lumry, R.: Fundamental problems in the physical chemistry of protein behaviour. In: Electron and Coupled Energy Transfer in Biological Systems. King, Klingenberg (eds.). New York: Marcel Dekker, 1971, Vol. I, Part A

Nickerson, K.W., Van Holde, K.E.: A comparison of molluscan and arthropod hemocyanin-I. Circular dichroism and absorption spectra. Comp. Biochem. Physiol. 39B, 855-872 (1971)

Pickett, S.M., Riggs, A.F., Larimer, J.L.: Lobster hemocyanin: properties of the minimum functional subunit and of aggregates. Science 151, 1005-1007 (1966)

Rossi Fanelli, A., Antonini, E.: Studies on the oxygen and carbon monoxide equilibria of human myoglobin. Arch. Biochem. Biophys. 77, 478-492 (1958)

Sullivan, B., Bonaventura, J., Bonaventura, C.: Function differences in the multiple hemocyanins of the horseshoe crab, *Limulus polyphemus* L. Proc. Natl. Acad. Sci. 71, 2558-2562 (1974)

Swartz, G.: On the kinetics of the helix-coil transition of polypeptides in solution. J. Mol. Biol. 11, 64-77 (1965)

Wyman, J.: Allosteric linkage. J. Am. Chem. Soc. 89, 2202-2218 (1967)

Kinetic Analysis of Oxygen-Binding of *Panulirus interruptus* Hemocyanin

H. A. Kuiper, M. Brunori, and E. Antonini

Introduction

Hemocyanin, occurring in the hemolymph of the spiny lobster, *Panulirus interruptus* (phylum Arthropoda) is a copper-containing respiratory protein. The undissociated protein is a hexamer containing six oxygen-binding sites, and has a molecular weight of 450,000. The subunits have a molecular weight of 75,000 and contain one oxygen-binding site each. Equilibrium oxygen-binding studies showed whole molecules to be cooperative, while the subunits bind oxygen noncooperatively with a low oxygen affinity compared to that of whole molecules (Kuiper et al., 1975). An extensive study of the oxygen equilibrium of the arthropod hemocyanin from *Callianassa californiensis* by Miller and Van Holde (1974) showed that oxygen binding could be described in terms of a two-state model according to Monod et al. (1965) with the presence of hybrid states during the allosteric transition of whole molecules.

Kinetic studies of oxygen binding by arthropod hemocyanins are restricted to an early one by Millikan (1933), and a recent study by Sullivan et al. (1974), who reported oxygen dissociation rate constants for undissociated *Limulus* hemocyanins and the different subunits.

This paper reports preliminary results of a study on the reaction kinetics of undissociated *P. interruptus* hemocyanin, under conditions of cooperative oxygen binding, and kinetic parameters of subunits, determined with the temperature-jump relaxation method and stopped-flow techniques.

Materials and Methods

Hemocyanin was isolated and stored lyophilized as described by Kuiper et al. (1975). For the range pH 7-9, 0.1 M Tris-HCl buffers, ionic strength 0.1 were used. For experiments above pH 9, 0.1 M ethanolamine and 0.04 M sodium tetraborate buffers made up to ionic strength 0.1 with sodium chloride were used. The hemocyanin concentration is given in terms of oxygen binding sites per liter, assuming 1 binding site per 75,000 molecular weight.

Stopped-flow measurements were carried out with a Gibson-Durrum stopped-flow apparatus. The dead time of the apparatus was approximately 3.5 ms. The oxygen dissociation reaction of hemocyanin was followed by mixing a solution of oxygenated protein with degassed buffer containing excess sodium dithionite. The oxygen combination reaction was followed by mixing a solution of deoxygenated hemocyanin with buffer containing oxygen at a known concentration (air-saturated buffer contains 270 μM oxygen). The reactions were monitored at wavelengths in the 337 nm and 570 nm absorption bands of oxygenated hemocyanin. The experiments were performed at 20°C. The dead time of the stopped-flow apparatus is not included in the various calculations of the kinetic constants.

Temperature-jump (T-jump) relaxation measurements were performed with an Eigen-De Maeyer instrument built by the Messanlagen Gesellschaft, Göttingen (Eigen and De Maeyer, 1963; Eigen, 1968). A temperature-jump cell with a 1 cm light path and a volume of 7 ml was used. A temperature increase of 4°C was obtained by discharging a condensor at 30 kV through the solution; the final temperature was 24°C. Changes in absorbance were followed around 360 and 570 nm under any given experimental conditions, the relaxation spectrum was analyzed using 3-5 traces at two sweep times each. Oxygenated protein solution was diluted with deoxygenated buffer in order to cover a range of concentrations of free binding sites and oxygen. The concentration of free binding sites was calculated from the total protein concentration and the oxygen-binding equilibrium curve. For experiments at very low oxygen saturations, the T-jump cell was filled with completely deoxygenated protein, and the oxygen concentration was raised by adding small amounts of a solution of oxygenated protein. The oxygen saturation levels at equilibrium were determined directly in the cell by measuring the absorbance in the 337 or 570 nm band using a Cary 14 recording spectrophotometer.

Results

Undissociated Cooperative Hemocyanin. At pH 9.6 in the presence of 10 mM $CaCl_2$, the protein binds oxygen cooperatively (Fig. 1). The Hill coefficient at 50% saturation is approximately 1.9.

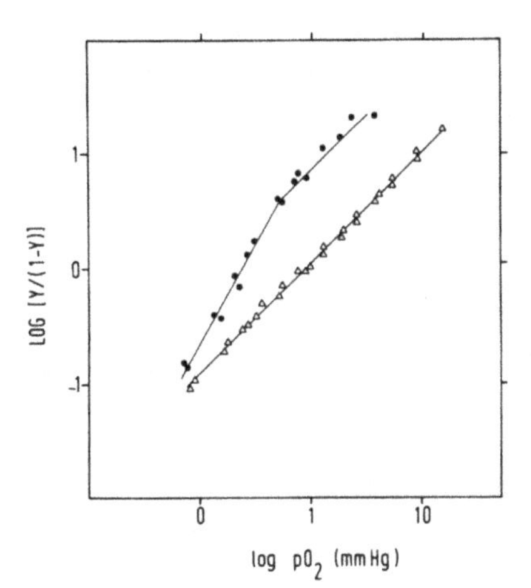

Fig. 1. Hill plots of oxygen equilibrium curves. Conditions: 0.04 M sodium tetraborate pH 9.6, ionic strength 0.1, 20°C. •: 10 mM $CaCl_2$ present; Δ: no calcium added. Y: fractional saturation with oxygen; pO_2: oxygen pressure (mm Hg)

Oxygen Combination as Measured with Stopped-Flow. The oxygen combination reaction is very fast. Upon mixing deoxygenated hemocyanin at a final concentration of 24 μM with buffer containing 33.8 μM oxygen, a 95% loss of the reaction in the dead time of the apparatus occurred as shown by the comparison of the observed ΔOD with the change in absorbance estimated from equilibrium data. This implies a very high combination rate constant in the order of 4-5 × 10^7/M/s.

Oxygen Dissociation as Measured with Stopped-Flow. Protein solutions with a concentration of 80-110 μM were mixed with degassed buffer + excess dithionite at several pH values. Upon lowering the pH, an increase in the first-order rate constant was observed, and at the same time an increasing percentage of the reaction was lost in the dead time of the

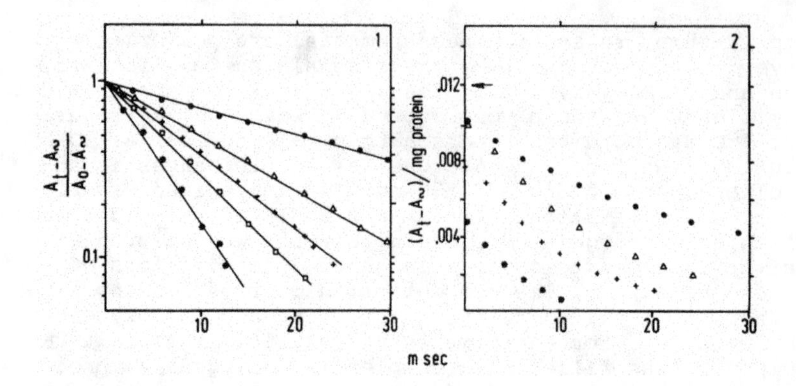

Fig. 2. 1,2. Time course of deoxygenation reaction at different pH values. Oxygenated undissociated hemocyanin was mixed with deoxygenated buffer containing excess of dithionite; the change in absorbance at 570 nm was followed
(1) Relative change in absorbance plotted as a function of time; A_o: absorbance observed at zero time; A_∞: absorbance observed after completion of the reaction; A_t: absorbance observed at time t. Conditions: pH 7-8.8, 0.1 M Tris-HCl + 10 mM $CaCl_2$, ionic strength 0.13; pH 9.25, 0.1 M ethanolamine + 10 mM $CaCl_2$, ionic strength 0.13. Experiments performed at 20°C. ●: pH 9.25, k_{off} = 33/s, Δ: pH 8.80, k_{off} = 80/s, +: pH 8.00, k_{off} = 102/s, □: pH 7.50, k_{off} = 120/s, *: pH 7.00, k_{off} = 192/s
(2) Change in absorbance per mg protein plotted as a function of time. *Arrow*: value for (A_o-A_∞)/mg protein as calculated from equilibrium data

apparatus (Fig. 2); the observed part of the reaction is apparently homogeneous. The increase in the oxygen-dissociation rate constant upon lowering the pH corresponds to the increase in P_{50} value as determined from the oxygen equilibrium (Fig. 3).

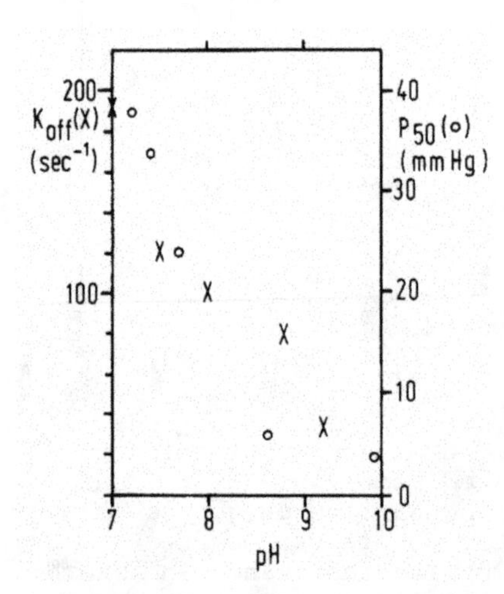

Fig. 3. Effect of pH on oxygen dissociation rate constant and P_{50} value of undissociated hemocyanin. Conditions: oxygen dissociation as described for Figure 2; oxygen equilibrium, pH 7.2-8.6, 0.05 M Tris-HCl + 10 mM $CaCl_2$, io onic strength 0.13; pH 9.68, 0.05 M ethanolamine + 10 mM $CaCl_2$, ionic strength 0.13. Temperature 20°C

Relaxation Kinetics. Temperature-jump experiments were performed at fractional saturation levels between 0.8 and 0.93; the protein concentration varied between 11 and 47 μM. The relaxation spectra are evidently heterogeneous (Fig. 4A). Data were analyzed in terms of the initial relaxation rate ($1/\tau^*$), which corresponds to one type of average of all elementary steps (Schwarz, 1968). According to the equation $(\tau^*)^{-1}$=

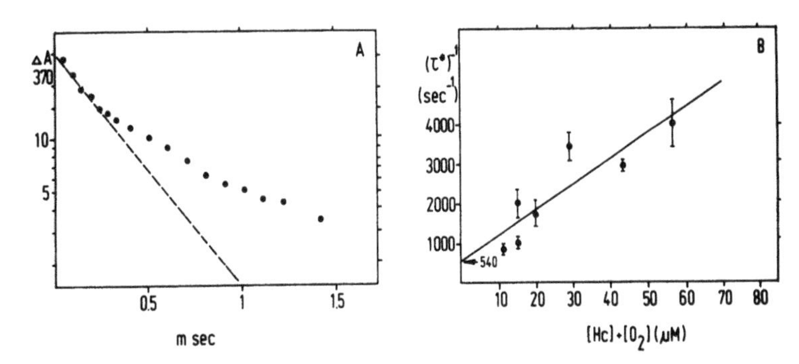

Fig. 4A and B. (A) Relaxation spectrum of reaction of hemocyanin with oxygen. Conditions: 0.04 M sodium tetraborate pH 9.6 + 10 mM $CaCl_2$, ionic strength 0.13; temperature 24°C (following jump); protein concentration = 23.4 μM binding equivalents, fractional saturation = 0.92. (B) Dependence of $(\tau^*)^{-1}$ on equilibrium concentration of free oxygen and free binding sites. Conditions as described under (A), except that protein concentration was varied from 11 to 47 μM binding equivalents. Fractional saturation levels varied from 0.80 to 0.95. *Drawn line* calculated with linear regression, corresponding to an apparent $k_{on} = 6.4 \times 10^7$/M/s and a k_{off} = 540/s

$k_{off} + k_{on}\{(O_2) + (Hc)\}$, which was derived for a bimolecular reaction (Eigen, 1968), the linear dependence of $1/\tau^*$ on the concentration of reactants (Fig. 4B) indicates that the observed overall process is a bimolecular one. This is presumably the binding of oxygen to hemocyanin. The apparent kinetic constants derived from this plot are: k_{off} = 540/s and $k_{on} = 6.4 \times 10^7$/M/s.

Dissociated Hemocyanin. At pH 9.6 in the absence of Ca^{2+}, the protein dissociates into subunits (molecular weight 75,000) containing one oxygen binding site.

Oxygen Dissociation as Measured with Stopped-Flow. The oxygen dissociation reaction shows heterogeneity (Fig. 5). A dissociation rate constant of 460/s was hardly measurable, while a loss of 70% of the reaction in the dead time of the apparatus indicates that the oxygen dissociation process is largely very fast. Increasing the dithionite concentration did not change the observed time course.

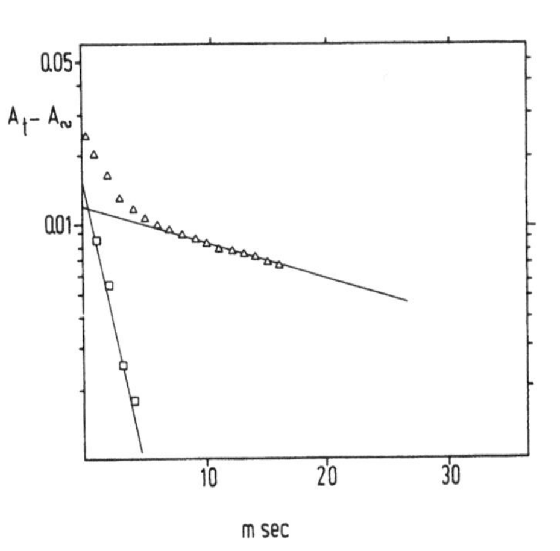

Fig. 5. Time course of deoxygenation reaction of oxygenated hemocyanin dissociated into subunits, measured at 570 nm and 20°C. Conditions: 0.1 M ethanolamine pH 9.2, ionic strength 0.10. Changes in absorbance were plotted as difference between observed absorbance at time t (A_t) and absorbance after completion of reaction (A_∞); □: fast phase, calculated by subtracting change in absorbance of slow phase (drawn line through last part of curve) from total absorbance change; Δ: k_{off} values as determined from these phases are respectively about 460 and 40/s

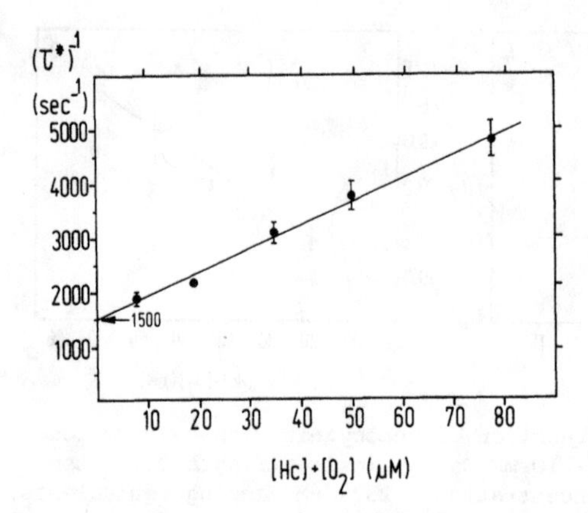

Fig. 6. Dependence of $(\tau^*)^{-1}$ on equilibrium concentration of free oxygen and free binding sites. Conditions: 0.04 M sodium tetraborate pH 9.6, ionic strength 0.10. Temperature 24°C (following jump), observation at 350 nm. Protein concentration was varied from 9 to 47 μM binding equivalents. Fractional saturation levels varied from 0.28 to 0.78. *Drawn line:* calculated with linear regression, corresponding to an apparent $k_{on} = 4.3 \times 10^7$/M/s and a $k_{off} = 1500$/s

Relaxation Kinetics. Temperature-jump relaxation experiments were performed at fractional saturation levels from 0.28 to 0.78; the protein concentration varied between 3.8 and 47 μM. The relaxation behavior is clearly heterogeneous and was analyzed in terms of initial relaxation rates. The data indicate that the relaxation behavior reflects a bimolecular reaction (Fig. 6). The apparent rate constants are: $k_{off} = 1500$/s and $K_{on} = 4.3 \times 10^7$/M/S.

Discussion

P. interruptus hemocyanin exhibits complex oxygen binding kinetics for both the undissociated and dissociated protein. The kinetic parameters of whole molecules and the dissociated subunits are summarized in Table 1.

Table 1. Kinetic parameters of oxygen binding to *Panulirus interruptus* hemocyanin

	Stopped-flow (20°C)		T-jump (24°C)			
	k_{on} (10^7/M/s)	k_{off} (s)	k_{on} (10^7/M/s)	k_{off} (s)	k_{off}/k_{on} (10^{-7} M)	k_{eq} (10^{-7} M)
					from t-jump data	from Fig. 1
Undissociated protein (pH 9.6 + 10 mM Ca^{2+})	4-5[a]	33[b]	6.4[c]	540[c]	84	19
Dissociated protein (pH 9.6)	-	460[d] and 40	4.3[e]	1500[e]	350	160

[a]Estimated from % loss of reaction in dead time of apparatus.
[b]Homogeneous as far as observable, with 15-20% loss of reaction in dead time.
[c]Heterogeneous, estimated error ± 1.2 × 10^7/M/s and ± 120/s respectively.
[d]70% of reaction lost in dead time.
[e]Heterogeneous, estimated error ± 0.5 × 10^7/M/s and ± 100/s respectively.

Undissociated hemocyanin shows, under conditions of cooperative oxygen binding, apparently homogeneous oxygen dissociation behavior as far as the reaction can be followed with stopped-flow measurements. At least 15% of the reaction is lost in the dead time of the apparatus, indicating that species may exist with higher dissociation rate constants. This is in agreement with T-jump measurements yielding a dissociation rate constant of 540/s. *Helix pomatia* hemocyanin (Van Driel et al., 1974) and *Limulus* hemocyanin (Sullivan et al., 1974) show a clearly autocatalytic oxygen dissociation behavior. Generally such behavior is understood to be the kinetic reflection of interactions between the oxygen-binding sites. The fact that no autocatalytic behavior is observed for *Panulirus hemocyanin* may be related to functional heterogeneity among the oxygen binding sites. This is consistent with T-jump measurements in which a clear heterogeneity in relaxation behavior at high oxygen-saturation levels is observed. The apparently homogeneous oxygen dissociation behavior may also be caused by a transition from a high oxygen affinity state to a lower one, which might be rate-limiting in the process of deoxygenation.

The pH dependence of the oxygen-dissociation constant may reflect the kinetic contribution to the alkaline Bohr effect observed in oxygen equilibrium studies (Kuiper et al., 1975). A similar effect has also been observed for undissociated *Limulus* hemocyanin (Sullivan et al., 1974).

In stopped-flow oxygen combination experiments, 80-95% of the reaction is lost in the dead time. The estimated second-order rate constant agrees well with the more quantitative results obtained with T-jump measurements.

T-jump relaxation measurements at high saturation levels reveal a heterogeneous behavior. At high saturation levels, both the high and low affinity states may be present; these may be characterized by different kinetic parameters. *H. pomatia* hemocyanin, on the other hand, shows an almost homogeneous relaxation behavior at high saturation levels, reflecting the binding of oxygen to the high oxygen affinity state (Van Driel et al., 1974). For the whole molecules we found a heterogeneous electrophoresis behavior (submitted for publication); this may correspond to the observed kinetic heterogeneity.

Dissociated hemocyanin shows heterogeneous oxygen dissociation behavior as measured with stopped-flow. About 70% of the reaction is lost in the dead time of the apparatus. Quantitatively, this agrees quite well with the observed k_{off} = 460/s.

T-jump relaxation kinetics also show heterogeneity; this may reflect differences between the subunits, since it is possible to separate the subunits into fractions with different electrophoretic properties (H.A. Kuiper, Ph.D. thesis, Univ. of Croningen, 1976). Sullivan et al. (1974) reported chromatographic heterogeneity of the subunits of *Limulus* hemocyanin; the oxygen equilibrium and the kinetics of the fractionated protein showed functional differences. Miller and Van Holde (1974) studied the oxygen binding properties of *Callianassa californiensis* hemocyanin: oxygen binding to the subunits, which is also much weaker than to whole molecules, could be described by the existence of a high and a lower oxygen affinity state of the protein. This may explain for *Panulirus* hemocyanin, too, the observed kinetic heterogeneity of the subunits. The low oxygen affinity of the subunits compared to that of whole molecules is caused by a larger oxygen dissociation constant of the former.

There is a discrepancy between the T-jump data and the stopped-flow dissociation results of both whole molecules and subunits. This may be caused by (1) fast processes that are not observed with stopped flow; for both aggregation states a certain percentage of the reaction is lost in the dead time of the stopped-flow apparatus; (2) rate-limiting steps which play an essential role under nonequilibrium conditions, for example when the oxygen dissociation rate constant of the high oxygen affinity state is much larger than the rate constant for the transition to a lower oxygen affinity state.

A further kinetic analysis is needed of whole molecules, unfractionated and fractionated subunits and hemocyanin reassociated from the various fractions of subunits for a complete quantitative description of the kinetic behavior of *P. interruptus* hemocyanin.

Acknowledgment. H.A.K. expresses his thanks to the E.M.B.O. for a short-term fellowship.

References

Driel, R. Van, Brunori, M., Antonini, E.: Kinetics of the co-operative reaction of *Helix pomatia* haemocyanin with oxygen. J. Mol. Biol. 89, 103-112 (1974)

Eigen, M.: New looks and outlooks on physical enzymology. Quart. Rev. Biophys. 1. 3-33 (1968)

Eigen, M., De Maeyer, L.: Relaxation methods. In: Techniques of Organic Chemistry. Friess, S., Lewis, E., Weissberger, A. (eds.). New York: Interscience, 1963, Vol. VIII, Part 2, pp. 895-1054

Kuiper, H.A., Gaastra, W., Beintema, J.J., van Bruggen, E.F.J., Schepman, A.M.H., Drenth, J.: Subunit composition, X-ray diffraction, amino acid analysis and oxygen binding behaviour of *Panulirus interruptus* hemocyanin. J. Mol. Biol. 99, 619-629 (1975)

Miller, K., Van Holde, K.E.: Oxygen binding by *Callianassa californiensis* hemocyanin. Biochemistry 13, 1668-1674 (1974)

Millikan, G.A.: The kinetics of blood pigments: haemocyanins and hemoglobin. J. Physiol. 79, 158-179 (1933)

Monod, J., Wyman, J., Changeux, J.P.: On the nature of allosteric transitions: a plausible model. J. Mol. Biol. 12, 88-118 (1965)

Schwarz, G.: Kinetic analysis by chemical relaxation methods. Rev. Mod. Phys. 40, 206-218 (1968)

Sullivan, B., Bonaventura, J., Bonaventura, C.: Functional differences in the multiple hemocyanins of the horseshoe crab, *Limulus polyphemus L.* Proc. Natl. Acad. Sci. 71, 2558-2562 (1974)

Hemocyanin of the Horseshoe Crab, *Limulus polyphemus*. A Temperature-Jump Study of the Oxygen Kinetics of the Isolated Components

C. Bonaventura, B. Sullivan, J. Bonaventura, and M. Brunori

Introdruction

The mechanism of oxygen binding to hemocyanin is a subject of considerable interest (Van Holde and Van Bruggen, 1971; Antonini and Brunori, 1974; Bonaventura et al., 1976). Comparative kinetic studies of oxygen binding are complicated by several factors. Oxygen binding by the copper atoms at hemocyanin's active site is often associated with marked homotropic and heterotropic effects. These effects are also evident in the oxygen kinetics, at least for the α-hemocyanin of *Helix pomatia* (Van Driel et al., 1974). Oxygen binding sometimes shifts the extent of subunit association. Moreover, molluscan and arthropod hemocyanins show wide ranges in structural and functional diversity (Van Holde and Van Bruggen, 1971) which increases the possibility of complications and makes it difficult to identify general features in the reaction mechanism.

It was previously reported that the high molecular weight hemocyanin of *Limulus polyphemus* (3.3×10^6 daltons) can be dissociated into subunits (70,000 daltons) and fractionated into five major chromatographic zones, denoted Hcy I-V (Sullivan et al., 1974). These components, which are distinct in their O_2 affinities, lack the homotropic effects of the parent molecule and do not undergo ligand-linked association-dissociation effects. It seemed, therefore, of interest to examine in detail the kinetics associated with oxygen binding to these various components. This paper is concerned with such a study carried out by the temperature-jump pulse relaxation method (Eigen and De Maeyer, 1963; Brunori and Schuster, 1969).

Materials and Methods

An unfractionated mixture of *Limulus* hemocyanin subunits (called whole stripped *Limulus* Hcy) can be obtained by dialysis of whole hemocyanin versus 0.05 M Tris, 0.01 M EDTA, pH 8.9. The five major zones obtained by DEAE-Sephadex chromatography of stripped hemocyanin are not all "pure" subunits. Electrophoresis suggests that there are at least eight structurally distinct polypeptides in the unfractionated mixture, and of the chromatographic zones only Hcy I and Hcy IV are electrophoretically homogeneous (Sullivan et al., 1976). A more complete description of the structural and functional differences between Hcy I-V is given elsewhere (Sullivan et al., 1974, 1976; Bonaventura et al., 1974; Chiang et al., Chap. 18, this vol.; Schutter et al., Chap. 3, this vol.).

Static spectra were obtained on a Cary 14 recording spectrophotometer and an Aminco DW-2A dual beam spectrophotometer. Temperature-jump studies were performed on an Eigen-De Maeyer temperature-jump instrument as described previously (Eigen and De Maeyer, 1963; Brunori and Schuster, 1969). Rates of oxygen dissociation were measured by the rapid mixing technique. Data collection and analysis were accomplished via an analog to digital converter (Aminco DASAR) coupled to a PDP 11 computer (Digital Corporation).

Results

Temperature-jump kinetics observed with whole stripped *Limulus* hemocyanin, stripped hemocyanin, and one of the *Limulus* hemocyanin components (Hcy IV) are shown in Figure 1. The approach to equilibrium is

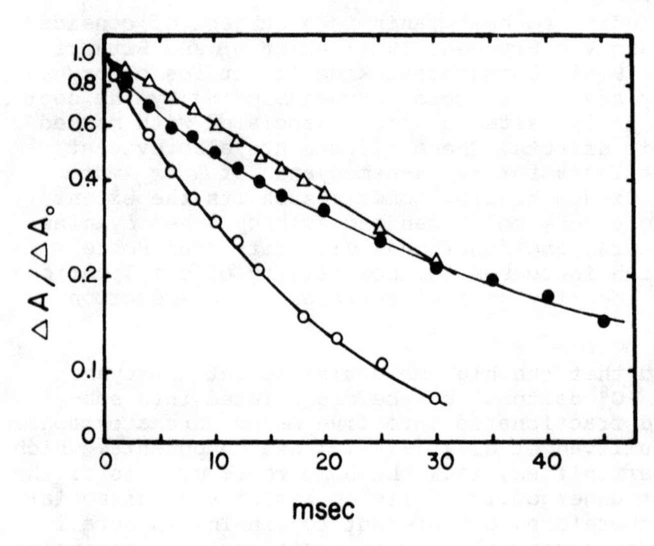

Fig. 1. Time courses for temperature-jump relaxation kinetics for whole *Limulus* hemocyanin (●), stripped *Limulus* hemocyanin (O) and Hcy IV (Δ). For all three preparations, 1.6 ml of air-equilibrated protein solution was mixed with 8.4 ml of deoxygenated 0.05 M Tris, pH 8.0 prior to a temperature-jump from 20° to 25°C. The resulting solution contains approximately 4 mg/ml of hemocyanin at an oxygen concentration of 4.3×10^{-5} M

somewhat heterogeneous for both whole *Limulus* hemocyanin and for stripped *Limulus* hemocyanin. In contrast, with Hcy III, as well as with the other components, simple kinetics are observed. The temperature-jump kinetics of Hcy I-V are roughly approximated by single exponential processes, (Fig. 2), with differences in the rates for the different components. Figure 3 shows the dependence of the observed reciprocal relaxation times on the sum of free oxygen and free binding sites for the five chromatographic zones of *Limulus* hemocyanin. There is a linear increase in k with increasing reactant concentration, as expected for bimolecular processes (Eigen and De Maeyer, 1963). The "on" constants calculated from Figure 3 are in rough agreement with those predicted from stopped-flow and equilibrium experiments with the five components as shown in Table 1. As predicted, the "on" constants do not show very

Table 1. Values of the O_2 rate constanst for *Limulus* hemocyanin components I-V

Components	k (/s)	k' (determined) (/M/s)	k' (predicted)[a] (/M/s)	K (M)
Hcy I	8.3	2.5×10^6	1.6×10^6	5.2×10^{-6}
II	11.7	2.2×10^6	2.33×10^6	5.0×10^{-6}
III	5.7	1.9×10^6	1.51×10^6	3.77×10^{-6}
IV	3.8	1.2×10^6	1.12×10^6	3.4×10^{-6}
V	31.0	2.7×10^6	1.73×10^6	17.87×10^{-6}

Conditions: 0.05 M Tris-glycine + 0.01 M EDTA, pH 8.9, 25°C.

[a] Values from equilibrium determinations and stopped-flow dissociation rate constants (i.e., K and k).

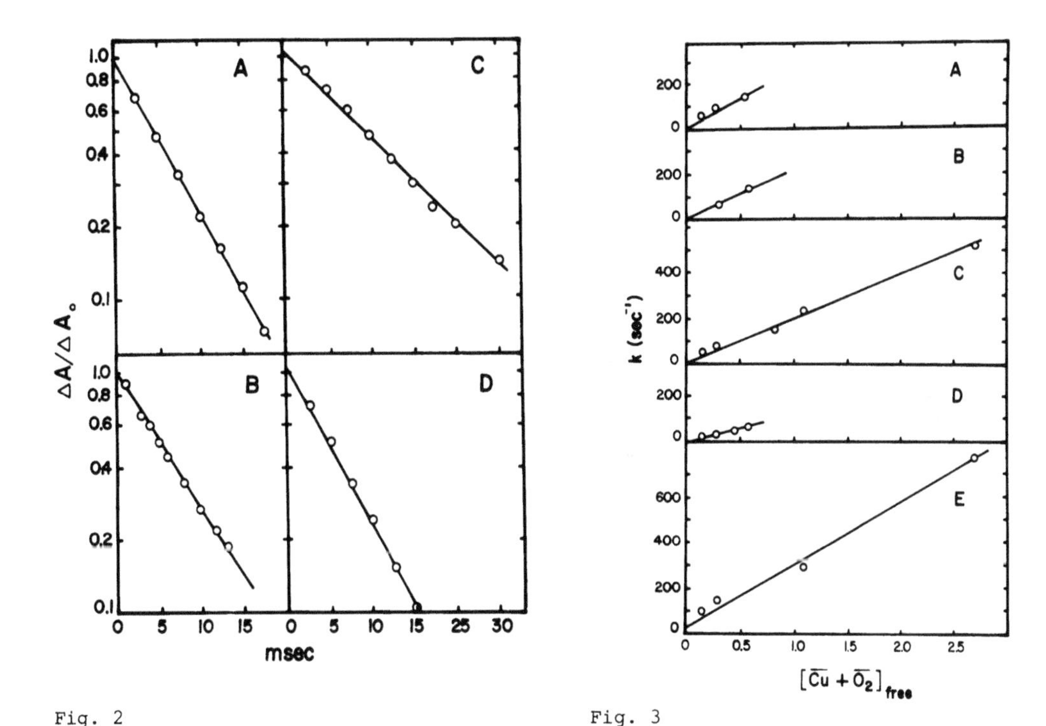

Fig. 2 Fig. 3

Fig. 2A-D. Time courses for temperature-jump relaxation kinetics of *Limulus* hemocyanin components. Temperature increased from 20° to 25°C during the jump. (A) Behavior of Hcy I at a protein concentration of 5.4×10^{-5} M. (B) Behavior of Hcy II at the same oxygen concentration. (C) Behavior of Hcy III at an oxygen concentration of 2.7×10^{-5} M. (D) Behavior of Hcy V, also at 2.7×10^{-5} M O_2. Proteins were all examined in 0.05 M Tris, pH 8.9, containing 0.01 M EDTA

Fig. 3A-E. Dependence of relaxation rate for the various hemocyanin components after a 20° to 25°C temperature-jump shown as a function of free ligand and free binding site concentration. (A)-(E) Zones I-V. Proteins were in 0.05 M Tris, pH 8.9, containing 0.01 M EDTA

much variation and the large differences in affinity between the components are largely due to differences in their rates of oxygen dissociation (Sullivan et al., 1974).

Although the temperature-jump kinetics for the isolated components are much simpler than for the whole hemocyanin or the mixture of stripped components, there is an unexpected kinetic feature that is apparent in the faster time range. Figure 4 shows the kinetic traces obtained in temperature-jump experiments with Hcy III and Hcy V at identical oxygen and protein concentrations. It is apparent that in addition to the slower phase, whose magnitude generally dominates the reaction, there is a fast phase whose optical density change is apposite to that associated with oxygen binding, which is a decrease of absorbance for an increase in temperature. Moreover, as shwon in Figure 4, the fast phase is much more conspicuous for Hcy III than for Hcy V.

Differences between the two phases became apparent when the wavelength dependence of the temperature-jump kinetics was examined. The slower phase has a wavelength dependence which follows the trend characteristic of the difference spectrum between deoxy- and oxyhemocyanin (max-

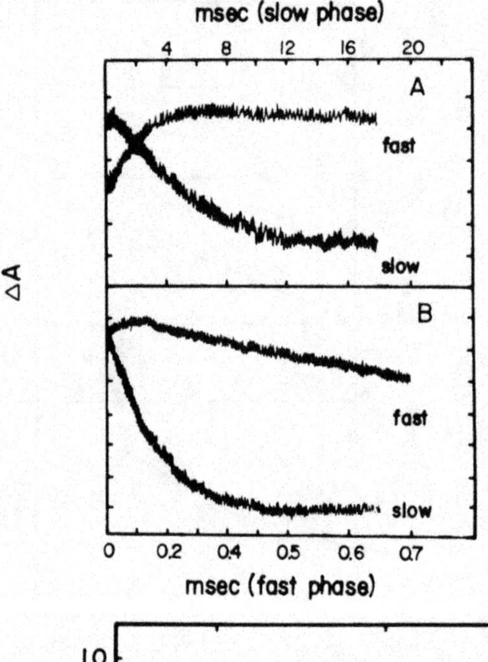

Fig. 4A and B. Drawings of oscilloscope traces [(A) for Hcy III; (B) for Hcy V] to illustrate fast as well as slow phase following a $5^{o}C$ temperature-jump. Both Hcy III and Hcy V were present at a concentration of 5.6 mg/ml. Oxygen concentration for both solutions: 1.08×10^{-4} M and saturation ∿ 90%. Both proteins were in 0.05 M Tris, pH 8.9, in presence of 0.01 M EDTA

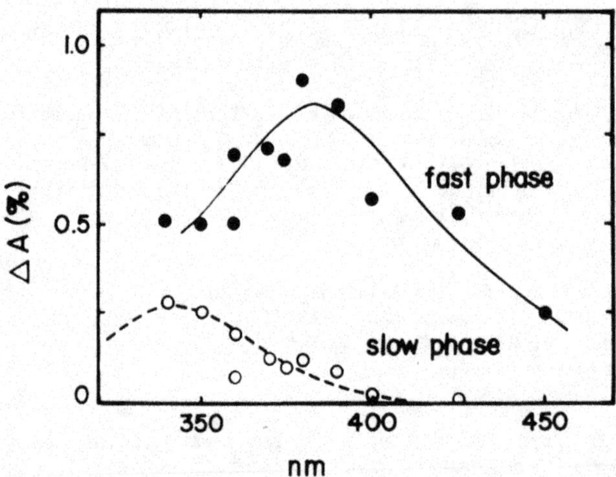

Fig. 5. Wavelength dependence of fast and slow phases of temperature-jump relaxation kinetics observed for Hcy III. ΔA(%) refers to absolute changes of absorbance irrespective of sign. Hemocyanin concentration approximately 5.6 mg/ml; solution saturated with oxygen. Temperature-jump from 20° to $25^{o}C$, in 0.05 M Tris, pH 8.9, containing 0.01 M EDTA. *Dashed line:* static difference spectrum for oxy-deoxy hemocyanin

imum at 340 nm). The faster phase, however, does not have the same wavelength dependence and shows a peak in the difference spectrum centered at about 380 nm (Fig. 5). Thus due to the different spectral characteristics the percentage absorbance change due to the fast and slow phases varies as a function of wavelength. The fast phase is spectrally distinct from the spectral change characteristic of oxygen binding. Its relaxation rate, for component III (10,000-14,000/s) does not vary appreciably with protein or oxygen concentration over the range shown in Figure 3C. This suggests that the process observed is not bimolecular. Similar rates were observed for all the other isolated components.

In oxygen equilibrium experiments with the isolated *Limulus* hemocyanin components it was found that the oxygen affinites of Hcy II and Hcy III decreased significantly upon the addition of NaCl (Sullivan et al., 1974). This effect, which saturated at about 2 M NaCl, was not observed with Hcy I, Hcy IV or Hcy V.

Temperature-jump experiments were carried out with Hcy II, Hcy III and Hcy IV in the presence and absence of 1 M NaCl in an attempt to learn more about the kinetic basis of the salt effects observed in equilibrium experiments. Hcy II and III show smaller relaxation times in the presence of salt, as was expected. Hcy IV also shows smaller values of τ in the presence of NaCl. This suggests that the lack of a NaCl effect in equilibrium experiments with Hcy IV may be due to compensation effects on the kinetic constants.

Discussion

The oxygen-binding properties of many molluscan and arthropodan hemocyanins have been studied (Van Holde and Van Bruggen, 1971; Antonini and Brunori, 1974; Bonaventura et al., 1976). Detailed kinetic studies have not been made previously, however, with preparations of isolated subunits. Functional heterogeneity of components in unfractionated hemocyanin preparations and variations in extent of aggregation must, in part, explain the heterogeneous kinetics observed for many hemocyanins. These complicating conditions seem to be absent in preparations of hemocyanin from *Limulus polyphemus* which have been treated with EDTA and purified by ion-exchange chromatography. Hcy I-V show differences in their oxygen affinities, and in their sensitivities to NaCl. These components thereby provide us with an opportunity to examine the kinetic interactions of oxygen with hemocyanin without the complications introduced by the presence of multiple interacting subunits.

The results of the temperature-jump studies with the isolated *Limulus* hemocyanin components show that, as far as the slow phase of the relaxation kinetics is concerned, these components behave as single-site molecules with relaxation times which follow the behavior expected for simple bimolecular processes. The differences between components in the oxygen binding rate constants could be directly determined, and follow the pattern which was calculated from complementary experiments using equilibrium data and stopped-flow techniques.

On the other hand, the fast phase of the relaxation kinetics is one for which we, as yet, have no adequate explanation. This fast phase is related to spectral changes which are expressed to different extents in the different hemocyanin components. Since all of the components probably have the same basic structure at the active site (two copper atoms which interact with a molecule of oxygen), we may presume that the differences between the components reflect differences in the immediate environment of the copper atoms at the active site. Although the *Limulus* hemocyanin components show differences in their oxygen affinities, there is no clear correlation between their oxygen affinities, and the extent of expression of the fast phase in their relaxation kinetics. Although no direct evidence for a kinetic coupling between the two resolved relaxation times is as yet available, a possible interpretation of the observation may be given along the lines proposed by Brunori (1969). On the basis of studies on the oxygen relaxation kinetics of octopus hemocyanin, a low affinity hemocyanin, he proposed that binding of oxygen to hemocyanin is a complex process even in the absence of site-site interactions.

The results which have been presented here complement experiments using the techniques of NMR spectroscopy, electron microscopy, and equilibrium and stopped-flow measurements on the five chromatographic zones of *Limulus* hemocyanin. The components show distinct structural and functional differences (Eigen and De Maeyer, 1963; Bonaventura et al., 1974; Sullivan et al., 1976; Chiang et al., Chap. 18, this vol.; Schutter et al., Chap. 3, this vol.). In future studies we hope that it will be possible to determine how the different properties of the *Limulus* components are expressed in the multisubunit aggregate found in vivo.

Acknowledgments. This work was supported in part by Grant HL 15460 from the National Institutes of Health, Grants BMS 73-01695 A01 and BMS 71-041432 AQ2 from the National Science Foundation, and Grant 866 from NATO. J. Bonaventura is an established investigator of the American Heart Association.

This paper represents number 4 on structure-function relationships in *Limulus* hemocyanin.

References

Antonini, E., Brunori, M.: Transport of oxygen: respiratory proteins. In: Molecular Oxygen in Biology. Hayaishi, O. (ed.). Amsterdam: North Holland, 1974, pp. 219-274

Bonaventura, C., Sullivan, B., Bonaventura, J., Bourne, S.: CO binding by hemocyanins of *Limulus polyphemus*, *Busycon carica* and *Callinectus sapidus*. Biochemistry 13, 4784-4789 (1974)

Bonaventura, J., Bonaventura, C., Sullivan, B.: Non-heme oxygen transport. In: Oxygen and Physiological Function. Jobsis, F. (ed.). Professional Information Library, Dallas, Texas (In press, 1976)

Brunori, M.: Kinetics of oxygen binding by *Octopus* hemocyanin. J. Mol. Biol. 46, 213-215 (1969)

Brunori, M., Schuster, T.M.: Temperature-jump studies of hemoglobin kinetics. J. Biol. Chem. 244, 4046-4051 (1969)

Eigen, M., De Maeyer, L.: In: Techniques of Organic Chemistry. Friess, S., Lewis, E., Wesberger, A. (eds.). New York: Interscience, 1963, Part 2, Vol. VIII, p. 895-1054

Sullivan, B., Bonaventura, C., Bonaventura, J.: Functional differences in the multiple hemocyanins of the horseshoe crab, *Limulus polyphemus*. I. Structural differentiation of the isolated components. J. Biol. Chem. (in press, 1976)

Van Driel, R., Brunori, M., Antonini, E.: Kinetics of the co-operative and non-cooperative reaction of *Helix pomatia* haemocyanin with oxygen. J. Mol. Biol. 89, 103-112 (1974)

Van Holde, K.E., Van Bruggen, E.F.J.: The hemocyanins. In: Biological Macromolecules Series. Timasheff, S.N., Fasman, G.D. (eds.). New York: Marcel Dekker, 1971, Vol. V, pp. 1-53

Evolutionary Studies on Hemocyanin

Evolutionary Studies on Hemocyanin

A. GHIRETTI-MAGALDI AND G. TAMINO

Introduction

The hemocyanins (Hcy) of Mollusca and Arthropoda have the same physiological role and show similarities in several physical and chemical properties such as the optical spectra, the multimeric state, the ability to dissociate in some experimental conditions, the isoelectric pH, etc.

Differences, however, are evident in their quaternary structure and in the content of copper and of carbohydrates, which have led several workers in the field to suggest that the Hcy of Mollusca and Arthropoda do not belong to a family of genetically related proteins but are the product of functionally convergent evolution. This idea has been readily accepted because it gives support to a recent hypothesis on the structure and size of the functional subunits and of the polypeptide chains of Hcy from the two phyla (Van Bruggen et al., Chap. 15, this vol.; Gielens et al., Chap. 11, this vol.; Bonaventura et al., Chap. 27, this vol.).

The existence of a genetic relationship (homology) between two proteins or groups of proteins is generally assessed by comparison and statistical analysis of their amino acid sequence. Since the primary structure of Hcy is not yet known, the homology of these proteins can be ascertained only on the basis of their amino acid composition. Analyses of many Hcy from Mollusca and Arthropoda have revealed great similarity also in amino acid composition.

Several methods have been suggested for discovering significant similarities in proteins from their amino acid composition (Metzger et al., 1968; Marchalonis and Weltman, 1971). All of them have been strongly criticized by Fitch (1973), who claims that homology among proteins can be established only by sequence comparison. Harris and Teller (1973), however, introduced the use of a function called "composition divergence" as a measure of the overall dissimilarity of the amino acid composition of two proteins, and produced clear evidence that the amino acid composition and the sequence homology are strictly correlated. They computed this value for a large number of proteins with known sequence and found a high correlation with the sequence homology. The same correlation and distribution of the data was obtained when the comparison between sequence and composition divergence was made on hypothetical proteins generated by computer on the basis of a model of cytochrome c evolution. Controls carried out on unrelated proteins provided further support for the hypothesis that the method can be used for predicting the sequence homology of unsequenced proteins.

Among other proteins, Harris and Teller (1973) also tested a group of Hcy analyzed by us (Ghiretti-Magaldi et al., 1966) and found low composition divergence values which suggest a sequence homology ranging from 54 to 96% for the Hcy of Mollusca and Arthropoda.

Since the presence of a given amino acid in a protein is genetically determined, there must be a correlation between amino acid frequency and the number of codons which specify for them. King and Jukes (1969) calculated the expected frequency of occurrence of a given amino acid from the redundancy of the genetic code and compared this value with the frequency observed using 53 polypeptides. They obtained a significant correlation between the expected and the observed values.

We have tested the possibility of introducing this information in Harris and Teller's formula simply by using the ratio of the molar fraction of the amino acid to the number of codons (Ghiretti-Magaldi et al., 1975). The modified formula from which a "corrected composition divergence" (D_c) is obtained, is the following:

$$D_c = \sqrt{\sum_{i=1}^{18} \left(\frac{X_{iA} - X_{iB}}{n_i} \right)^2}$$

where X_{iA} and X_{iB} are the molar fractions of the amino acid i in the proteins A and B respectively and n_i is the number of codons for the same amino acid.

When tested on a group of 15 comparisons between two proteins which are homologous by sequence, a correlation coefficient of 0.90 was obtained between D_c and sequence homology. The homologies for two bacterial azurins and two plastocyanins were 63 ± 10% and 87 ± 10%, respectively when estimated from the linear regression graph of the D_c values, and 70% and 78% from sequence alignment. A certain degree of homology was also detected between azurins and plastocyanins using the D_c values. This has been recently confirmed by Ryden and Lundgren (1976) on the basis of the sequence alignment.

The composition divergence method was applied to a group of representative Hcy from Mollusca and Arthropoda. The D_c values obtained indicated a strong homology among Hcy from the different phyla. The calculated homologies were highly consistent with the phylogenetic relations derived from morphological and paleontological evidence.

These results stimulated us to compare as many Hcy compositions as available for two main purposes: to test the resolving power of the method for the construction of a phylogenetic tree and to inspect the possibility that convergence phenomena might have taken place during the evolutionary history of Hcy.

Experimental Section

The corrected composition divergence formula has been used for all comparisons. This equation is independent of both molecular weight and order of comparison of the two proteins. As stated by Harris and Teller: "This function was made independent of molecular weight in order to eliminate any problem that might be caused by compositions from products of a gene duplication or by compositions based on polymeric forms of the same protein subunit". As the Hcy from all classes containing this respiratory pigment, except the *Placophora*, had been previously characterized, most of the compositions were taken from the literature (Ghiretti-Magaldi et al., 1966; Witters and Lontie, 1968; Van Holde and Van Bruggen, 1971; Sundara Rayulu, 1974; Hall and Wood, 1975). The Hcy of *Haliotis*, *Sepia*, *Carcinus*, *Portunus*, *Squilla* and *Androctonus* were recently analyzed in our laboratory and the composition of the spider's Hcy was kindly sent to us by Dr. B. Linzen (Univ. Munich).

Results

The results obtained by comparing the Hcy composition of 29 species of Mollusca and Arthropoda have been assembled in a two-entrance table or a composition divergence matrix (Fig. 1). For practical purposes the D_c values in this figure have been multiplied by 10^4. The triangles delimited by the lines indicate the intra-phylum or intra-class com-

	H. tuberculata	*M. trunculus*	*M. brandaris*	*N. antiqua*	*B. undatum*	*C. gracilis*	*P. leopoldvillensis*	*C. neptuni*	*L. stagnalis*	*H. pomatia α*	*H. pomatia β*	*A. fulica*	*O. vulgaris*	*O. macropus*	*E. moschata*	*S. officinalis*	*C. maenas*	*P. depurator*	*E. spinifrons*	*C. sapidus*	*C. magister*	*H. vulgaris*	*P. vulgaris*	*S. mantis*	*L. polyphemus*	*A. australis*	*C. solei*	*D. californica*	*E. hellus*	*S. longicornis*
Haliotis tuberculata	–																													
Murex trunculus	101	–																												
Murex brandaris	95	41	–																											
Neptunea antiqua	82	61	78	–																										
Buccinum undatum	87	93	107	63	–																									
Colus gracilis	146	129	155	100	107	–																								
Pila leopoldvillensis	309	334	330	302	297	316	–																							
Cymbium neptuni	259	258	266	229	216	219	107	–																						
Lymnaea stagnalis	241	252	261	241	236	252	290	279	–																					
Helix pomatia α	308	322	313	297	277	327	114	195	275	–																				
Helix pomatia β	250	304	294	278	260	309	114	192	261	50	–																			
Achatina fulica	272	313	306	283	266	316	87	71	247	71	52	–																		
Octopus vulgaris	143	116	133	110	121	152	333	258	189	333	244	314	–																	
Octopus macropus	109	96	123	86	96	118	305	246	232	305	286	293	72	–																
Eledone moschata	173	81	89	73	98	125	353	250	247	353	334	297	63	65	–															
Sepia officinalis	152	139	165	97	123	101	345	250	251	345	328	327	141	146	147	–														
Carcinus maenas	120	97	108	100	106	145	326	263	255	326	302	308	179	136	126	138	–													
Portunus depurator	132	112	129	108	114	132	332	255	247	332	309	311	163	147	146	123	55	–												
Eriphia spinifrons	149	89	98	104	124	154	351	272	309	351	331	340	154	137	126	149	78	93	–											
Callinectes sapidus	128	89	111	83	83	122	319	242	257	319	293	301	125	101	105	131	68	79	94	–										
Cancer magister	167	130	154	136	161	160	386	302	300	386	360	413	167	166	151	153	101	104	112	118	–									
Homarus vulgaris	129	87	93	106	112	164	394	294	271	394	316	327	151	129	125	150	84	122	83	64	124	–								
Palinurus vulgaris	187	134	153	152	158	175	378	308	334	378	370	380	174	234	154	167	124	100	98	126	125	92	–							
Squilla mantis	223	139	179	155	154	154	375	289	278	375	354	361	166	143	160	115	124	129	138	132	120	110	33	–						
Limulus polyphemus	204	203	211	249	245	293	404	363	344	404	383	444	236	197	214	251	215	231	214	183	245	158	193	190	–					
Androctonus australis	267	266	247	270	242	302	357	342	297	357	357	368	210	243	270	262	138	123	149	131	153	150	167	115	146	–				
Cupiennius solei	196	208	219	170	149	172	240	176	215	240	221	213	193	186	196	201	204	208	245	182	245	210	252	233	310	289	–			
Dugesiella californica	146	168	179	133	110	141	297	247	223	297	275	272	148	143	144	157	112	181	191	143	177	175	207	177	271	265	125	–		
Eurypelma hellus	154	181	176	138	114	145	295	244	224	295	271	269	148	146	145	174	160	163	195	148	183	180	213	204	274	282	123	18	–	
Scutigera longicornis	239	271	284	265	264	239	459	352	444	459	449	459	255	266	272	277	312	301	299	284	309	284	256	306	325	382	329	316	319	–

Fig. 1. Matrix of composition divergence values for all Hcy studied. All values have been multiplied by 10^4

parisons and the rectangles the inter-phyla or inter-class ones. The results obtained from the intra-mollusks and intra-arthropods and Mollusca/Arthropoda comparisons are summarized in Table 1.

Table 1. Average composition divergence values for comparisons between molluscan, arthropod and molluscan/arthropod species

Comparison	$D_c$av.	sd	se	D_c max	D_c min	N. compar
Mollusca/Mollusca	.0209	.0094	.0008	.0359	.0041	136
Arthropoda/Arthropoda	.0184	.0080	.0008	.0382	.0018	105
Mollusca/Arthropoda	.0229	.0094	.0006	.0459	.0083	255

$D_c$av.: D_c average
sd: standard deviation
se: standard error
D_c max: maximal D_c value
D_c min: minimal D_c value
N. compar: number of comparisons

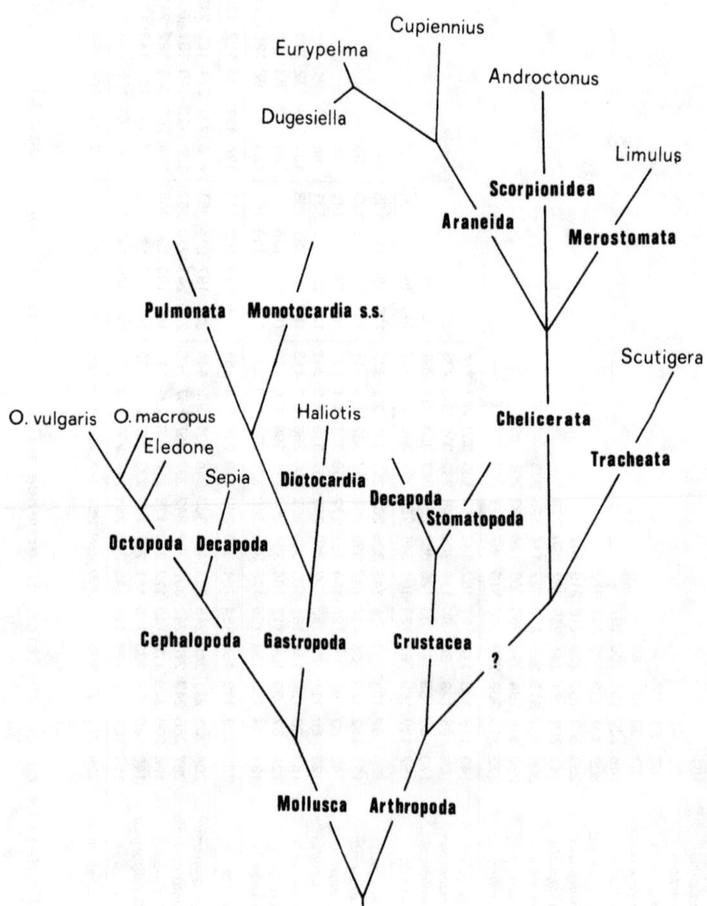

Fig. 2. Dendrogram showing phylogenetic relationships between Mollusca and Arthropoda. Lengths of branches are proportional to values given in Table 2

The data from the matrix of Figure 1 have been used to construct the phylogenetic tree of Figure 2. In this tree the position of the branches is arbitrary and there is no correlation with the evolutionary time. The length of the branches was made proportional to the composition divergence. These lengths were calculated from the arithmetical mean of the D_C values for a given taxonomic group. The values are reported in Table 2 with the standard deviation. When more than two branches stem from one point, as in the case of Chelicerata, the branch

Table 2. Length of branches of dendrogram in Figure 2

Comparison	$D_C$av.	sd	D_C max	D_C min	N. compar.
Mollusca					
Cephalopoda/Gastropoda	.0205	.0092	.0353	.0073	52
Octopoda/Decapoda	.0145	-	.0147	.0141	3
Octopus/Eledone	.0064	-	.0065	.0063	2
O. vulgaris/O. macropus	.0072	-	-	-	1
Pros. Monotoc./Diotoc.	.0195	.0089	.0309	.0082	12
Monotoc.s.s./Pulmonata	.0263	.0075	.0359	.0071	32
Arthropoda					
Crustacea/other Arthropoda	.0224	.0058	.0382	.0112	50
Crustacea/Decapoda/Stomatopoda	.0135	.0025	.0194	.0110	8
Chelicerata/Tracheata	.0334	.0026	.0382	.0316	5
Araneida/other Chelicerata	.0282	.0014	.0310	.0265	6
Scorpionidea/other Chelicerata	.0287	-	.0310	.0147	4
Merostomata/other Chelicerata	.0250	-	.0310	.0123	4
Cupiennius/Dugesiella + Eurypelma	.0124	-	.0125	.0123	2
Dugesiella/Eurypelma	.0018	-	-	-	1

lengths are proportional to the mean D_C calculated from the comparisons of the species of one branch with all those of the other branches. Thus, for instance, the length of the *Androctonus* branch includes the comparisons Scorpionidea/Merostomata + Araneida.

Since the Gastropoda and the Crustacea Decapoda were represented by several species, we attempted the construction of detailed trees for both groups. These trees (Figs. 3 and 4) have been constructed by a method independent of taxonomic considerations. The single species were clustered together in small groups or in pairs on the basis of the minimal D_C values. The clusters were included in a group of higher order when the mean calculated from the squares of the D_C values was significantly lower than that obtained from the entire set of data.

Discussion and Conclusions

The data presented here bring additional evidence that Hcy belong to a single family of phylogenetically related proteins, although the evidence cannot be taken as proof because we are using a comparison not based on the primary structure.

It is true that if one compares, by whatever method based on amino acid composition, shark immunoglobulin with bovine pepsin, lobster glyceraldehyde-3-phosphate dehydrogenase and tobacco mosaic virus coat protein, one can get some low divergence values simply by chance (Fitch, 1973), but it is also true that the value of composition di-

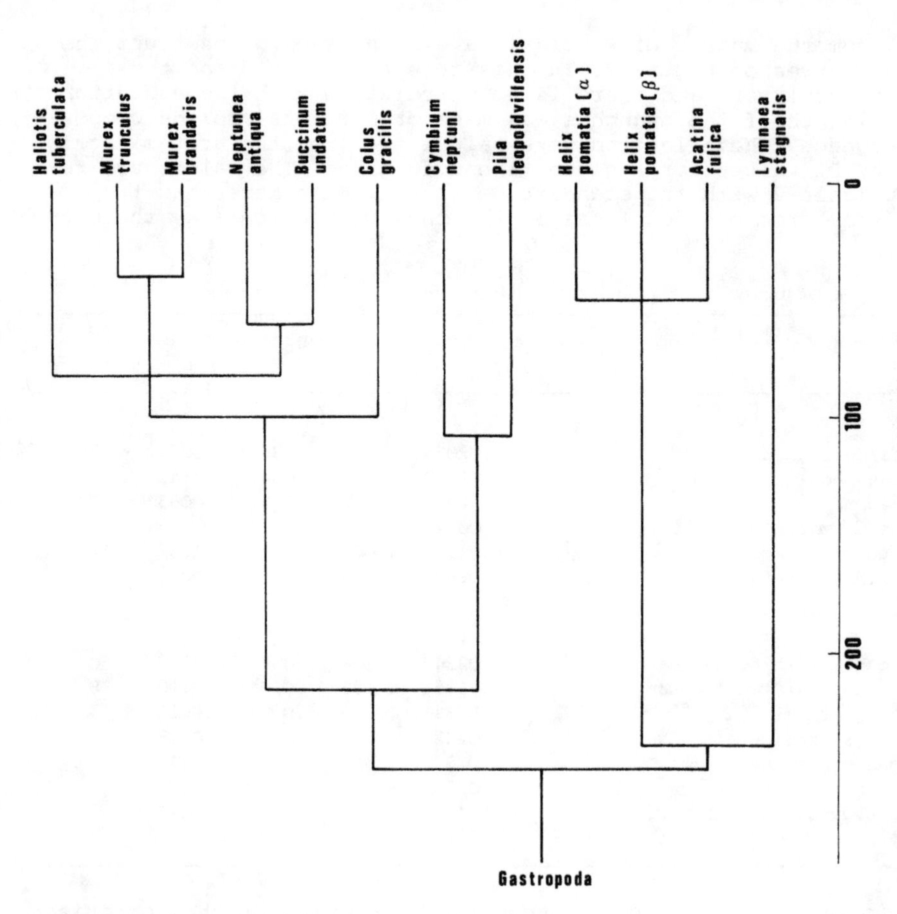

Fig. 3. Dendrogram showing relationships between Gastropod species. Scale indicates actual D_C values multiplied by 10^4 (see text for explanation)

vergence cannot be used for all proteins at random, and that function must be taken into account (Harris and Teller, 1973). The composition divergence can be used to estimate the degree of homology only when functionally related proteins are being compared.

The values of Table 1 give an estimated homology of 70 ± 10% for molluscan species, of 75 ± 10% for Arthropoda and of 60 ± 10% between Mollusca and Arthropoda. Also the distribution of the D_C values does not change appreciably when Hcy of the two different phyla are compared.

The phylogenetic tree of Figure 2 has been constructed by fitting the observed D_C values into the taxonomic information. No inconsistency is evident when the composition divergence is compared with the other data on the evolution of Mollusca and Arthropoda.

The high divergence values found among Gastropoda clearly reflect the adaptation of these animals to different environments.

When comparing Arthropoda Hcy we found the necessity of connecting the Tracheata (*Scutigera*) and the Chelicerata to the Crustacea by an additional branch, indicated in Figure 2 by a question mark. Actually this branch has been suggested by many taxonomists but has not yet received

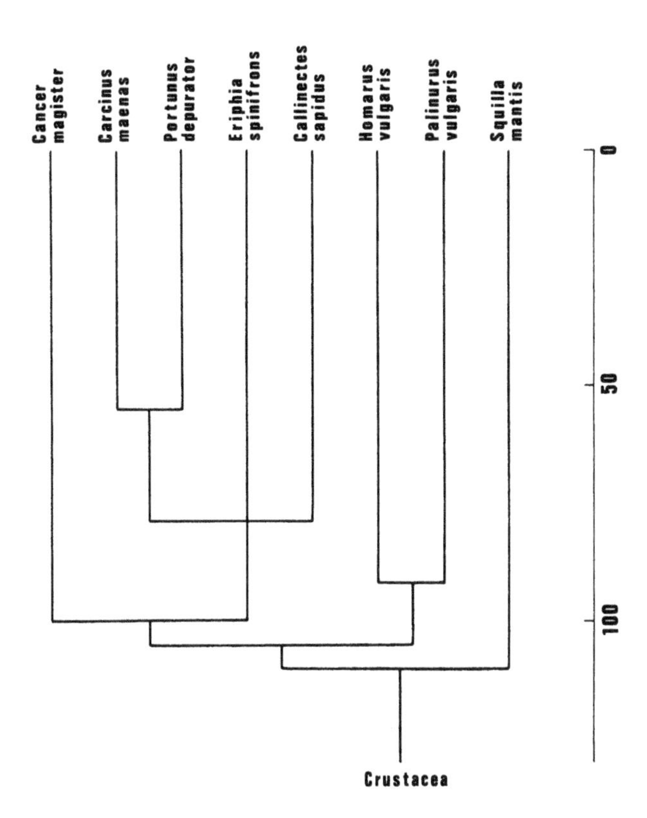

Fig. 4. Relationships between Crustacean species. Scale as in Figure 3 (see text for explanation)

a satisfactory denomination (Omodeo, personal communication). The two tarantulas have been indicated in the figure with the different generic names originally used: *Dugesiella* and *Eurypelma*. We were later informed that they have now been included in the same genus and are called *Dugesiella* (Linzen, personal communication). The new systematic arrangement is in close agreement with the D_C values obtained by us.

The trees of Figures 3 and 4 have been constructed only on the basis of the composition divergence, by a method which is independent of taxonomic considerations. These trees are consistent when compared to the taxonomy of the species concerned. For the gastropods we find *Cymbium*, a marine neogastropod, too close to *Pila*, which is an amphibian mesogastropod. The position of *Pila*, close to terrestrial pulmonates, can be an indication of functional convergence. The isolated position of *Lymnaea* with respect to the other pulmonates confirms the data which suggest that the fresh-water forms are derived as post-adaptation of the terrestrial ones (Ghiretti and Ghiretti-Magaldi, 1975).

As for the Crustacea Decapoda, *Cancer magister* is too distant from the other crabs. For this case, however, we must consider that we are dealing with very low D_C values for which resolution, as already stated (Ghiretti-Magaldi et al., 1975), is not very good. Moreover the 3% overall error intrinsic in the amino acid analyses is bound to affect the low D_C values more than the higher ones. If we consider that these trees are based on only one character, we may conclude that the method has surprisingly high resolution.

This first attempt to reconstruct the phylogenetic relations existing between species of Mollusca and Arthropoda on the basis of the amino acid composition of Hcy indicates that the method, as suggested by

Harris and Teller and as modified by us, may prove a valuable tool for discovering non-casual similarities among functionally related proteins.

The precise phylogenetic history of Hcy requires sequence comparisons. For these, however, we must be sure that the polypeptide chains we are comparing are the true fundamental units of the protein, otherwise a correct alignment cannot be achieved. As the data on polypeptide chains in Hcy are still contradictory, one can predict that measuring homology by sequence still has far to come.

It is important that the first approach to the problem of the genetic relationships among Hcy has given consistent results; the hypothesis of a polyphyletic origin by convergent evolution can be considered a rather unlikely event, especially when considering high order taxonomic groups such as the phyla.

Acknowledgments. We wish to thank Prof. P. Omodeo and Dr. S. Minelli for helpful discussion and criticism, Prof. B. Linzen for sending to us the composition of spider's Hcy and Mr. D. Cervellin who did the amino acid analyses.

References

Fitch, W.M.: Aspects of molecular evolution. Ann. Rev. Gen. 7, 343-380 (1973)

Ghiretti, F., Ghiretti-Magaldi, A.: Respiration. In: Pulmonates. Fretter, V., Peake, J. (eds.). London: Academic Press, 1975

Ghiretti-Magaldi, A., Nuzzolo, C., Ghiretti, F.: Chemical studies on hemocyanins. I. Aminoacid composition. Biochemistry 5, 1943-1951 (1966)

Ghiretti-Magaldi, A., Tamino, G., Salvato, B.: The monophyletic origin of hemocyanins on the basis of the amino acid composition. Structural implications. Boll. Zool. 42, 167-179 (1975)

Hall, R.L., Wood, E.J.: The carbohydrate content of gastropod hemocyanin. Biochem. Soc. Trans. 4, 307-309 (1975)

Harris, C.E., Teller, D.C.: Estimation of primary sequence homology from aminoacid composition of evolutionary related proteins. J. Theor. Biol. 38, 347-362 (1973)

Holde, K.E. Van, Van Bruggen, E.F.J.: Hemocyanins. In: Subunits in Biological Systems. Timasheff, N.S., Fasman, C.D. (eds.). New York: M. Dekker, 1971, Vol. V, Part A, pp. 1-55

King, G.L., Jukes, T.H.: Non-darwinian evolution. Science 164, 788-797 (1969)

Marchalonis, J.J., Weltman, J.K.: Relatedness among proteins: A new method of estimation and its application to immunoglobulins. Comp. Biochem. Physiol. 38B, 609-625 (1971)

Metzger, H., Shapiro, M.B., Mossiman, J.E., Vinton, J.E.: Assessment of compositional relatedness between proteins. Nature (London) 219, 1166-1168 (1968)

Ryden, L., Lundgren, J.O.: Homology relationships among the small blue proteins. Nature (London) 261, 344-346 (1976)

Sundara Rayulu, G.: A comparative study of the organic components of the hemolymph of a Millipede *Cyngalobolus bugnioni* and a centipede *Scutigera longicornis* (Myriapoda). Symp. Zool. Soc. London 32, 347-364 (1974)

Witters, R., Lontie, R.: Stability regions and aminoacid composition of Gastropod hemocyanins. In: Physiology and Biochemistry of Hemocyanins. Ghiretti, F. (ed.). New York: Academic Press, 1968

Physiology of Hemocyanin

Haemocyanin-Producing Cells in Gastropod Molluscs

T. SMINIA

Introduction

The respiratory pigment haemocyanin (Hcy) is found in only two phyla of the animal kingdom, in arthropods and in molluscs. Several studies have indicated that this blood pigment is not cell-bound as is haemoglobin in the red blood cells of vertebrates, but occurs free in the blood plasma (Scheer, 1967; Ghiretti, 1968). Many physiological and biochemical studies have provided information in the respiratory function, the chemical characteristics and the fine structure of Hcy (Redmond, 1971; Ghiretti and Ghiretti-Magaldi, 1972). In contrast, relatively little is known about the biosynthesis of Hcy or about the cells which produce this blood pigment. In fact it was only recently that evidence was obtained on the sites of the synthesis of this respiratory pigment. There is only one morphological study on Hcy synthesis in arthropods, that of Fahrenbach (1970) on the horseshoe crab *Limulus polyphemus*, which shows that Hcy is formed in a special type of blood cell, the cyanoblast. As far as the molluscs are concerned, cephalopods and gastropods have been studied. Dilly and Messenger (1972) have suggested, on morphological grounds, that in the cephalopod *Octopus* sp. Hcy is produced in special organs, the paired branchial glands. The idea that these (branchial) glands synthesize Hcy is supported by a morphological study by Schipp et al. (1973) on three other cephalopod species and by experimental evidence on the synthesis of Hcy (Messenger et al., 1974). In gastropods, morphological studies have been performed on five opisthobranch species (Schmekel and Weischer, 1973), two species of terrestrial pulmonates (Reger, 1973; Skelding and Newell, 1975) and on one freshwater pulmonate species (Sminia, 1972; Sminia et al., 1972; Sminia and Boer, 1973). The studies suggest that in the opisthobranch Hcy is formed in a blood gland. In both terrestrial and freshwater pulmonates, on the other hand, Hcy is apparently synthesized in so-called pore cells (Sminia and Boer, 1973; Skelding and Newell, 1975), which are scattered throughout the connective tissue.

Pore cells have also been found in prosobranch and opisthobranch gastropods (Nicaise et al., 1966; Schmekel, 1972; Sminia, unpublished), as well as in bivalves (Cheng and Rifkin, 1970; Ruddell and Wellings, 1971). It is, however, not yet known whether pore cells in these molluscs also produce Hcy.

The present paper deals with the fine structure and the function of Hcy-producing cells in gastropods. On the basis of morphological data conclusions will be drawn about the synthesis and release of Hcy by these cells. In addition, the morphological evidence will be discussed in relation to similar evidence on Hcy-producing cells in cephalopods and arthropods.

Blood Gland and Pore Cells

The blood gland of opisthobranchs is a leaf-shaped organ, situated above the cerebro-pleural and pedal ganglia; it communicates with a branch of the aorta (Schmekel and Weischer, 1973). The gland has no ducts. It is composed of gland cells, arranged in rows or clusters which are embedded in the connective tissue. The nuclei of the cells are relatively small. In the cytoplasm numerous free ribosomes and an occasional cistern of the granular endoplasmic reticulum (GER) are found; the Golgi apparatus is poorly developed. The most prominent feature of the blood gland cell is that its plasma membrane has numerous deep invaginations. These are commonly filled with globules of about 250-300 Å in diameter or with strands composed of the same units. This material, which is not found in or attached to cell organelles (ribosomes, GER and Golgi apparatus) involved in the synthesis, storage and transport of proteinaceous materials, is comparable in structure to globular and linearly polymerized Hcy.

On the basis of these morphological observations it has been assumed that the blood gland synthesizes Hcy. It is obvious, however, that this evidence is not conclusive: Hcy molecules have not been found inside the cells. Also the fact that high levels of copper are present in the blood gland (Schmekel and Weischer, 1973) cannot be taken as proof of the assumption that the gland produces Hcy: it may merely indicate that the gland contains large amounts of the pigment.

It is not known whether or not pore cells are present in the species in which the blood gland was investigated. As mentioned above, pore cells have been found in other opisthobranchs (Nicaise et al., 1966; Baleydier et al., 1969; Schmekel, 1972), but the possible involvement of these cells in Hcy synthesis has not been reported.

The most comprehensive studies on pore cells have been made on pulmonate snails (Plummer, 1966; Wondrak, 1969; Sminia, 1972; Wolburg-Buchholz, 1972a, b; Dyer and Cowden, 1973; Nicolas, 1973; Richardot, 1975; Skelding and Newell, 1975). These cells occur scattered throughout the connective tissue; their number varies greatly, however, in the various parts of the connective tissue. Pore cells may be round, ovoid, elongated or branched; they vary considerably in size (diameter 20-60 μ). The cells normally have a round nucleus (diameter up to 12 μ) which contains a conspicuous nucleolus (Fig. 1). When examined with the light microscope, some cells have a transparent cytoplasm, apparently due to large fields of glycogen, while others are entirely filled with grnaules of different size and shape, some of which are lysosomes (see below). The plasma membrane of pore cells shows many vesicular and tubular invaginations, many of which are bridged by cytoplasmic tongues (Figs. 2, 3, 4). In cross-sections of these areas the impression is gained that the cells have perforated plasma membrane, which has led to the designation "pore cells" (Plummer, 1966) or "cellule â sillons" (Nicolas, 1973). The substructure and functional significance of the pores will be discussed below. The invaginations which are not bridged by cytoplasmic tongues have a smooth membrane in contrast to the bridged ones, which bear a spiny coat at the intracellular side (Fig. 4). Another difference between the two types of invagination is that Hcy molecules have only been found in those invaginations which have wide openings (Fig. 2).

Electron microscope studies on the connective tissue of various gastropods have shown that many cell types which were originally described

---

Fig. 3. Part of peripheral cytoplasm of a pore cell of *L. stagnalis*. *BM*: basement membrane; *GER*: granular endoplasmic reticulum; *I*: invagination bridged by cytoplasmic tongues (*T*); *M*: mitochondrion; *P*: pore; *V*: vesicle connected with or derived from the invaginations (× 20,000)

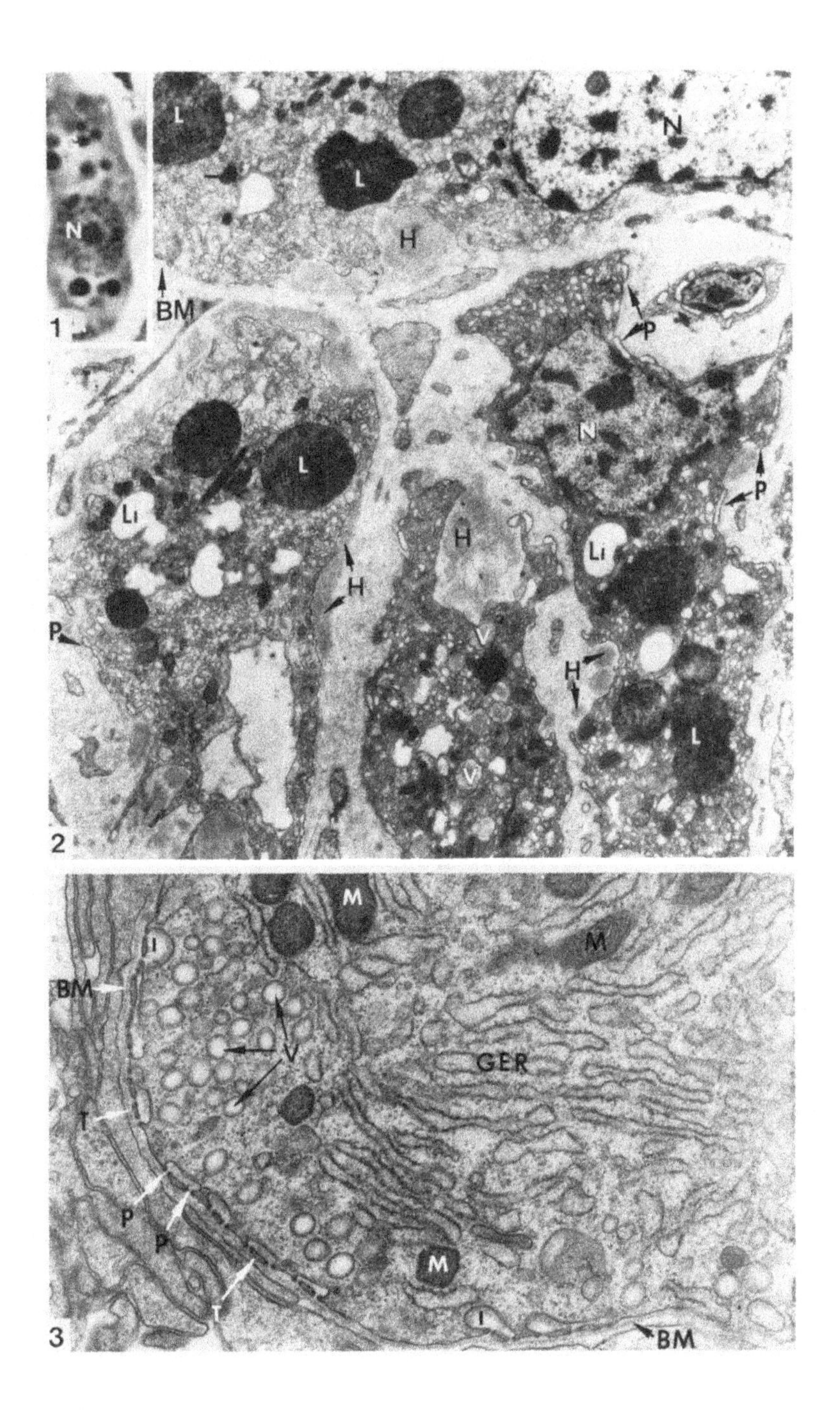

Fig. 1. Light micrograph of a pore cell of *L. stagnalis* with conspicuous nucleus and nucleolus (*N*). *G*: cytoplasmic granules (× 2250)

Fig. 2. Electron micrograph of pore cells of *L. stagnalis*. Large amounts of Hcy (*H*) occur in close association with cells. Each cell is surrounded by a basement membrane (*BM*). *L*: lysosome; *Li*: lipid droplet; *N*:nucleus; *P*: pore; *V*: vacuole containing Hcy molecules (× 8000)

Fig. 3. Legend see p. 280

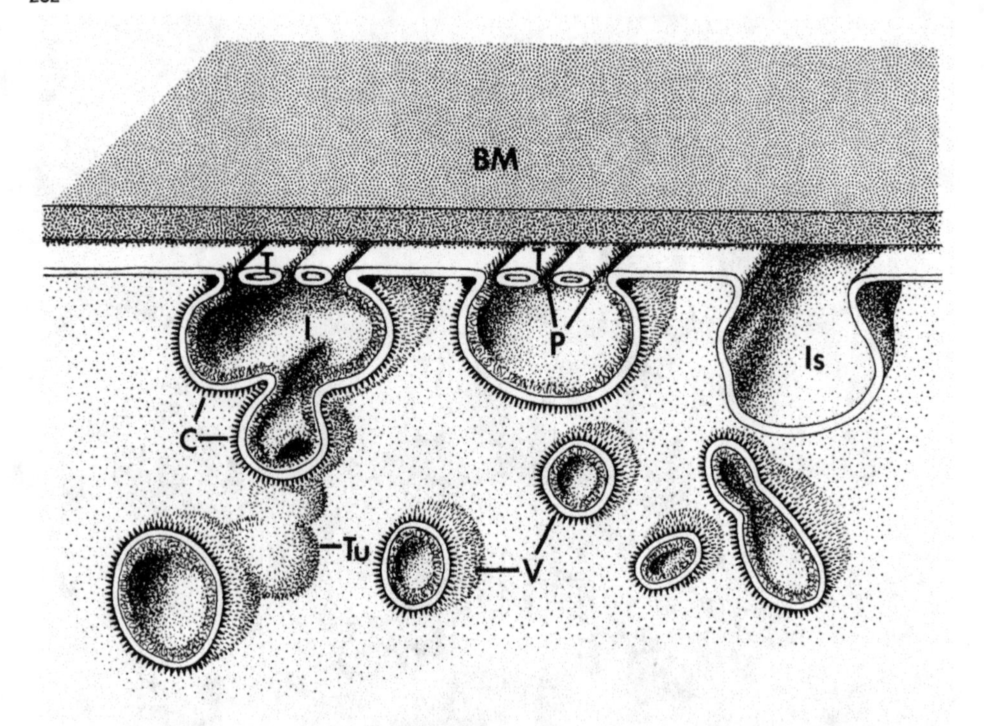

Fig. 4. Schematic drawing of pore system of a pore cell. *BM*: basement membrane; *C*: spiny coat at cytoplasmic side of tubular (*Tu*) and vesicular (*V*) invaginations of plasma membrane; *I*: invagination bridged by cytoplasmic tongues (*T*); *Is*: invagination not bridged by cytoplasmic tongues; *p*: pore

under various names on the basis of light microscope investigations - the cells may very greatly in appearance, hence different functions were attributed to them - are in fact pore cells (Sminia, 1972; Richardot, 1975).

A prominent feature of the pore cells is that the GER and the Golgi apparatus are strongly developed (Baleydier et al., 1969; Sminia, 1972; Nicolas, 1973; Richardot, 1975; Skelding and Newell, 1975). The cisterns of the GER are usually very dilated and contain granular and/or crystalline material (Sminia, 1972; Sminia and Boer, 1973; Nicolas, 1973; Skelding and Newell, 1975). The granules present in the GER of pore cells of the pulmonates *Lymnaea stagnalis* and *Arion hortensis* have the characteristic shape and dimensions (cylinders with a diameter of about 250 Å and a height of about 300 Å) of Hcy molecules (Fig. 5). Apparently the crystals also consist of Hcy. They are composed of dense arrays of hollow cylinders, which have a diameter of about 250 Å and a center-to-center space of about 320 Å. The walls of the cylinders, which are about 75 Å thick, consist of 12 circularly arranged subunits (Figs. 6a, b, c; Stang-Voss and Staubesand, 1971). Hcy molecules were also found in vacuoles in the peripheral parts of the cytoplasm (Sminia, 1972; Skelding and Newell, 1975). It has been suggested that the vacuoles are derived from the GER. Hcy molecules were not observed in vacuoles originating from the Golgi apparatus, which indicates that this organelle is not involved in the synthesis, packaging or transport of Hcy. The Golgi apparatus of pore cells is often very active (Sminia, 1972; Dyer and Cowden, 1973; Skelding and Newell, 1975): it forms many vacuoles, of which at least some are considered lysosomes on the basis

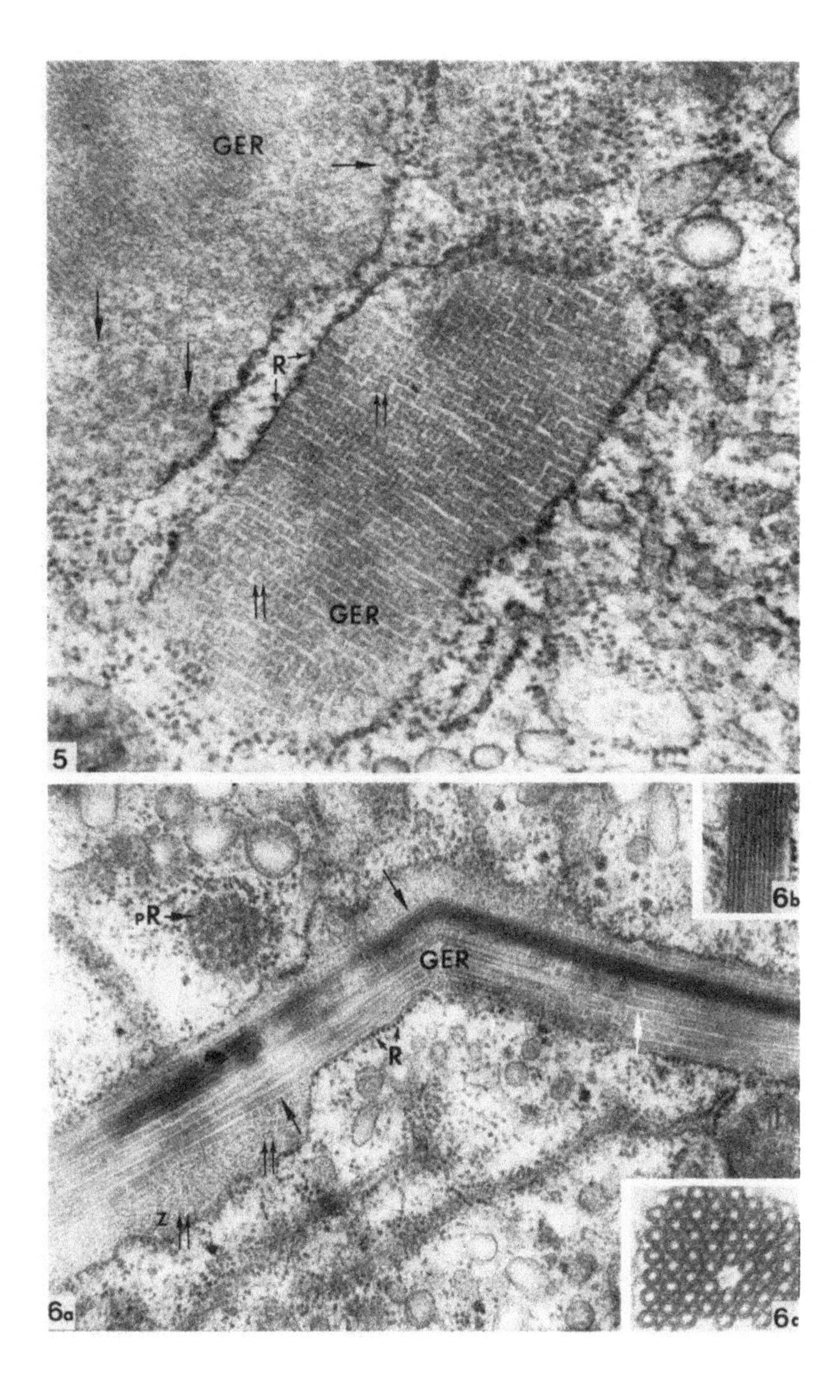

Fig. 5. Dilated part of granular endoplasmic reticulum (*GER*) of a pore cell of *L. stagnalis* filled with Hcy molecules. Note circular (*arrows*) and rectangular profiles (*double arrows*). *R*: ribosomes attached to ER membrances (× 70,000)

Fig. 6a-c. (a) Part of cytoplasm of a pore cell of *L. stagnalis* showing a cistern of the granular endoplasmic reticulum (*GER*), which contains closely packed (*arrows*) and separate (*double arrows*) Hcy molecules. *R*: free ribosomes; *pR*: cluster of (poly) ribosomes. (× 55,000). (b) Longitudinal section of Hcy crystal, showing that it is composed of densely packed hollow cylinders. (× 60,000). (c) Cross section of a Hcy crys-

of morphological and histochemical criteria (Sminia, 1972; Nicolas, 1973). These lysosomes obviously play a role in the digestion of material which is selectively ingested by pore cells (see below).

The assumption that pore cells produce Hcy is further sustained by observations made with an analytical electron microscope, which revealed the presence of a relatively large amount of copper at those sites in the pore cells where Hcy is supposed to be present, i.e. in the GER and the invaginations of the plasma membrane (Sminia et al., 1972).

Reger (1973) has described cells in the connective tissues of the slug *Limax* sp. under the name interstitial cells; these cells possess many Hcy molecules in their well-developed GER. Although in the electron micrographs of the cells presented by Reger no pores can be seen in the plasma membrane, these interstitial cells are probably identical to pore cells. This can be inferred from their distribution and general ultrastructure, which are similar to those of pore cells.

It is interesting to mention that in planorbid snails which have haemoglobin as blood pigment it is the pore cells which synthesize this pigment. This conclusion, which was drawn both on morphological and histochemical observations (Sminia et al., 1972), sustains the assumption that, in general, pore cells are blood pigment producing cells.

Another peculiar feature of pore cells is that each cell is individually surrounded by a basement membrane, which separates it from the other connective tissue elements (Figs. 3, 4; Nicolas, 1973; Richardot, 1975; Skelding and Newell, 1975). This observation suggests that pore cells have a non-mesenchymal origin. In this respect it is interesting to mention that in 1-3-week-old embryos of a number of gastropods a large group of cells has been found in the head region dorsal to the cerebral ganglia. It has been demonstrated that these cells, which were originally described under the name nuchal cells (Block, 1938; Raven, 1966), are morphologically identical to pore cells (Regondaud, 1972). There are, however, no definite data on the origin of this group of cells. According to some authors (Block, 1938) these cells are of mesodermal origin; others are, on the other hand, of the opinion that they are derived from the ectoderm (Fol, 1880; Erlanger, 1891).

In a preliminary study on *L. stagnalis* the conclusion of Regondaud - this author worked with the same species - that nuchal cells are pore cells, was confirmed (Sminia, unpublished). In this study it was furthermore shown that at an age of three weeks the group of nuchal cells dissociates into separate cells which subsequently spread over the connective tissue.

Many authors have paid special attention to the intriguing pore system of the pore cells (Nicaise, 1966; Plummer, 1966; Wolburg-Buchholz, 1972a; Nicolas, 1973). The distance between the cytoplasmic tongues is 200-300 Å. Recently Boer and Sminia (1976) have shown that two diaphragms, forming a three-dimensional sieve structure with holes of about 200-220 Å, are present in the slits between the cytoplasmic tongues of the pore cells of *L. stagnalis*. This sieve shows structural similarities to the sieve present between the pedicels of podocytes which forms the ultrafilter in the urinary systems of both vertebrates (Rodewald and Karnovsky, 1974) and invertebrates (Boer and Sminia, 1976). It has been shown by injection experiments that pore cells have a great, but selective, endocytosis capacity (Cuenot, 1899; Buchholz et al., 1971; Sminia, 1972; Wolburg-Buchholz and Nolte, 1973). Boer and Sminia (1976), using colloidal gold suspensions with particles of different dimensions, have shown that the size of the holes of the pore cells' sieves matches that of particles which can be ingested by this type of cell.

In addition to their involvement in Hcy synthesis, several other functions have been attributed to pore cells. First, on the basis of their endocytosis capacity pore cells have been regarded as fixed macrophages (Wolburg-Buchholz, 1972b), thus as cells involved in the cellular defence mechanism of snails.

However, since pore cells ingest material very selectively they cannot be considered true macrophages. Secondly, pore cells have been considered as glycogen-storing cells since they may contain - especially in terrestrial pulmonates - large amounts of this polysaccharide (Dyer and Cowden, 1973). This function is probably almost only a subsidiary one since gastropods have connective tissue cells specialized for the storage of carbohydrates (Sminia, 1972; Richardot, 1975). Another function which has been ascribed to pore cells is their possible involvement in the synthesis and release of collagen connective tissue fibrils, i.e. they have been considered to be fibroblasts (Plummer, 1966; Nicaise et al., 1966; Nicolas, 1973). However, gastropods possess another cell type, structurally identical to the fibroblasts of vertebrates, which has been identified as a fibroblast (Plummer, 1966; Sminia, 1972). Although the ultrastructure of pore cells is in accordance with the assumption that they are protein-producing cells, it seems peculiar that gastropods would have two types of fibroblast (Plummer, 1966).

Pore Cells: Haemocyanin Synthesis and Release

The present data indicate that in gastropods Hcy is synthesized in pore cells, which are scattered throughout the connective tissue. In these cells high numbers of Hcy molecules have been found in the lumen of the GER. Apparently Hcy is synthesized on the ribosomes attached to the ER membranes and subsequently stored inside the cisterns. It is still unknown how Hcy is released from the pore cells. Hcy molecules have never been observed inside vacuoles formed by the Golgi apparatus, the organelle normally involved in the synthesis and release of secretion products. Probably Hcy is directly released from the GER via openings which originate by fusion of the GER membranes with the plasma membrane. The idea is supported by the observation that in the peripheral parts of the cytoplasm vacuoles and smooth surfaced cisterns, crowded with Hcy molecules, have been found by Sminia (1972) and by Skelding and Newell (1975). According to the latter authors the smooth cisterns are continuous with the ribosome-bearing parts of the GER; presumably GER cisterns filled with Hcy loose their ribosomes and then fuse with the plasma membrane, thus forming those invaginations of the plasma membrane which are not bridged by cytoplasmic tongues. The fact that these invaginations are often filled with Hcy molecules is in accordance with the hypothesis that at these sites Hcy is released from the cell. Obviously they cannot reach the blood via the invaginations bridged by cytoplasmic tongues; the pores of the slit membranes between the tongues are smaller (200-220 Å) than the Hcy molecules (250 × 300 Å).

Haemocyanin-Producing Cells in Invertebrates: A Comparison

The pore cells of gastropods will be compared to the branchial gland of cephalopod molluscs and to the cyanoblast of arthropods (*L. polyphemus*). (The "blood gland" of opisthobranchs is not taken into account, since it is still doubtful, as mentioned, whether or not this gland produces haemocyanin). It is evident that in molluscs Hcy is produced in "fixed" cells, which are probably of ectodermal origin, while in arthropods Hcy is synthesized in "free" cells which are obviously derived from the mesoderm. The different types of cell are shown diagrammatically in Figures 7a, b and c.

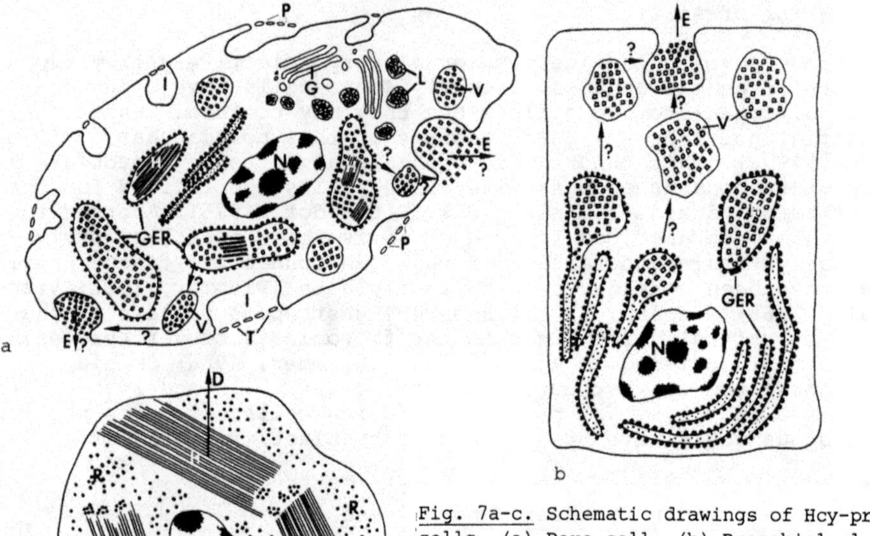

Fig. 7a-c. Schematic drawings of Hcy-producing cells. (a) Pore cell. (b) Branchial gland cell. (c) Cyanoblast. *Arrows* in Fig. 7a, b indicate possible route of Hcy from site of synthesis (ribosomes bound to ER membranes; *GER*: granular endoplasmic reticulum) to site of release (Hcy is presumably liberated by exocytosis, *E*). *Arrows* in Figure 7c point to way of release of Hcy crystals (synthesis takes place at free ribosomes) which occurs by disruption of the cell (*D*). *G*: Golgi apparatus; *H*: Hcy molecules or crystals; *I*: invaginations of the plasma membrane bridged by cytoplasmic tongues (*T*); *L*: lysosome; *N*: nucleus; *P*: pore; *R*: ribosomes; *V*: vacuole filled with Hcy molecules

Comparing the processes of synthesis, storage and release the following remarks can be made:

1. The synthesis of Hcy in branchial gland cells (cephalopods) and pore cells (gastropods) takes place on ribosomes attached to ER cisterns; in cyanoblasts (arthropods) synthesis occurs at free (poly)ribosomes. In the first two cell types Hcy is stored inside the GER cisterns, in the latter cell type it occurs free in the cytoplasm.

2. In both cyanoblasts and pore cells Hcy molecules are stored in an aggregated form, as large crystals. In branchial gland cells, on the other hand, Hcy molecules apparently do not aggregate.

3. Hcy is liberated from the cyanoblast by disruption of the cell (holocrine secretion), while in branchial gland cells and pore cells, Hcy is probably released by exocytosis (merocrine secretion).

The conclusion that the branchial gland of cephalopods synthesizes Hcy is based on morphological as well as on experimental evidence. Only morphological evidence on Hcy synthesis is, however, available for cyanoblasts and pore cells. In our laboratory immuno-enzyme-histochemical experiments are in progress to further elucidate the role of pore cells in Hcy synthesis.

Acknowledgments. The author thanks Dr. H.H. Boer for his valuable criticism during the preparation of the manuscript, Professor Dr. J. Lever for reading the manuscript, Mr. G.W.H. van der Berg for drawing the diagrams and Miss Benita Plesch for correcting the English text.

References

Baleydier, C., Nicaise, G., Pavans de Ceccatty, M.: Etat fibroblastique et différenciation fibrocytaire des cellules conjonctives de *Glossodoris* (Gasteropode Opisthobranche). Comp. Rend. Acad. Sci. Paris 269, 175-178 (1969)
Bloch, S.: Beiträge zur Kenntnis der Ontogenese von Süßwasserpulmonaten mit besonderer Berücksichtigung der Mitteldarmdrüse. Rev. Suisse Zool. 45, 157-220 (1938)
Boer, H.H., Sminia, T.: Sieve structure of slit diaphragms of podocytes and pore cells of gastropod molluscs. Cell. Tiss. Res. (in press, 1976)
Buchholz, K., Kuhlmann, D., Nolte, A.: Aufnahme von Trypanblau und Ferritin in die Blasenzellen des Bindegewebes von *Helix pomatia* and *Cepaea nemoralis* (Stylommatophora, Pulmonata). Z. Zellforsch. 113, 203-215 (1971)
Cheng, T.C., Rifkin, E.: Cellular reactions in marine molluscs in response to helminth parasitism. A symposium on diseases of fishes and shellfishes. Am. Fish. Soc. Spec. Publ. 5, 443-496 (1970)
Cuenot, L.: L'excretion chez les mollusques. Arch. Biol. (Liège) 16, 4-96 (1899)
Dilly, P.N., Messenger, J.B.: The branchial gland: a site of haemocyanin synthesis in *Octopus*. Z. Zellforsch. 132, 193-201 (1972)
Dyer, R.F., Cowden, R.R.: Electron microscopy of the esophageal ganglion complex of the gastropod pulmonate *Triodopsis divesta*. I. Ultrastructure of the epineurium. J. Morphol. 139, 125-154 (1973)
Erlanger, R. von: Zur Entwicklung von *Paludina vivipara*. II. Morphol. J. 17, 636-680 (1891)
Fahrenbach, W.H.: The cyanoblast: haemocyanin formation in *Limulus polyphemus*. J. Cell Biol. 44, 445-453 (1970)
Fol, H.: Etudes sur le développement des mollusques. III. Sur le développement des gasteropods pulmones. Arch. Zool. Exptl. Gen. 8, 103-232 (1880)
Ghiretti, F.: Physiology and Biochemistry of Haemocyanins. London-New York: Academic Press, 1968
Ghiretti, F., Ghiretti-Magaldi, A.: Respiratory proteins in molluscs. In: Chemical Zoology. Florkin, M., Scheer, B.T. (eds.). New York-London: Academic Press, 1972, Vol. VII, pp. 201-214
Messenger, J.B., Muzii, E.O., Nardi, G., Steinberg, H.: Haemocyanin synthesis and the branchial gland of *Octopus*. Nature (London) 250, 154-155 (1974)
Nicaise, G., Garrone, R., Pavans de Ceccatty, M.: Aspects membranaires du fibroblaste, au cours de la genèse du collagène chez *Glossodoris* (Gasteropode: Opisthobranche). Comp. Rend. Acad. Sci. Paris 262, 2248-2250 (1966)
Nicolas, M.T.: Cellules à sillons et genèse des macromolecules extra-cellulaire dans la gaine conjonctive periganglionaire d'*Helix aspersa* (Gasteropode Pulmone). Thesis: Univ. Claude Bernard, Lyon, France, 1973
Plummer, J.M.: Collagen formation in Achatinidae associated with a specific cell type. Proc. Malac. Soc. London 37, 189-198 (1966)
Raven, C.P.: Morphogenesis: The Analysis of Molluscan Development. New York: Pergamon Press, 1966
Redmond, J.R.: Blood respiratory pigments-Arthropoda. In: Chemical Zoology. Florkin, M., Scheer, B.T. (eds.). New York-London: Academic Press, 1971, Vol. VI, pp. 119-144
Reger, J.F.: A fine structure study on haemocyanin formation in the slug *Limax* sp. J. Ultrastruct. Res. 43, 377-387 (1973)
Regonaud, J.: Observation ultrastructurale des cellules nucales de l'embryon de *Lymnaea stagnalis* L. (Gasteropode Pulmone Basommatophore). Comp. Rend. Acad. Sci. Paris 275, 679-682 (1972)

Richardot, M.: Determinisme de la formation du septum chez *Ferrissia wautieri* (Mirolli). Données écologiques, biologiques et physiologiques. Thesis: Univ. Claude Bernard, Lyon, France, 1975

Rodewald, R., Karnowsky, M.J.: Porous structure of the glomerular slit diaphragm in the rat and the mouse. J. Cell Biol. 60, 423-433 (1974)

Ruddell, C.L., Wellings, S.R.: The ultrastructure of the oyster brown cell, a cell with a fenestrated plasma membrane. Z. Zellforsch. 120, 17-28 (1971)

Scheer, B.T.: Animal Physiology. New York-London-Sydney: John Wiley, 1967, pp. 211-214

Schipp, R., Hohn, P., Ginkel, G.: Elektronenmikroskopische und histochemische Untersuchungen zur Funktion der Branchialdrüse (Parabranchialdrüse) der Cephalopoda. Z. Zellforsch. 139, 253-269 (1973)

Schmekel, L.: Feinstruktur der Spezialzellen von normal ernährten und hingernden Aeolidiern (Gastr. Nudibranchia). Z. Zellforsch. 124, 419-432 (1972)

Schmekel, L., Weischer, M.: Die Blutdrüse der Doridoidea (Gastropoda, Opisthobranchia) als Ort möglicher Haemocyanin-Synthese. Z. Morphol. Tiere 76, 261-284 (1973)

Skelding, J.M., Newell, P.F.: On the functions of pore cells in the connective tissue of terrestrial pulmonate molluscs. Cell Tiss. Res. 156, 381-390 (1975)

Sminia, T.: Structure and function of blood and connective tissue cells of the freshwater pulmonate *Lymnaea stagnalis* studied by electron microscopy and enzyme histochemistry. Z. Zellforsch. 130, 497-526 (1972)

Sminia, T., Boer, H.H., Niemantsverdriet, A.: Haemoglobin producing cells in freshwater snails. Z. Zellforsch. 135, 563-568 (1972)

Sminia, T., Boer, H.H.: Haemocyanin production in pore cells of the freshwater snail *Lymnaea stagnalis*. Z. Zellforsch. 145, 443-445 (1973)

Stang-Voss, C., Staubesand, J.: Mikrotubulare Formationen in Zisternen des endoplasmatischen Retikulums. Electronenmikroskopische Untersuchungen an Bindegewebszellen von *Lymnaea stagnalis* L. (Pulmonata). Z. Zellforsch. 115, 69-78 (1971)

Wolburg-Buchholz, K.: Blasenzellen im Bindegewebe des Schlundrings von *Cepaea nemoralis* L. (Gastropoda, Stylommatophora). I. Feinstruktur der Zellen. Z. Zellforsch. 128, 100-114 (1972a)

Wolburg-Buchholz, K.: Blasenzellen im Bindegewebe des Schlundrings von *Cepaea nemoralis* L. (Gasteropoda, Stylommatophora). II. Aufnahme und Speicherung von Ferritin. Z. Zellforsch. 130, 262-278 (1972b)

Wolburg-Buchholz, K., Nolte, A.: Vergleichende Untersuchungen an Amoebozyten und Blasenzellen von *Cepaea nemoralis* L. (Gastropoda, Stylommatophora). Unterschiedliche Endozytosefähigkeit der Zellen. Z. Zellforsch. 137, 281-292 (1973)

Wondrak, G.: Die Ultrastruktur der Zellen aus dem interstitiellen Bindegewebe von *Arion rufus* (L)., Pulmonata, Gastropoda. Z. Zellforsch. 95, 249-262 (1969)

Subject Index

MOLECULAR BIOLOGY

BIOCHEMISTRY

AND BIOPHYSICS

Editors:
A. Kleinzeller, G.F. Springer, H.G. Wittmann

Volume 23
M. Luckner, L. Nover, H. Böhm
Secondary Metabolism and Cell Differentiation
52 figures, 7 tables. VI, 130 pages. 1977
ISBN 3-540-08081-3

The volume is based on the concept that the biosynthesis of secondary natural substances is a widely found characteristic of cell specialization in almost all organisms. As in other specialization processes, the formation of the enzymes involved is dependent on the stage of development of the respective cell. For this reason the central theme of the book, together with chapters on coordinated enzyme synthesis and possible effectors of gene expression in this field of metabolism, is the analysis of differentiating programs, including the formation of enzymes of secondary metabolism. Particular aspects of the formation of secondary natural substances in plant cell culture are treated in a separate chapter.

Volume 24
Chemical Relaxation in Molecular Biology
Editors: I. Pecht, R. Rigler
141 figures, 47 tables. Approx. 470 pages. 1977
ISBN 3-540-08173-9

The purpose of this monograph is to give a representative cross section of the current research activities dedicated to the analysis of elementary steps in biological reactions. This covers the range of the following topics: hydrogen-bond formation, nucleotide base pairing, protein folding, isomerisation of protein and nucleic acid conformations, interactions between protein and proteins, nucleic acid and proteins, enzymes and substrates, antibody and haptens or ionic transport through membranes. A common denominator in these studies is the search for an understanding of the laws that govern the dynamic behaviour of living systems.

Springer-Verlag Berlin-Heidelberg-New York

PROGRESS

IN MOLECULAR AND

SUBCELLULAR BIOLOGY

Editors:
F.E. Hahn, H. Kersten, W. Kersten, W. Szybalski
Advisors: T.T. Puck, G.F. Springer, K. Wallenfels
Managing Editor: F.E. Hahn

Volume 4
88 figures, 27 tables. XI, 251 pages. 1976
ISBN 3-540-07487-2

The new volume of this well-established series contains seven articles which range from biophysical studies on the conformations of DNA, through papers on gene expression in bacterial viruses and cellular slime molds, to inhibitors of nucleic acid syntheses as antitumor agents and the role of adenosine as a regulator of coronary blood flow. This volume continues the tradition of the series to combine carefully selected articles in the field of biomedicine with those in which life scientists are particulary interested.
Contents: S. Bram: The Polymorphism of DNA. - H.J. Witmer: Regulation of Bacteriophage T4 Gene Expression. - W. Lotz: Defective Bacteriophages: The Phage Tail-like Particles. - M. Sussman: The Genesis of Multicellular Organization and the Control of Gene Expression in Dictyostelium discoideum. - D. Oesterhelt: Isoprenoids and Bacteriorhodopsin in Halobacteria. - P. Chandra, L.K. Steel, U. Ebener, M. Woltersdorf, H. Laube, G. Will: Inhibitors of DNA Synthesis in RNA Tumor Viruses: Biological Implications and Their Mode of Action. - R.A. Olsson, R.E. Patterson: Adenosine as a Physiological Regulator of Coronary Blood Flow.

Volume 5
56 figures. XIII, 176 pages. 1977
ISBN 3-540-08192-5

Volume 5 of this series presents a collection of five progress reports concerning recent developments in molecular biology outside the classical area of molecular genetics. It continues the emphasis on medical pharmacological aspects of molecular biology which has characterized this series throughout its existence. The text covers the following topics in detail. F.E. Hahn: The Double Helix Revisited, Watson and Olby. C. Reiss, T. Arpa-Gabarro: Thermal Transition Spectroscopy: A New Tool for Submolecular Investigation of Biologic Macromolecules. I. Schuster: Interactions of Drugs with Liver Microsomes. W. Flamenbaum, J.H. Schwartz, R.J. Hamburger, J.S. Kaufman: The Pathogenesis of Experimental Acute Renal Failure: The Role of Membrane Dysfunction. P. Gund: Three-Dimensional Pharmacophoric Pattern Searching. N.W. Gabel: Chemical Evolution: A terrestrial Reassessment.

Springer-Verlag Berlin-Heidelberg-New York